Springer Series on Polymer and Composite Materials

Series Editor

Susheel Kalia, Army Cadet College Wing, Indian Military Academy, Dehradun, India

The "Springer Series on Polymer and Composite Materials" publishes monographs and edited works in the areas of Polymer Science and Composite Materials. These compound classes form the basis for the development of many new materials for various applications. The series covers biomaterials, nanomaterials, polymeric nanofibers, and electrospun materials, polymer hybrids, composite materials from macro- to nano-scale, and many more; from fundamentals, over the synthesis and development of the new materials, to their applications. The authored or edited books in this series address researchers and professionals, academic and industrial chemists involved in the areas of Polymer Science and the development of new Materials. They cover aspects such as the chemistry, physics, characterization, and material science of Polymers, and Polymer and Composite Materials. The books in this series can serve a growing demand for concise and comprehensive treatments of specific topics in this rapidly growing field. The series will be interesting for researchers working in this field and cover the latest advances in polymers and composite materials. Potential topics include, but are not limited to:

Fibers and Polymers:

- Lignocellulosic biomass and natural fibers
- Polymer nanofibers
- Polysaccharides and their derivatives
- Conducting polymers
- Surface functionalization of polymers
- Bio-inspired and stimuli-responsive polymers
- Shape-memory and self-healing polymers
- Hydrogels
- Rubber
- Polymeric foams
- Biodegradation and recycling of polymers

Bio- and Nano- Composites:-

- Fiber-reinforced composites including both long and short fibers
- Wood-based composites
- Polymer blends
- Hybrid materials (organic-inorganic)
- Nanocomposite hydrogels
- Mechanical behavior of composites
- The Interface and Interphase in polymer composites
- Biodegradation and recycling of polymer composites
- Applications of composite materials

More information about this series at http://www.springer.com/series/13173

Rasel Das
Editor

Two-Dimensional (2D) Nanomaterials in Separation Science

Springer

Editor
Rasel Das
Department of Chemistry
State University of New York at Stony Brook
New York, NY, USA

ISSN 2364-1878 ISSN 2364-1886 (electronic)
Springer Series on Polymer and Composite Materials
ISBN 978-3-030-72456-6 ISBN 978-3-030-72457-3 (eBook)
https://doi.org/10.1007/978-3-030-72457-3

This Springer imprint is published by the registered company Springer Nature Switzerland AG
The registered company address is: Gewerbestrasse 11, 6330 Cham, Switzerland

Preface

"Two-Dimensional (2D) Nanomaterials in Separation Science" is a monograph book, which offers help in two ways—(a) understanding the fundamentals of 2D materials, especially graphene-family members, hexagonal boron nitride, molybdenum disulfide, tungsten disulfide, metal organic framework, zeolite, MXene nanosheets, etc. and (b) shaping your thoughts and understanding on the uses of these 2D nanomaterials in separation science including the membrane filtration and adsorption. This book starts with a special chapter, focusing on novel pollutants, such as nanoparticles and nanoplastics, which have recently become emerging water pollutants of global importance. The authors of individual chapters cover both simulation and experimental evidences of 2D nanosheet-based separation technologies which are often viewed as the more difficult concepts, especially to isolate complex solutes from wastewater. The authors give insight mechanistic details in simple enough language to make them more understandable. At first, they scrutinize the literatures, and based on that enough discussions on 2D nanosheets synthesis; characterizations, membrane fabrication, and demonstration of these membranes in isolating solutes are corroborated in detail. Researchers are often burdened with the task of finding research gaps, and the enormity of such knowledge gaps is given in the last part of each chapter which might help the readers to overcome any stagnancy in this research field. The authors also give their perspectives and concentrated on concepts which could be useful to young researchers and industrial R&D scientists. My rationale is that new research breakthrough is considerably difficult to find in the ever-expanding literatures, but if the given challenges and perspectives are appropriately mastered, which are implemented in my book, one can bring new state-of-the-art technologies in the field of separation science.

New York, USA — Rasel Das

Contents

Introduction

Rasel Das

Abstract This chapter defines the two-dimensional (2D) nanomaterials that have been used in separation science. Then, chapter summaries can be found at the end of this chapter.

Keywords Water pollutants · Water purification · Separation science

1 Background

Industrialization, urbanization, population growth, etc., have severely contaminated freshwater resources. It is due to the addition of macro-, micro- [1] and nano-pollutants [2] into the water bodies. Among these, nanoplastics and nanoparticles have recently become emerging water pollutants of global importance. These pollutants have detrimental effects not only on human health, but also on aquatic flora and fauna [3]. Huge investment is necessary each year due to these polluted water consequences. Therefore, water purification using state-of-the-art technologies is necessary for an unlimited safe water supply [4]. Nowadays, these technologies need novel materials, especially two-dimensional (2D) nanomaterials (NM), due to their atomic-scale thickness, nano-sized pores, high surface area, stability, and high purification ability [5]. It has been around 160 years that scientists have studied layered materials. However, very recently, they have tried to figure out the actual uses of these materials for various technological applications [6]. Most of the bulk materials, for example, three-dimensional (3D) graphite, when thinned into nanoscale, for example, graphene, and these thinnest NMs are called 2D nanosheets at their physical limit as shown in Fig. 1.

Since 1859, many occasional attempts were taken to study graphene, but it is widely synthesized and used after 2004 when Professor Sir Andre Geim and Professor

R. Das (✉)
Department of Biochemistry and Biotechnology, University of Science and Technology Chittagong (USTC), Foy's Lake, Chattogram 4202, Bangladesh
e-mail: raseldas@daad-alumni.de

R. Das (ed.), *Two-Dimensional (2D) Nanomaterials in Separation Science*, Springer Series on Polymer and Composite Materials,
https://doi.org/10.1007/978-3-030-72457-3_1

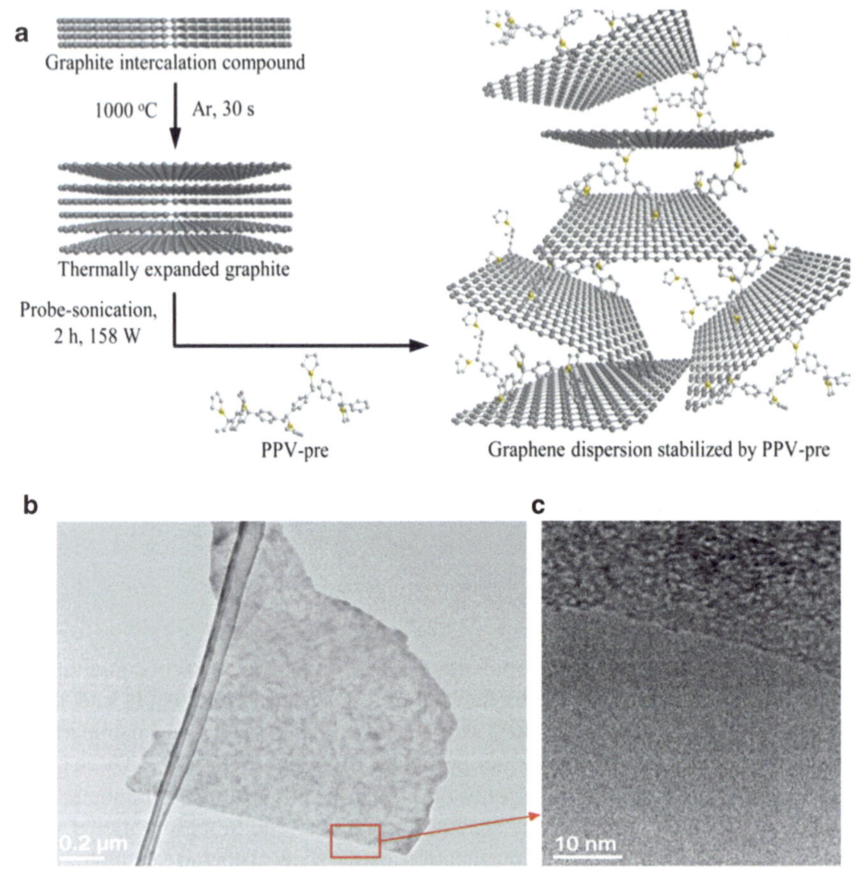

Fig. 1 **a** Exfoliation of 3D graphite to few-layer 2D graphene nanosheets using poly (p-phenylenevinylene) precursor (PPV-Pre), **b** transmission electron microscope (TEM) and **c** high resolution (HR)TEM images of a graphene-PPV-pre single-layer sheet. Figures are reproduced with permission from the American Chemical Society [7]

Sir Kostya Novoselov (from The University of Manchester) discovered a monolayer thick sp^2-hybridized 2D carbon sheet using a piece of graphite and scotch tape. Graphene has excellent electrical, thermal, optical and mechanical properties [8], which have long been the interest in the field of material science, condensed physics, synthetic chemistry, etc. It has been claimed that the graphene hardness is 200 and 30 times higher than steel and diamonds, respectively [9]. Due to its impermeability to water and gas, graphene has been widely studied in separation science by making suitable pores on it [10]. Atomic-scale thickness could ensure its high liquid permeability which is several times higher than most of the commercial nanofiltration (NF) membranes. It increases cost efficiency and energy loss. There are three types of graphene, such as raw or pristine graphene, graphene oxide (GO)

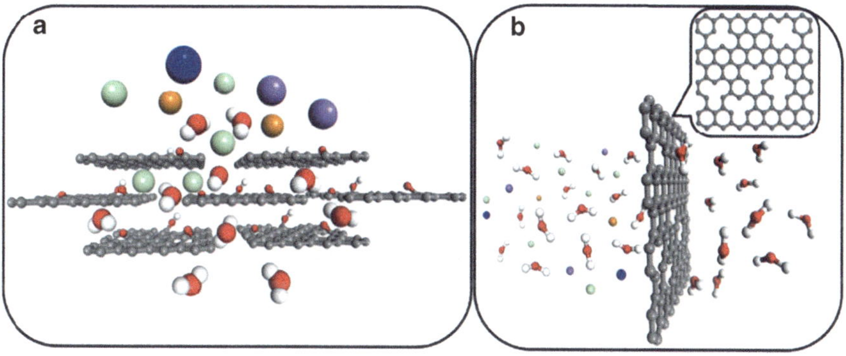

Fig. 2 **a** Stacking of GO nanosheets in the lamellar membrane and **b** nanoporous graphene membrane. Here C: black, H: white, O: red, K^+: blue, Na^+: purple, Mg^{2+}: orange and Cl^-: cyan. Figure is reproduced with permission from the Royal Society of Chemistry [11]

and reduced GO (rGO) which have been used for membrane fabrications. The free-standing graphene-family-based membranes mainly appear in two forms, (a) lamellar membranes, formed by stacking of GO/rGO, and (b) nanoporous graphene membrane where pore is generated using oxygen plasma, ion bombardment, etc. (Fig. 2).

As shown in Fig. 2a, water molecules typically permeate through the fabricated nanochannel between each GO/rGO layer/sheets, while hydrated salt ions/pollutants are blocked. On the other hand, water molecules could pass through the artificially created desired nanopores in the graphene (called nanoporous membrane) to hinder the permeation of large hydrated salt ions (Fig. 2b). Therefore, they have the potentiality for size-selective transport through the graphene nanopores. These membranes significantly outperform the commercial polymeric membranes in terms of high water flux and solute selectivity [12]. However, most of the GO-based membranes have a tendency to be swelled in presence of water, compromising with solute selectivity and have shown poor lifelong stability.

After tremendous efforts for graphene production, very recently, researchers are motivated to synthesize its iso-structural form called hexagonal boron nitride (h-BN). Similar to graphene, h-BN has good mechanical and thermal conductivity. In contrast, h-BN has high thermal and chemical stabilities and a wide bandgap (5.5–5.9 eV) [14]. H-BN is also a layered 2D material, consisting of equal numbers of boron (B) and nitrogen (N) atoms in an alternative way in a hexagonal lattice. Compared to carbon in graphite, B-atoms are situated directly above or below corresponding N atoms as shown in Fig. 3a, and the interlayer distance is almost 0.33 nm. Compared to C-C bond in graphene, B-N bond in h-BN is partially ionic and therefore, would have good interaction property for solutes. Despite having these leapfrogging properties, the application of h-BN in separation science is sparsely reported. Recently, Chen et al. fabricated amino-functionalized h-BN lamellar membrane as depicted in Fig. 3b–d which retains solutes based on size-exclusion (molecular sieving) mechanism [13].

Very recently, transition metal disulfides (TMDS) that mainly include tungsten disulfide (WS_2) and molybdenum disulfide (MoS_2) have found tremendous research

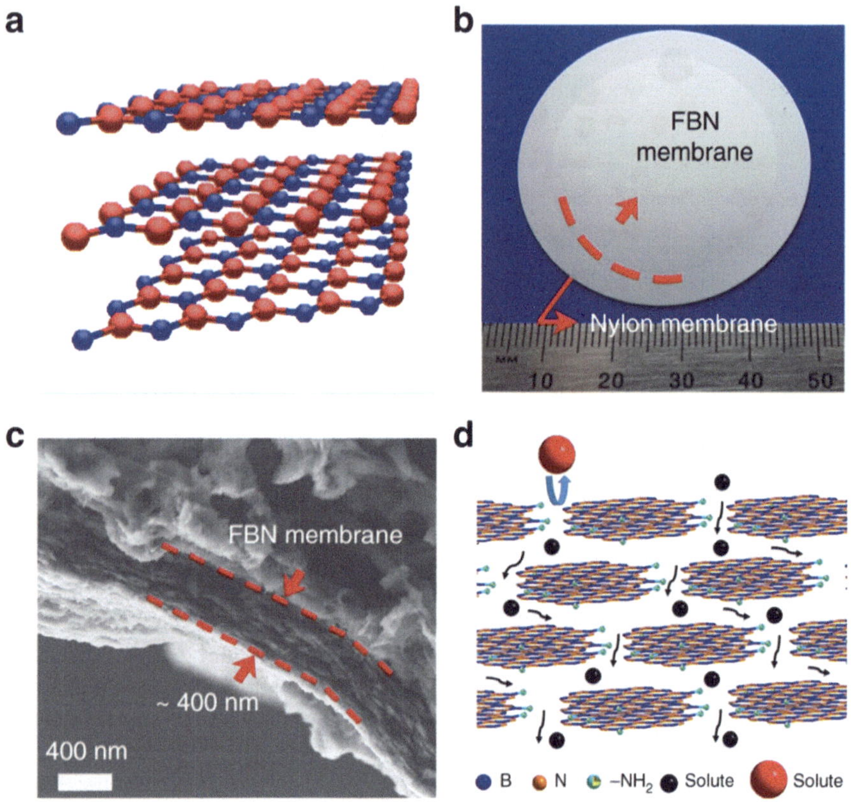

Fig. 3 **a** Illustration of 2D h-BN nanosheets, **b** h-BN film deposited onto a nylon membrane support, **c** cross-section SEM image of a typical h-BN membrane and **d** mechanism of action for solute separation. Figures b-d are reproduced with permission from Springer-Nature [13]

interests due to their chemical and thermal stabilities, low-cost material and excellent processibility. However, using these 2D TMDS for solutes separation is rarely reported in the literature. Methods like chemical vapor deposition, mechanical exfoliation, and other chemical strategies are used for the large-scale synthesis of their 2D nanosheets. For example, a rolling-out method has been used to synthesize WS_2 nanosheets, having a 100 nm lateral dimension from 1D $W_{18}O_{49}$ using surfactant [15]. Ball milling and heating methods have also been used by mixing WO_3 and S and heated the powders at 600 °C in Ar [16]. On the other hand, MoS_2 consists of two types of atoms, that is, molybdenum (Mo) and sulfur (S). The thickness of monolayer MoS_2 is around 1.0 nm. Compared to steel, it is a strong material (Young's modulus: 270 ± 100 GPa). Because of its flexibility, both Mo and S are suitable to create nanopore even with desired functionality. Figure 4 shows simulated MoS_2 membrane with nanoscale pore sizes which enhances water permeation and has good solute retentions.

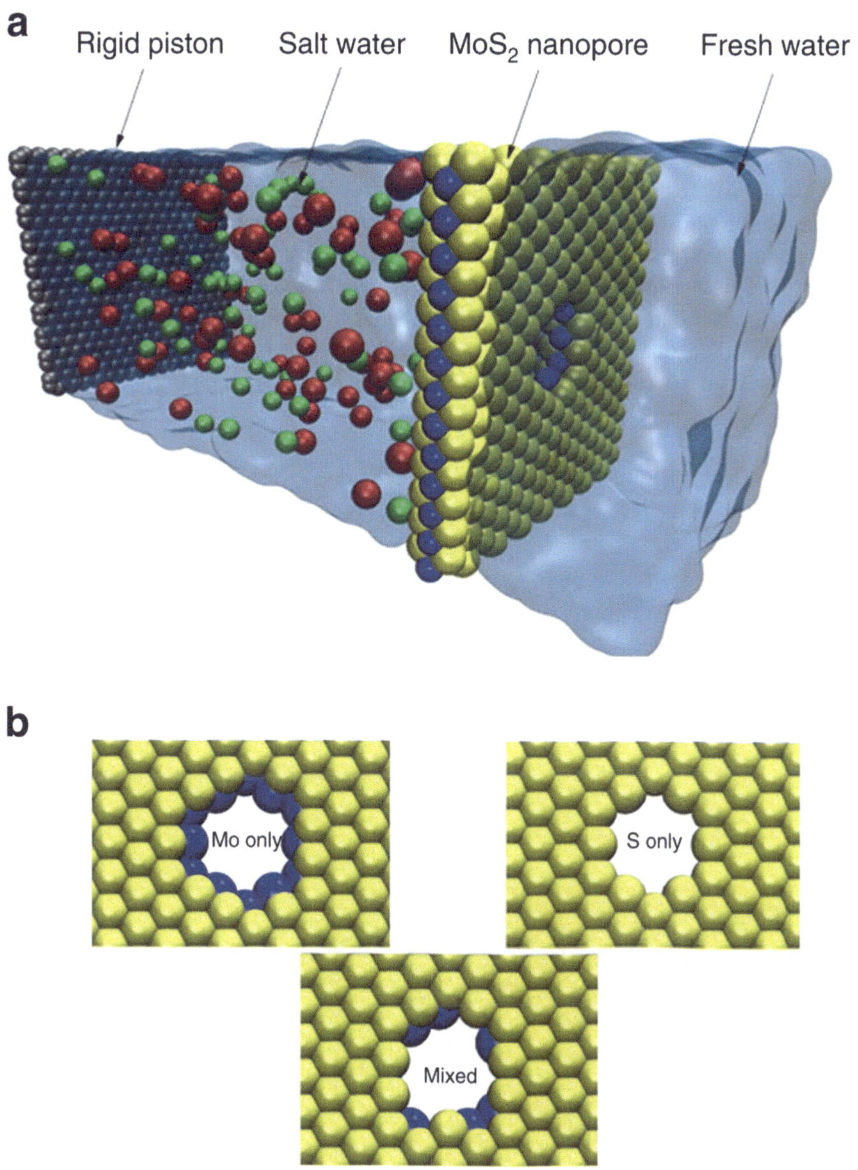

Fig. 4 **a** Illustration of the simulation box consisting of a MoS_2 sheet (molybdenum in blue, and sulfur in yellow), water (transparent blue), ions (in red and green) and a graphene sheet (in gray). **b** Left: Mo only pore type. Right: S only pore type. Bottom: mixed (Mo and S) pore type. Figures are reproduced with permission from the Nature-Springer [17]

Metal-organic framework (MOF)—as the name implied, it is a combination of metal ions/clusters and the organic linker. These combinations prepare secondary building units that are then self-associated to polymeric architectures (structured frameworks) as shown in Fig. 5. The main properties of MOF are their high crystallinity, very high surface area (>6000 m^2/g) and have ultrahigh porosity (<90% of its volume) [18]. Similar to a sponge, MOF is capable to host and release many solutes into/from their pores. It is considered as one of the fastest-growing class of materials, and >20000 MOFs have been synthesized in the last twenty years. Because of its variable structural units (i.e., metal ions and organic components), many researchers are now interested to prepare these materials for many potential applications, such as separation science, clean energy, gas storage, etc. Depending on the nature of coordination between organic ligands and metal nodes, the overall

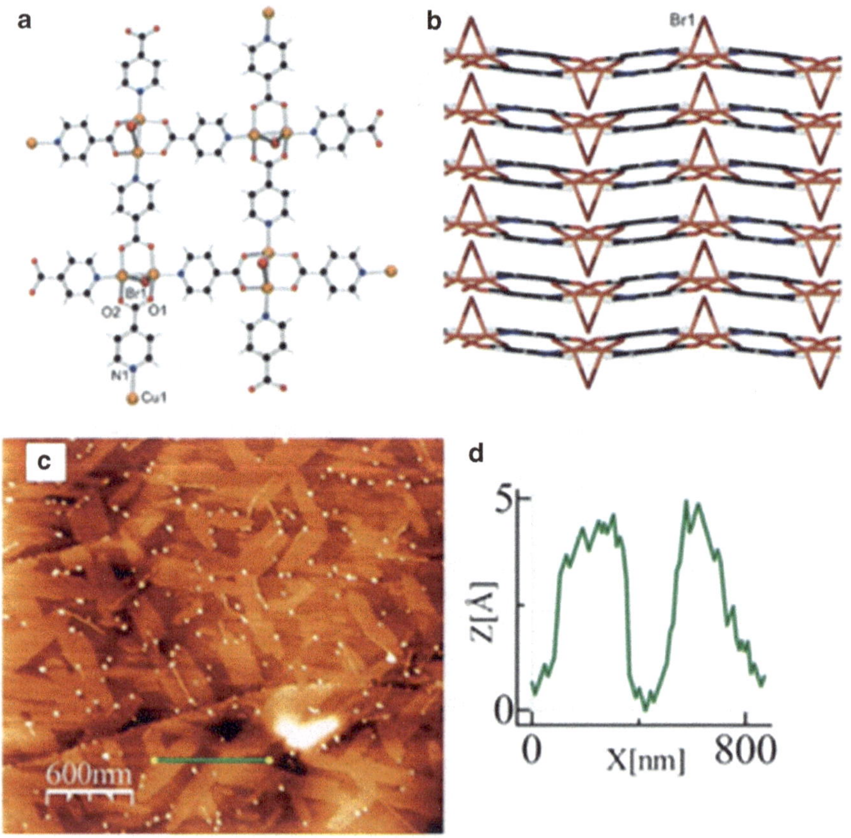

Fig. 5 **a** Illustration of monolayer $Cu_2Br(IN)_2$ framework, **b** Superposition (stacking) of layers along the a-axis, **c** AFM topography image of $Cu_2Br(IN)_2$ nanosheets deposited on highly oriented pyrolytic graphite support, **d** Height profile across the green line in (**c**). Figures are reproduced with permission from the Royal Society of Chemistry [20]

shapes (3D) of MOF could be octahedra, irregular polyhedral, dodecahedra, etc. These 3D shapes have two major drawbacks, such as the requirement of activation treatment and poor solvents dispersibility. Therefore, recent studies are focusing on 2D MOF nanosheets synthesis either from 3D bulky MOFs using exfoliation methods or others. Compared with 3D MOFs, 2D MOF nanosheets have ultrathin thickness, accessible reactive/active sites, high surface area, etc., which are unique properties to develop industrial-grade separation techniques. Because of thin thickness and tunable pore size, the selectivity and permeability of 2D MOFs nanosheets are completely suitable for fabricating membranes with unprecedented advantages [19].

Zeolite is popular for acting as a host framework towards various solutes. It is mainly due to their microporous structure, high aspect ratio, thinner nanosheets (short diffusion distance), hydrophilicity, etc. [21]. It mainly consists of 'corner-sharing TO_4 tetrahedra where T means tetrahedrally coordinated framework atoms, for example, Si and Al or other heteroatoms [22]. Roth et al. reported that 213 different types of zeolite are possible to synthesize and/or naturally available by tuning a range of reaction parameters, such as pressure, time, temperature, reactant types and structure-directing agents as indicated by the yellow color in Fig. 6. Among these structural variations, nearly 200 types belong to 3D structure and only 10 cases can be prepared before the condensation of monolayers to 3D zeolite as revealed in the left and right sides of Fig. 6, respectively [22]. Figure 7 reveals a few 2D zeolite nanosheets synthesis methods including hydrothermal, surfactant-templated, and partial zeolite hydrolysis.

MXenes (general formula: $M_{n+1} X_n$) are transition metal carbides or nitrides. They are the new member of the 2D materials family which could be obtained by etching of A atomic layer from MAX ternary ($M_{n+1}AX_n$) where M is an early transition metal (e.g., Ti, Zr, Hf, W, V, Nb, Ta, Cr, Sc, etc.), A is a non-metal element (i.e., Si, Al, Ga, etc.) and X is C or/and N. Yury's research group reported the MXene (i.e., Ti_3C_2) using HF as etchant at first in 2011 from Ti_3AlC_2 [23]. This Ti_3C_2 is commonly expressed by $Ti_3C_2T_X$, where T_X depicts surface functional groups, such as $=$O, $-$OH, $-$F, etc. Subsequently, various MXenes are reported in the literatures including V_4C_3TX, $Nb_4C_3T_X$, $Zr_3C_2T_X$, $Ta_4C_3T_X$, $Ti_4N_3T_X$, V_2CT_X, Ti_2CT_X and Cr_2CT_X, etc. Li et al. reported halides-terminated MXenes by using $ZnCl_2$ salts as the etchant and Zn-based MAX as a precursor as shown in Fig. 8. Such 2D nanosheets are used to fabricate membrane which shows extremely short transport pathway and many nanochannels. These properties ensure a very high water transport (>1000 L m^{-2} h^{-1} bar^{-1}) and solute rejection (over 90 %) [24]. Therefore, MXenes nanosheets are promising fashioned 2D materials for fabricating advanced water purification membranes.

2 Conclusions

Based on these preceding materials, this book consists of eight chapters, excluding this introductory chapter. These chapters bring meaningful information, focusing

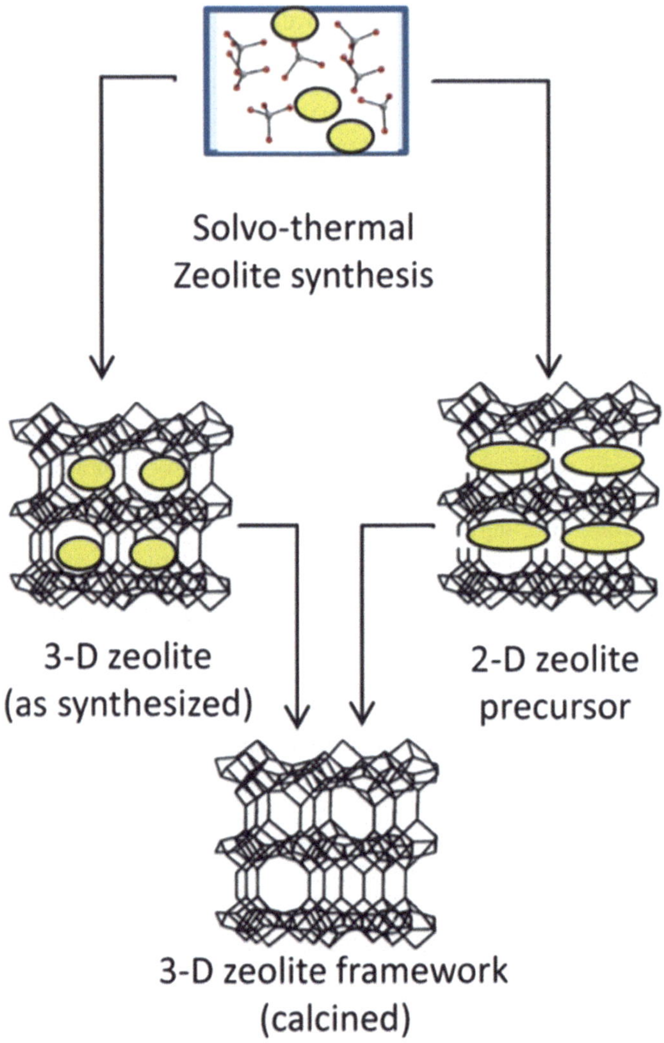

Fig. 6 Schematic illustration of zeolite synthesis using the solvothermal route. Figure is reproduced with permission from the American Chemical Society [22]

on recent threats to water pollutions and highlighting novel 2D materials-based methods to purify contaminated water bodies. Chapter 2 underlines major contaminants of emerging concerns, especially pharmaceuticals and personal care products. Two broad categories of novel pollutants including nanoparticles and nanoplastics are corroborated in detail. The authors show pollutants discharge routes from domestic/agriculture/industry into freshwater resources, and also highlight the exposure risks of these pollutants to humans, plants and animals. They summarize some treatments and tackling techniques (classical and advanced) of these

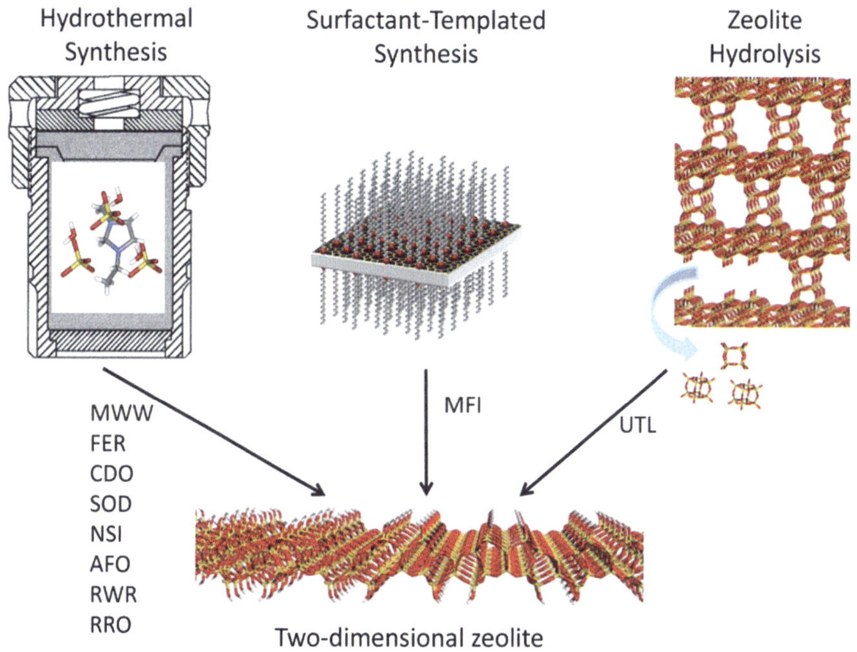

Fig. 7 Schematic illustration of different reaction pathways that results in 2D zeolites. The atoms, such as Si, O, N, and H are indicated using yellow, red, blue, and white, respectively (except surfactant route). Figure is reproduced with permission from the American Chemical Society [22]

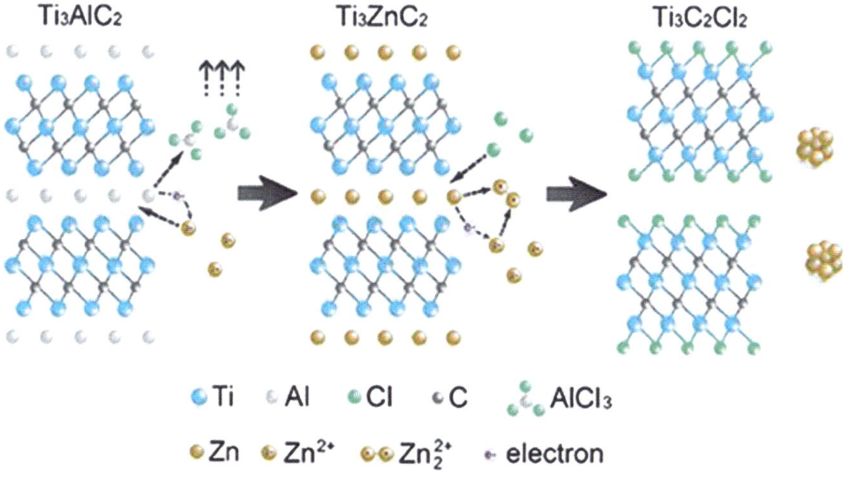

Fig. 8 Schematic of the molten salt etching and delamination process. Figure is reproduced with permission from the American Chemical Society [25]

pollutants where challenges are given to design future research works. Free-standing nanoporous graphene-based membrane for water purification is outlined in Chap. 3. Drawbacks and advantages of some state-of-the-art membrane technologies are given. Membrane testing methods, ideal membrane properties, as well as prospects are drawn in this chapter. Before giving experimental working details, the chapter authors focus on simulation studies that could help to fabricate an efficient nanoporous graphene membrane. Their mechanical stability and large-scale membrane synthesis are underlined. A few characterization tools of nanoporous graphene membranes are summarized. In the last part of this chapter, they show whether such membrane is feasible for large-scale application and also mention some research gaps to overcome any stagnancy in this research field. Chapter 4 mainly focuses on recent advances in GO and/or rGO based 2D- and 3D-membrane technologies. At first, the preparation of GO and rGO is outlined. Next, the experimental works on lamellar and 3D membranes, while the computational studies are discussed in detail. Some graphene-based desalination membrane and graphdiyne films are highlighted. The authors describe some research gaps and draw their perspectives to design future research works. Chapter 5 demonstrates the uses of GO and rGO-based composite membranes (i.e., reinforcing these materials into other polymeric and/or NMs) for water purification. At the beginning of this chapter, the authors describe the commonly used methods (e.g., solvent processing, *in situ* polymerization, melt processing, etc.) in order to synthesize the composites. Then, they summarize some membrane characterization methods for surface roughness, morphological study, contact angle analysis, thermal stability investigation and mechanical property evaluation. After that, the experimental evidence of these membranes for solutes separation is given in detail. In the last part of the chapter, some research gaps are given based on systematic literature studies. Reverse osmosis and electrodialysis-based nanoporous graphene membranes are reviewed in Chap. 6. It reveals some unique fabrication methods of this membrane and discloses the separation mechanisms of solutes. Both the representative theoretical and experimental works in this field are deeply analyzed. Although using porous graphene membranes in the field of electrodialysis is less highlighted in the literature, this chapter compiles all the essential information in one place. Chapter 7 beautifully presents h-BN based separation technologies, especially membrane and sorption. At first, the authors describe all the possible routes for h-BN synthesis. Then, they demonstrate a computational study that indicates the feasibility of such technology for efficient water purification. Experimental evidence of h-BN based solute separation is discussed in detail. Lastly, they compare the performances of h-BN with GO-based membranes which might be benefitted for general readers. Research gaps are also given at the end of this chapter. Chapter 8 is based on MoS_2 and WS_2 based separation methods. At the beginning of the chapter, the authors discuss these 2D nanosheets synthesis methods. Next, they show whether the free-standing TMDS is feasible for filtration technology. What is more interesting about this chapter is to add a section on a comparative study among TMDS and 1D/2D materials. Research gaps and the author's perspectives are given at the end of the chapter. The final Chap. 9 mainly covers three very recently used and fascinating 2D materials, such as MOF, MXenes, and zeolite nanosheets in desalting and heavy metal ions removal. These materials have been significantly spotlighted for their tremendous possibility in separation science. The synthesis methods of these materials are fundamentally important and described in detail. Both the theoretical and experimental works, demonstrating the feasibility of such materials to fabricate future state-of-the-art technologies are given. An interesting section focusing on the intercalation of these 2D nanosheets into relatively classical GO membranes to develop an ultra-grade water purification method is underlined. Readers will find some research gaps and the author's perspective in the last part of this chapter.

References and Future Readings

1. Das R, Abd Hamid SB, Annuar MSM (2016) Highly efficient and stable novel nanobiohybrid catalyst to avert 3, 4-dihydroxybenzoic acid pollutant in water. Sci Rep 6:1–11
2. Das R, Leo BF, Murphy F (2018) The toxic truth about carbon nanotubes in water purification: a perspective view. Nanoscale Res Lett 13:183
3. Das R, Vecitis CD, Schulze A, Cao B, Ismail AF, Lu X et al (2017) Recent advances in nanomaterials for water protection and monitoring. Chem Soc Rev 46:6946–7020
4. Pan C-Y, Xu G-R, Xu K, Zhao H-L, Wu Y-Q, Su H-C, et al. (2019) Electrospun nanofibrous membranes in membrane distillation: recent developments and future perspectives. Sep Purifn Technol 2019:221:44–63
5. Xu G-R, Xu J-M, Su H-C, Liu X-Y, Zhao H-L, Feng H-J et al (2019) Two-dimensional (2D) nanoporous membranes with sub-nanopores in reverse osmosis desalination: Latest developments and future directions. Desalination 451:18–34
6. Bhimanapati GR, Lin Z, Meunier V, Jung Y, Cha J, Das S et al (2015) Recent advances in two-dimensional materials beyond graphene. ACS Nano 9:11509–11539
7. Wee B-H, Wu T-F, Hong J-D (2017) Facile and scalable synthesis method for high-quality few-layer graphene through solution-based exfoliation of graphite. ACS Appl Mater Inter 9:4548–4557
8. Allen MJ, Tung VC, Kaner RB (2010) Honeycomb carbon: a review of graphene. Chem l Rev 110:132–45
9. Novoselov KS, Geim A (2007) The rise of graphene. Nat Mater 6:183–191
10. Das R (2017) Advanced membrane materials for desalination: carbon nanotube and graphene. In Inorganic Pollutants in Wastewater: methods of analysis, removal and treatment 16:322
11. You Y, Sahajwalla V, Yoshimura M, Joshi RK (2016) Graphene and graphene oxide for desalination. Nanoscale 8:117–119
12. Thebo KH, Qian X, Zhang Q, Chen L, Cheng H-M, Ren W (2018) Highly stable graphene-oxide-based membranes with superior permeability. Nat Commun 9:1–8
13. Chen C, Wang J, Liu D, Yang C, Liu Y, Ruoff RS et al (2018) Functionalized boron nitride membranes with ultrafast solvent transport performance for molecular separation. Nature Commun 9:1–8
14. Sainsbury T, Satti A, May P, Wang Z, McGovern I, Gun'ko YK et al (2012) Oxygen radical functionalization of boron nitride nanosheets. J Am Chem Soc 134:18758-71
15. Jw S, Yw J, Sw P, Nah H, Moon T, Park B et al (2007) Two-dimensional nanosheet crystals. Ange Chem Inter Edt 46:8828–8831
16. Wu Z, Fang B, Bonakdarpour A, Sun A, Wilkinson DP, Wang D (2012) WS2 nanosheets as a highly efficient electrocatalyst for hydrogen evolution reaction. Appl Catal B: Environ 125:59–66
17. Heiranian M, Farimani AB, Aluru NR (2015) Water desalination with a single-layer MoS 2 nanopore. Nat. Commun. 6:1–6

18. Zhou H-C, Long JR, Yaghi OM (2012) Introduction to metal–organic frameworks. ACS Publications
19. Liu J, Yu H, Wang L, Deng Z, Nazir A, Haq F (2018) Two-dimensional metal-organic frameworks nanosheets: synthesis strategies and applications. Inorg Chim Acta 483:550–564
20. Amo-Ochoa P, Welte L, González-Prieto R, Miguel PJS, Gómez-García CJ, Mateo-Martí E et al (2010) Single layers of a multifunctional laminar Cu (I, II) coordination polymer. Chem Commun 46:3262–3264
21. Jamali SH, Vlugt TJ, Lin L-C (2017) Atomistic understanding of zeolite nanosheets for water desalination. J Phys Chem C. 2017;121:11273–80
22. Roth WJ, Nachtigall P, Morris RE, Cejka J (2014) Two-dimensional zeolites: current status and perspectives. Chem Rev 114:4807–4837
23. Naguib M, Kurtoglu M, Presser V, Lu J, Niu J, Heon M et al (2011) Two-dimensional nanocrystals produced by exfoliation of Ti3AlC2. Adv Mater 23:4248–4253
24. Ding L, Wei Y, Wang Y, Chen H, Caro J, Wang H (2017) A two-dimensional lamellar membrane: MXene nanosheet stacks. Ang Chem 129:1851–1855
25. Li M, Lu J, Luo K, Li Y, Chang K, Chen K et al (2019) Element replacement approach by reaction with Lewis acidic molten salts to synthesize nanolaminated MAX phases and MXenes. J Am Chem Soc 141:4730–4737

Emerging Water Pollutants and Wastewater Treatments

Adejumoke Abosede Inyinbor, Olugbenga Solomon Bello, Oluwasogo Adewumi Dada, and Toyin Adedayo Oreofe

Abstract The environment is continuously threatened with various kind of pollutants, which have detrimental effects not only on human health, but also on ecosystems. In general, the water body is an easy route to wastewater disposal where many contaminants of emerging concerns, such as pharmaceuticals and personal care products etc. are present. Very recently, other classes of novel water pollutants occur which include nanoplastics and nanoparticles. Here, we discuss both classical and newly emerging water pollutants categories with their possible consequences. In addition, treatments and tackling techniques of these water pollutants are summarized. Ten different areas of concern to solve these problems and challenges of emerging water pollutants are highlighted.

Keywords Emerging contaminants · Classical water pollutants · Nanopollutants · Nanoplastics · Nanoparticles

1 Introduction

Water pollution is one class of other types of pollution that has attracted so much attention. The importance of water to humanity, animals and plants can never be underestimated, and water pollution threatens clean water availability. Water pollution can occur as a result of the release of waste product and contaminants into surface water, runoff into the river, drainage systems leaching into groundwater, liquid spills, eutrophication, littering and wastewater discharges into water bodies.

A. A. Inyinbor (✉) · O. S. Bello · O. A. Dada
Department of Physical Sciences, Landmark University, P.M.B 1001 Omu Aran, Nigeria
e-mail: inyinbor.adejumoke@landmarkuniversity.edu.ng

O. S. Bello
Department of Pure and Applied Chemistry, Ladoke Akintola University of Technology, P.M.B 4000 Ogbomosho, Nigeria

T. A. Oreofe
Department of Chemical Engineering, Landmark University, P.M.B 1001 Omu Aran, Nigeria

R. Das (ed.), *Two-Dimensional (2D) Nanomaterials in Separation Science*,
Springer Series on Polymer and Composite Materials,
https://doi.org/10.1007/978-3-030-72457-3_2

Water quality deterioration is not the only result obtained from water pollution, but this act also threatens human health and the balance of aquatic ecosystems, economic development and social prosperity [1].

Pollutants threatening the water sanctity are of a wide range. These include heavy metals and dyes of many kinds, domestic wastes, such as oil, salts, soaps, detergents, surfactants, nutrients, organic matter and chemical oxygen demand that is able to produce energy as high as 130 kJ/g-COD [2]. Wastewater from various industrial activities also contain pollutants of a wide range. Since water bodies are considered a convenient route of wastewater disposal, their accompany pollutants becomes a threat to water bodies. Indiscriminate disposal of solid wastes, as well as runoff from agricultural activities, contribute serious pollution to the water environment. Discharges of grey and/or grey water coupled with sewage disposal into water bodies also create great threats to the water environment. These threats from various quarters have left about 67% of the world population in water scarcity at least once in a month while 1.6 billion people currently face economic water shortage [3, 4].

In recent times, another category of pollutants classed the contaminants of emerging concerns (CECs) have become a subject of concerns to environmentalists. CECs may have been released into the environment over a period of time without traces or notice until new technologies for their determination were developed. Other categories of CECs were recently identified as threatening to human and the ecosystem [5]. Many CECs are also biodegradation products of some chemicals. CECs include pharmaceutical and personal care product wastes (PPCPs) such as antibiotics, antihypertensives, analgesics, anticonvulsant, and antidepressant amongst others; Endocrine disruptive Chemicals (ECs), such as hormones, flame retardants, artificial sweetener, plasticizers and insects' repellants [6].

CECs have been established as a great threat to the aquatic environment and human health. Their high toxicity, non-degradation characteristics and sneaky behavior make them a serious risk in the environment. Pharmaceuticals, such as antibiotics that have very wide usage and are not completely absorbed in the body may enter the water environment through various routes (Fig. 1). Their impact in the water environment ranges from accumulation in water and sediment through bioaccumulation in aquatic organisms to antibiotics resistant bacteria and gene breeding. High production and usage of various CECs result in the continuous and ready release of the same into water bodies [7].

Categories of other pollutants, such as the classical water pollutants, nanoparticles and nanoplastics amongst others, have also been of great threat to water availability. Hence, in addition to discussing selected pollutants types, this chapter highlights their associated risks, methods of screening and tackling various pollutants.

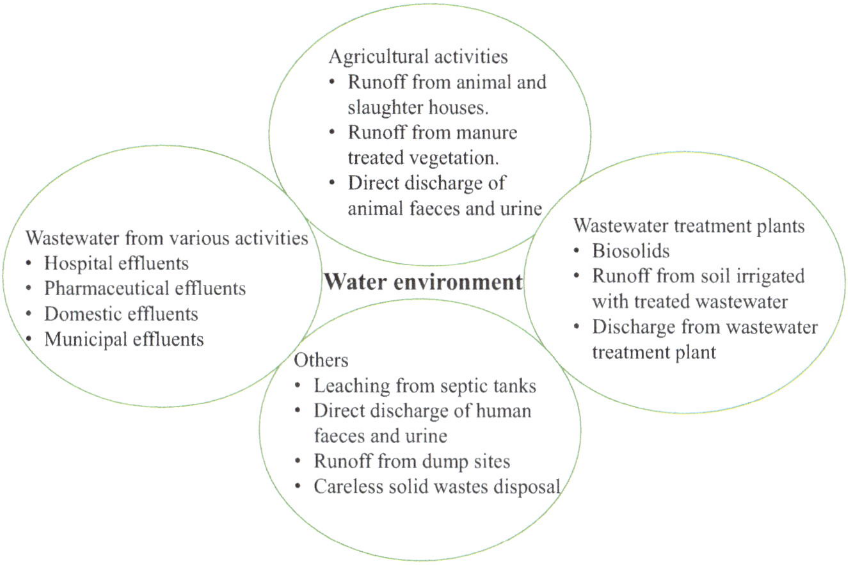

Fig. 1 Possible source for antibiotics in the water environment

2 Classical Water Pollutants

Pollutants are agents of pollution. The range of pollutants entering the water body is enormous and these come from various sources. A broad division is the biological pollutants and the chemical pollutants. Classical chemical pollutants have been grouped into two distinct categories; the macro pollutants which occur in very minute concentration and the natural organic pollutants [8]. Here, common water pollutants will be discussed with a focus on nutrients and inorganic pollutants.

2.1 Nutrient as Water Pollutants: Sources, Effects and Clean Up

Ecological systems greatly depend on nutrients, such as nitrogen (N) and phosphorus (P) for healthy survival. However, great deterioration has been seen due to the excessive discharge of nutrients into water bodies [9]. A high concentration of nutrients may be discharged into water bodies from identifiable or discrete sources, such as animal houses and farmyards or municipal wastewater treatment plants. Other sources discharging low concentrations of nutrient into water bodies may be forest plantations, pastures and arable lands [10]. Another significant source of nutrient into water bodies is from desert dust. Dust from deserts travels long distance and are mostly deposited on water shores hence enriching water bodies with nutrients [11].

Phytoplankton bloom is one way in which nutrients negatively affects water bodies. This renders water unfit as it causes turbidity, lack of transparency, large thick scum spread on the water surface, and releases toxins into the water. All of these threaten water quality as well as a public health [12]. Phytoplankton is known to release biotoxins into the water while its high reproduction causes serious algae bloom. The negative impact of algae bloom includes huge dissolved oxygen consumption, hence choking and deaths of aquatic organisms [13]. Algae bloom also threats freshwater availability due to unpleasant color and odors impacted on water bodies [12]. Removal of nutrients is therefore expedient to ensure water safety and security.

Many techniques of nutrients control cum algae bloom control have been in practice. A seaweed whose growth thrives on nutrients and sunlight competes with algae and phytoplankton for their basic growth needs, hence an effective control measure for algae bloom cum nutrients removal. In addition, seaweeds also have the tendency to release algae growth-inhibiting chemicals as well as shield off sunlight from algae and phytoplankton thus controlling their growth [14]. Algae bloom may be controlled via certain physical and chemical parameters, such as pH and temperature control, transparency and illumination control as well as nutrients control. Natural occurrences including turbulence [15] and climate change [16], as well as the deliberate diversion of water [17] amongst others, serve as control measures for an algae bloom. While removal of nutrient from water is found economically unfriendly, phytoremediation serves as a cheap emerging alternative [18].

2.2 *Inorganic Pollutants: The Heavy Metal Menace*

Inorganic pollutant types are wide, which range from acids through bases and to heavy metals. They also include oxyanions, halides, cations as well as radioactive materials, some of which occur in natural water in low concentration. However, various anthropogenic activities have left water bodies with a high concentration of inorganic pollutants [19]. Nutrients which have been detailed in Sect. 2.1 are also generally grouped with inorganic pollutants. In this section, we will focus on heavy metal as inorganic pollutant representative.

Heavy metals have become a global challenge as it threatens all components of the environment. Although the minute concentration of heavy metals exists in the natural environment, various metal-based industries, as well as metal utilizing activities, have continued to increase heavy metal concentration in the environment [20]. Here, we make a study around some heavy metals that are also classed priority pollutants, such as arsenic, cadmium, lead and mercury.

(a) **Arsenic**

Arsenic which is commonly found associated with ores becomes a great challenge to the water environment globally [21]. Arsenic which is grouped as a class 1 carcinogen is found associated with more than 300 minerals. The great use of these arsenic-containing minerals in metallurgical processes has made arsenic a common water pollutant [22]. Arsenic carcinogenic activities depend on duration and dosage; it as well focuses on different tissues and organs of the body including the lungs, skin and bladder. Arsenic has been reported in varying concentrations in drinking water in the United State [23]. Physicochemical processes, such as adsorption and ion exchange as well as bioremediation have been used for arsenic removal from the environment [24].

(b) **Cadmium**

Cadmium enters the environment through natural sources, such as phosphate rocks while numerous anthropogenic activities basically release cadmium into water bodies. They are highly toxic, persist in the environment and are referred to as probable human carcinogens [25]. Cadmium attacks the bones, liver, renal systems and lungs, causing shortness of breath and hypertension amidst other diseases. Major human exposure route to cadmium has been traced to food irrigated with wastewater followed by a nonfood material (tobacco). Cadmium accumulation is higher in women than in men, this is closely associated with hemoglobin deficiency [26]. The concentration of cadmium has been set to be less than or equal to 3.0 $\mu g/L$ in water and less than or equal to 5.0 $\mu g/L$ in wastewater to be discharged in the environment [27, 28]. Various conventional methods, such as precipitation, ion exchange, coagulation, filtration, solvent extraction and adsorption have been used for cadmium removal from water. Adsorption is a preferred technique because of its simplicity of operation.

(c) **Lead**

The major exposure route to lead has been from leaded paints, although leaded gasoline and industrial release of lead are also prominent. Danger associated with lead exposure is enormous; this is aggravated in children as it damages cognitive and behavioral development [29]. High levels of lead exposure experienced in children have been recently traced to lead presence in water. This is attributed to water distribution channel lines constructed with leaded materials [30]. Lead poisoning may result in diverse neurological disorders, and long-term exposure can form a basis for brain tumor. Adsorption has been reported as a preferred method of lead removal compared with other conventional water treatment techniques [31].

(d) **Mercury**

Mercury does not exist with gold in the raw ore, but that is extensively used in gold extraction. The highest level of mercury pollution has been reported to emanate from artisanal gold mining [32, 33]. Other sources of mercury pollution are cement production, waste disposal, coal combustion, chemical synthesis, ferrous and nonferrous metal manufacturing as well as chlor-alkali industries [34]. Mercury is known to affect the cardiovascular system, renal system as well as causes damage to the central nervous system. Children and pregnant women are highly prone to the effects of mercury amongst ages and categories. The prominent impact of mercury in pregnancy may result in underweight babies as well as lethargic growth and development in children [35]. The major ways of mercury removal are adsorption of Hg (II) and reduction of Hg (II) to Hg (0) which subsequently volatilizes into the air [36].

3 New Anthropogenic Water Pollutants: Focus on Nanoparticles and Nanoplastics

Water is an essential component of human life. All metabolisms in human and aquatic organisms are water dependent. Among the 17 sustainable development goals (SDG), access to clean and safe water has been signposted as *sine qua non*. However, one of the main sources of water pollution leading to a barrier to accessibility and availability of safe water has been accredited to several unregulated and unguided anthropogenic activities [37]. This section seeks to explore the new anthropogenic water pollutants majoring in nanoparticles and nanoplastics.

3.1 Nanotechnology and Nanoparticles

Nanotechnology derived its origin from different Science and Engineering fields where new materials are structured by molecular alteration [38, 39]. Nanoparticles (NPs) are materials structured with a size range of 1–100 nm, possessing at least one dimension within that size. The two main approaches involved in the synthesis of several categories of NPs are 'Top-down and Bottom-up'. The bottom-up approach is applied by building materials by the principle of molecular recognition, while top-down is a destructive approach that involves breaking down larger entities into the micro size and later to nanosize [40–42]. A major advantage of NPs among other materials lies on their sizes. Based on morphology, physical and chemical properties, NPs are broadly classified into natural and synthetic forms. The natural NPs exist in the natural environment, including water bodies. They have their sources from (i) humic substances, biomolecules, and viruses (ii) volcanic ash, metal oxides/hydroxides, and clay minerals and (iii) microbial activities. On the other hand, the synthetic NPs are the product of different anthropogenic activities and they are broadly classified into 'Incidental and Engineered NPs'. Incidental NPs are accidentally released by various indirect anthropogenic activities, such as combustion and corrosion processes, automobile exhaust, industrial and mining wastes [43]. Engineered NPs originate from different anthropogenic activities intentionally tailored towards large production of materials, such as cosmetics, computer chips and accessories, solar cells and home appliances. Based on their precursors, they are further categorized into six classes as follows and shown in Fig. 2.

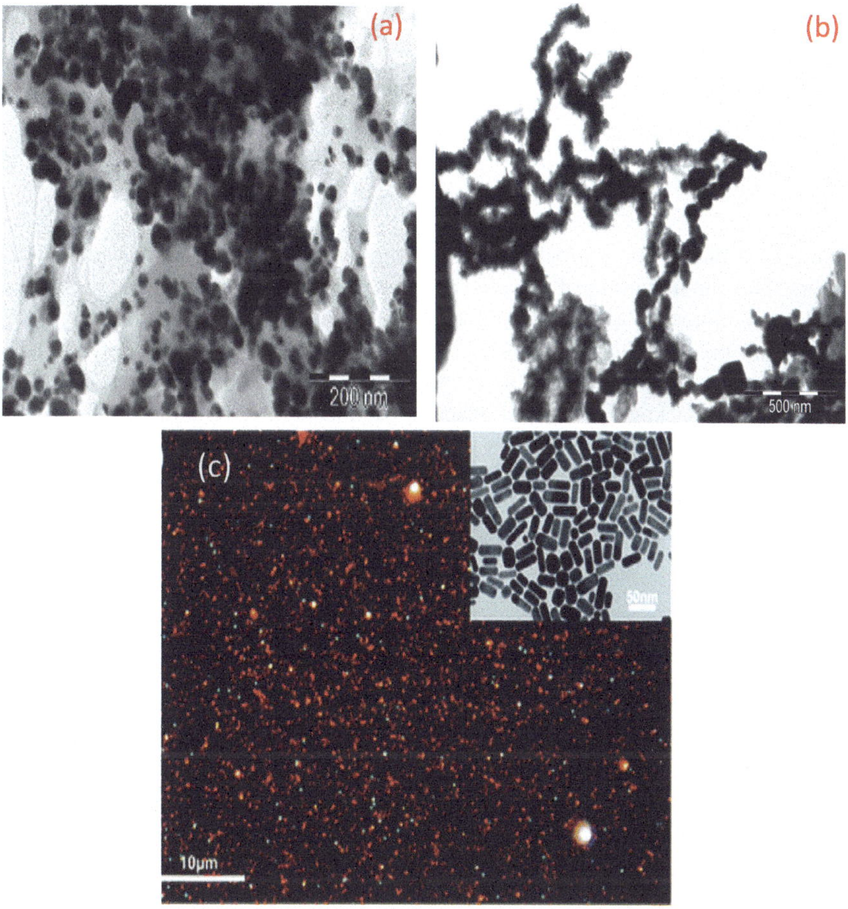

Fig. 2 **a** TEM image of AgNPs. Reproduced with permission from Elsevier [55, 56], **b** TEM image of CS-nZVI [57] and **c** TEM of Au NPs. Reproduced with permission from the Royal Society of Chemistry [58]

- Carbon-based (fullerene and carbon nanotube etc.) [44, 45].
- Metallic and metal oxide NPs characterized with good surface plasmon resource (SPR), magnetic and optoelectrical properties with distinctive micrographs.
- Ceramic NPs formed from a mixture of metals and metalloid oxides with good relevance in imaging and photocatalysis based on unique properties [46, 47].
- Polymeric NPs are those made of different polymeric organic molecules. They could be nano-capsules or nanosphere depending on the synthetic method. They find wide application in drug development [48].
- Liquid-based NPs are organic-based containing liquid moieties and mostly spherical in nature with a vast biomedical application as a result of size advantage (10–1000 nm) [49–51].

- Semiconductor NPs are nanomaterials with quantum dot effect, luminescence and optoelectronic properties. Due to their extremely small sizes (1–20 nm), they find applications in electronic and bio-sensing [49, 52–54].

3.2 *Plastics and Nanoplastics*

Plastics are man-made petrochemical products with vast applications in various sectors. The classifications of recycled and unrecycled plastics based on the Society of Plastics Industries (SPI) are polycarbonate (PC), polyvinyl chloride (PVC), low-density polyethylene (LDPE), polypropylene (PP), high-density polyethylene (HDPE), and polystyrene (PS). The relevance of plastics could be accessed in food and beverages, packaging, textile, consumer products, transportation, electrical and electronics, and industrial machinery [59]. Despite the global application and advantages derived from plastics, their indiscriminate disposal in the environment has been a global challenge for more than five decades. Presented in Fig. 3a–d are pictures of menace created by the indiscriminate disposal of plastics.

Fig. 3 Different landfill sites with plastics waste around the world [60]

Size is one of the major parameters for categorizing materials into macro and nanoscale. It is the fragmentation of larger plastics that led to the formation of micro and nanoplastics. Recent studies have shown that macro plastics are those with size greater than five millimeters (>5 mm), microplastics have a size range between one micro meter (1 μm) to five millimeters (5 mm), while nanoplastics have sizes less than one micro meter (0.1 μm). Nanoplastics are therefore defined as the product of polymer fragmentation that exhibits colloidal properties with the size range from between 1 and approximately (1 μm = 1000 nm) [61, 62]. Both nanoparticles (NPs) and nanoplastics (NaPs) derive their sources through several anthropogenic activities. They enter the water body via primary and secondary routes. The primary routes, as well as primary NaPs, are concerned with the intentionally produced products, such as pharmaceutical products, paints, and personal care products like cosmetics. On the other hand, secondary routes, as well as secondary NaPs, refer to the source that has to do with the degradation and fragmentation of larger particles. They successfully enter the water body by improper and unselective disposal of waste disintegrated by natural agents, such as the sun, waves and winds. Generally, both NPs and NaPs are released during production, uses and guided or unguided disposal of NPs containing products [48, 63–65].

NPs find relevance in environmental remediation and structuring of environmental sensors for pollutant detection [37, 66, 67], drug delivery and antibacterial applications [55, 68], engineering electronics and nanocrystalline materials [69, 70] energy generation from photoelectrochemical applications [71], and in mechanical industries for the building of coating, lubricant, and adhesives [55, 72, 73].

Despite the vast applications of NPs, their toxicity arising from their usage is hazardous to the water body. Figure 4 presents various routes of nanoplastics into the water environment. Similarly, NaPs toxicity arising from the post-consumer indiscriminate waste disposal has been reported. As earlier established, discharges of NaPs can be launched from microplastic and macroplastic because of fragmentation and degradation. Direct emission from merchandise and applications consists of electronics, magnetics, optoelectronics, waterborne paints, biomedical products, adhesives, coatings, and 3D—printing [74]. Both NPs and NaPs cause gastrointestinal disorders, liver, reproductive and neurotoxicity when they are ingested by human or aquatic organisms [75, 76]. They have the capacity to blend with heavy metals, polyaromatic hydrocarbon (PAH), and polychlorinated biphenyl (PCBs). NaPs have the capacity to transport contaminant. Using molecular simulation, [77], investigated nano-sized polystyrene reporting cellular functions could be affected when nano-sized polystyrene pervades into the lipid membranes. Bioaccumulation of NPs and NaPs affects plants, vertebrates and invertebrate aquatic organisms. It also alters the life cycles of bacteria and microbes [77–80].

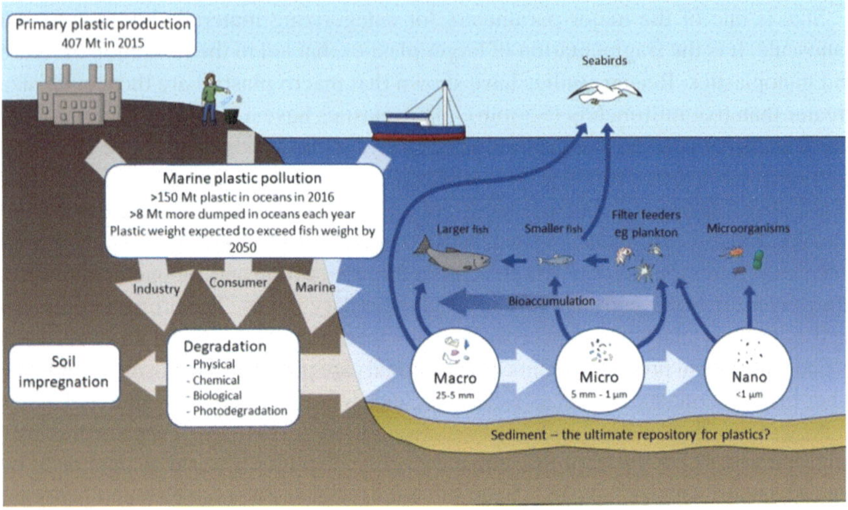

Fig. 4 Sources and route of the release of nanoplastics into the water body. Reproduced with permission from Elsevier [61]

4 Environmental Fate of Water Pollutants

Water is used on daily basis for a different purpose in human activities, such as industrial, agricultural and domestic activities amongst others [81]. Good quality water is required in achieving the desired objective of this specific activity. These processes, however, introduce substances known as pollutants either through point source (direct discharge) or non-point source (indirect discharge) into the water environment. Some of these sources include surface runoffs, oil spills, leaching into groundwater amongst others [82, 83]. These pollutants have characteristics of nature which reflect in their toxicological, bio-accumulative and persistency in the water environment [84]. The build-up of pollutants in water results in low water quality amongst others, such as deleterious human health, harmful effect on aquatic animals and formation of complexes in the ecosystem [85].

4.1 Industrial Water Pollutants

Industrialization and population growth in most economies had made the principal source of water pollution to be generated from the industries [86]. Industrial processes require the use of a large volume of water and subsequently lead to the discharge of a large volume of waste into the environment. The wastewater discharged from industries is also known as effluents. Industrial effluent is unique to each industry depending on the type of product produced. Hence, the variations of water pollutants in each water body and subsequently the variation in its effect on the environment [87]. Apart from industrial effluents, industrial waste can also exist in solid and gaseous form and these can also pollute the water bodies through rainfalls, leaching of the groundwater and surface runoffs. Some of these industrial water pollutants include personal care products, pesticides, fungicides, pharmaceuticals, phosphates, lead, surfactants, metals, caustic soda, asbestos, abrasives, mercury and cadmium amongst others [88–90].

4.2 Municipal Water Pollutants

Municipal water pollutants are related to waste generated from household or domestic activities. They could be found in solid, semisolid or liquid forms. Prior to the industrial revolution, domestic wastes are disposed by individual residents in a landfill site. At the emergence of industrialization, the mode of disposal changed to the use of vehicles and thus emission of CO_2 into the environment increased [91]. The constituents of municipal waste had changed drastically with industrialization, enormous use of personal care products and detergents had also contributed to the release of pollutants into water bodies [92]. Moora et al. [93] classified municipal wastes as domestic harmful materials (such as insecticides containers, medications, and shoe polish), inert waste substances (demolition, construction, debris, rocks, and dirts materials), bio-degradable constituents (green wastes, food, paper, wood) and fossil organic materials (plastics, metals, bottles glass). Municipal waste materials from landfills and incinerators are transported into water bodies through groundwater leaching, surface runoffs and rainfall.

4.3 Agricultural Waste Pollutants

The substances disposed into the environment through agricultural activities are termed agricultural wastes. Some of the agricultural production activities are seed planting, livestock breeding, meadowland, grazing land, dairy farming, horticulture and fruit plantation [94]. A number of waste are generated from this process some of which include products processing waste, rural domestic waste, materials eroded from farmlands, poultry manure, livestock manure, vegetation, inorganic mineral and salt from irrigation. It can also be classified as biodegradable, environmentally friendly and reproducible [95, 96]. The pollutants are disposed into the water bodies by these activities, thereby polluting the environment. Some of these pollutants are found in rivers, lakes and coastal oceans through irrigation and surface runoffs. A long-term accumulation of such pollutants leads to environmental hazards in plants, animals and humans. The severity of water pollution depends on the type of agricultural activity, soil type and groundwater level [97].

5 Exposure Risks and Biocompatibility

Pollutants persist or pseudo-persist in nature to degradation and continuous accumulation poise environmental risks [98]. Deterioration of water quality is not the only resultant effects of water pollutants, but the detrimental threats to human health, animal ecological system and plant habitats [99]. The interwoven interaction of the ecosystem through the food web-enhanced the environmental risk of these pollutants [100]. Hence, the probability of transference from plants to animals and to humans.

5.1 Effect on Human

Pollutants in water find its path into the food web through diverse means and these include diet and direct discharge into the water environment amongst others [101]. This is prominent in developing economies in that there is little or no awareness of the existence of these pollutants in water bodies. Such an economy either uses conventional water treatment techniques or no treatment at all before the discharge of water to the environment for use. The conventional treatment technique removes the large pollutants, while the trace level contaminants are still predominant in the water bodies. Accumulation and assimilation of these substances in the body system introduce different health challenges [102]. There is a wide range of health challenge traceable to ingestion of pollutants some of these includes: leukemia, cancer, brain tumor, dysmenorrhea, infertility and abortion amongst others [103, 104].

5.2 Effect on Plants

The effect of water pollutants on plants cannot be overlooked as both man and animal feeds on plants directly or indirectly through the food chain. The pathway of these pollutants is through irrigation, leaching and surface runoff [103]. However, plants uptake both vital and non-vital nutrients from the soil for its growth. Accumulations of the non-vital nutrients are detrimental to plants, humans and animals [105]. The ingestion of these pollutants via the plant root forms the basis of its transference to the food chain. Subsequently, accidental ingestion of unwashed raw plants contributes to human and animal exposure [106].

5.3 Effect on Animals

Animals are directly affected by water pollutants as it serves as habitat for aquatic animals and source of diet on plants for terrestrial animals. Water quality that aids

Fig. 5 Heap of dead fish found on the Atlantic shore. Adapted from https://www.bloomberg.com/news/articles/2020-05-15/heavy-metal-pollution-causes-mass-fish-deaths-in-nigeria

the growth and well-being of these animals reduce, and hence there is a pronounced reduction in the species richness and corresponding loss of habitat. Exposure of fish of different species to a class of water pollutants results in health effects, such as altered mating performance, sexual growth and the existence of intersex species [107]. This may result in aquatic isolation and population decline. Water pollution may also cause the death of aquatic organisms depending on pollutants concentration. A typical example of the effect of industrial pollution was reported in Nigeria. Figure 5 shows a heap of dead fish on the Atlantic shore. The laboratory analysis of some of the dead fishes revealed the presence of an unusual amount of metals traceable to discharge on the waterways.

6 What New Strategies Are Needed for Tackling New Wastewater Pollutants?

6.1 Reverse Osmosis

This is a technique of contaminant removal with the use of semi-permeable utilizing pressure as the physical parameter. This operation is generally carried out using pressure higher than the osmotic pressure of the dissolved contaminants. It is effective in removing molecular contaminants whose diameter or hydrated shells are larger than that of water. The technique makes use of an osmotic membrane made of cellulose in acetone whose pore size is quite small. This operation involves the separation of the sample solution from a solution of a high ionic strength using a membrane permeable to water alone. Propelling of water into the filtrate chamber with the aid of the pressure is needed in reverse osmosis-based filtration system [108]. Generally, different researchers have reported the use of membrane filtration in reverse osmosis for dyes, heavy metal ions, salt removal, treatment and recycling of textile wastewater, paper industries wastewater, secondary waste removal and recovering of protein from casein [109–112]. Reverse osmosis has found relevance in the purification of portable water contaminated with different salts, heavy metals, dyes and it is also gaining relevance in the industries. Reverse osmosis was used in the reclamation of municipal wastewater where secondary effluent was treated [113]. Effective sequestration of hazardous cationic and anionic pollutants has been reported using reverse osmosis and some functionalized adsorbents as fillers [114]. Al-Bastaki [115] also investigated the uptake of methyl orange dye together with Na_2SO_4 salt from artificial wastewater utilizing reverse osmosis [115]. Despite the application of the method, the disadvantage is that it is expensive to purify a large volume of water.

6.2 Electrodialysis

This is an electrochemical process involving the separation of salt from water by ion transfer. This technique is only effective in the removal of ionizable species, for instance, common salt (NaCl) which exists in solution as Na^+ and Cl^-. Some species, such as silica, which does not ionize are unaffected by electrodialysis, but can be removed by reverse osmosis. The knowledge of electrolysis plays out vividly in electrodialysis. Here, however, filters or membranes are used interspaced and alternately between the electrodes. Positively charged filters will allow the free flow of negative ions while hindering cations. In the same manner, negatively charged filters allow the free flow of positives ions and hinders anions. Series of a compartment within the tank stores the separated ions. Hence, cells with increased salts concentration are in alternate arrangement with cells depleting the salts. Desalinated water then flows for collection. Electrodialysis is a good technique in brackish water treatment, but has not been employed in well explored for desalination of seawater for economic reasons [116].

6.3 Ion Exchange

Production of potable water from wastewater using ion exchange has been explored in recent times as an alternative to conventional treatment techniques. This process is simple and does not require energy for its operation [117]. The ion exchange attracts pollutants with the opposite charge for ease of removal and treatment of wastewater. For instance, an anion pollutant attracts a cationic ion exchange material. Various natural and synthetic materials can be used to exchange ions thus softening hard waters and treatment of polluted water [118]. Biomasses are a form of natural ion exchanger, biodegradable in nature and originate from animal residues, agricultural-based residue, microbes and wood amongst others. Biomasses are readily available and easy of disposal due to its high level of biodegradability. Clays with high iron and aluminum content are good ion exchange adsorbent. It can be easily acquired, and it is characterized by good hydrodynamic properties [119]. Resins, one of the foremost ion exchangers used in wastewater treatment is polymeric in nature with diverse functional groups that attracts ions from the solution. Resins are either cationic or anionic in nature with a relatively high recovery efficiency rate. It is simple to operate; it can be recycled and reused. Resins are non-bio-degradable, and commercial resins are very costly; this is a major drawback for its use in ion exchange [120].

6.4 Catalytic Process

The catalytic process is one of the advanced treatment techniques employed for wastewater treatment. It operates through the improvement of the advanced oxidation process with a catalyst. There are two different classes of catalysts employed for wastewater treatment which include heterogeneous and homogeneous catalysts [121]. However, the regeneration and reusability of catalysts are paramount due to its high cost. These characteristics make heterogeneous catalysts more viable than homogeneous catalysts due to ease of separation from reaction mixture [122]. Also, it attracts more attention due to its high removal efficiency rate, high oxidative capacity, no intermediate pollutants and strong mineralization rate [123]. Some of these catalytic processes include catalytic ozonation, photocatalyst, catalytic wet air oxidation, etc.

In the catalytic ozonation process, hydroxyl radicals (OH) are formed through an interfacial reaction mechanism that reacts with a variety of organic compounds in wastewater. Degradation of micropollutants is enhanced with processes established on the utilization of hydroxyl radicals [124]. Ozone conversion into hydroxyl radicals through a series of chain reactions is enhanced with the catalyst materials which in turn attracts the ozone molecules [122]. Also, larger toxic organic compounds in wastewater are degraded with catalytic wet air oxidation technique. Air/oxygen over a catalyst oxidized these larger compounds either completely into carbon dioxide and water or partially into smaller weight organic compounds. These smaller-weight organic compounds are further mineralized into carbon dioxide and water but required more energy to achieve this [125] Catalyst materials include Fe-based catalyst [126], horseradish/H_2O_2 [127], ZNPO molecular sieves [128] among others.

The main objective of the catalytic process is to reduce the toxicity of wastewater effluent either by conversion to biodegradable intermediate or complete conversion to carbon dioxide and water.

6.5 Adsorption

Adsorption is an aspect of surface chemistry that involves the sequestration of contaminants (adsorbate) by a surface or interface of the adsorbent. It finds applications in the following areas: removal and separation of substances from gas and solution phases; purification of impurities from the pharmaceutical and chemical products; removal of metals in its ionic form from solution; purification of the process effluent for the control of water pollution; removal of odor, taste, and color from municipal water supplies and removal of impurities from oil, dry-cleaning solvents [129–131]. Its simplicity and easy operations have given it much attraction [132].

On the basis of the types of the interface between the adsorbate and adsorbents, it can be classified into physisorption and chemisorption. Physisorption involves heterogeneous and reversible adsorption mechanism characterized by weak van der Waal interphase physical forces, a non-specific multilayered process taking place all over the adsorbent and it is dependent on the surface area, temperature pressure and adsorbate nature with a lower energy of activation between 20 and 40 kJ mol^{-1}. Whereas chemisorption is a homogeneous and irreversible adsorption mechanism characterized by strong chemical forces between adsorbate-adsorbent interphase, monolayer in nature with high specificity process occurring at the center of the adsorbent and high activation energy between 40–400 kJ mol^{-1}. Chemisorption is also influenced by the nature of the adsorbate, temperature, and surface area [133–135]. Two major types of adsorption techniques are batch and dynamics (Column). Patel [136] reported five types of adsorption techniques employed for contaminants removal. The first is batch in nature and the remaining four are dynamics involving the use of column. They are also generally referred to as continuous fixed bed, moving bed, fluidized, and pulsed bed. Of all these five techniques, the most explored by researchers based on ease of operation, cost effectiveness, efficiency and amount of adsorbent used is batch technique [136]. There are seven generally operational factors influencing adsorption, namely pH, concentration, contact time, temperature, adsorbent dosage, stirring speed and ionic strength.

A number of researchers have worked on various categories of adsorbents classified as (i) natural adsorbents e.g., agro wastes, plant residues, biological adsorbents and other eco-friendly materials used directly with modification (ii) synthetic adsorbents, such as porous materials like nanomaterials, metal-organic frameworks (MOFs), and other artificial adsorbents that are polymeric in nature, (iii) semi-synthetic adsorbents: these are prepared by modification of natural adsorbents and incorporation of functional materials to enhance their performances [137, 138].

Adsorption is found very effective for the removal of categories of pollutants, such as heavy metals, dyes, pharmaceuticals and other organic pollutants [139–144]. The expensive commercial activated carbon has however made adsorption techniques economically unfriendly. Focus on economically friendly adsorbents as an alternative to commercial activated carbon has been explored in recent times [145–151].

7 Conventional Wastewater Treatment Processes

Many wastewater treatments basically use a combination of biological, chemical, and physical, processes. In addition, the removal of solid wastes, such as inorganic and organic solids and sometimes essential element from wastewater may be included. Listing the different forms of available treatment in the order in which they are executed: preliminary, primary, secondary and tertiary otherwise referred to as advanced wastewater treatment [152]. Furthermore, a disinfection method may be included in order to make the wastewater free from some pathogenic microbes such as bacteria and viruses.

(a) **Preliminary treatment**

This primarily involves the uptake of large particle and coarse solids materials that may be present in wastewater or discharged from industries. This is necessary so as to enhance the efficiency of the subsequent water treatment units and to avoid blockage of pipes of some of the equipment. The preliminary treatment process can be grouped into the following: coarse screening, grit removal, commination of large objects. Preliminary treatment is relatively cheap to operate, that is, manually cleared screens to require either little or no maintenance of equipment. The automated cleared screen requires a lower workforce than manually cleared screen. Similarly, it is easy to operate. Grit removal reduces damages and downstream processes of maintenance. Screening, handling and disposal are eliminated which may help in reducing odors, flies associated with screening. However, a manually cleaned screen requires frequent scraping in order to prevent clogging blocking and elevated backwater level that causes accumulation of a solid on the surface. It is relatively costly to operate. The efficiency of solid capturing by downstream units may be reduced. High equipment maintenance costs. There may be the accumulation of rag on air diffuse. There may be an increase in the plastic build-up in the digestion tank. During the digestion process, solids from comminutors and grinders will not decompose.

Note: Coarse Screening is seldom supplemented by the use of comminutors. This usually reduces the particles with large size to a smaller size. The size reduction will enhance the removal of the particles in other treatment processes in sludge form. Treatment of wastewater at the preliminary stage often requires the inclusion of devices measuring flow; usually, standing-wave flumes are always included at the preliminary treatment stage.

(b) **Primary treatment**

This treatment involves the use of sedimentation and skimming process for the uptake of organic and inorganic solid that is capable of settling. The percentages of materials removed during primary treatment are: suspended solids (SS): 55–75%, biochemical oxygen demand (BOD): 25–50% fat and oil: 60–65%. During these stages, it should be noted that some organic compounds containing phosphorus, nitrogen, and heavy metals connected with solids wastes are also sequestrated in the process of primary sedimentation. This treatment doesn't interfere with the constituents that are dissolved and those that are colloidal in nature [144]. The effluent that is generated as a result of primary sedimentation is known as the primary effluent. The least wastewater irrigation pre-treatment required is the primary treatment. The level of treatment depends on the objectives of the need. Primary treatment may be described as sufficient for wastewater remediation if the water is intended for use in irrigation of crops not consumable by man or for wetting vineyards. Secondary treatments are required in countries where problems associated with food storage are usually avoided. This is applicable to non-food crop irrigation as well. Small portions of the primary effluent may be used for irrigation in cases where off-line storage is available.

Anaerobic digestion is a type of biological treatment that is usually adopted for primary sludge treatment in large sewage treatment plant (7600 m^3/d) in the United States. The volume of wastewater that requires ultimate disposal is reduced by the metabolic action of anaerobic and facultative bacteria on it [2]. This results in the stability of the sludge and improvement of its dewatering properties. Digestion usually takes place in anaerobic digesters. These are tanks with a cover that has its depth ranging from 7 to 14 m. The duration of wastewater in a digester takes about 10 days and 60 days for a high—rate digester and standard—rate digester, respectively. This enhances the mixing and heating of the wastewater. There is the possibility of generating gas during digestion. The gas generated usually has a methane content of about 60–65%. This can be harnessed as an energy source. Sludge can be processed in various ways in the small sewage. This includes land application, in-process storage, direct application to sludge drying beds, and aerobic digestion.

(c) **Secondary treatment**

Effluents are usually subjected to further treatments. The third phase of wastewater treatment is characterized by the secondary treatment process. This stage enhances the removal of suspended solids as well as organic matters. Secondary treatment of wastewater is usually subsequent to primary treatment. Removal of colloidal dissolved and biodegradable organic matters usually occur during this treatment phase. Biological treatment that occurs in the presence of oxygen is referred to as aerobic biological treatment. Wastewater is usually subjected to the action of aerobic microbes (majorly bacteria). The bacteria usually initiate the metabolism of the organic matter present in the wastewater [153]. Hence, inorganic end–products, such as CO_2, NH_3, and H_2O as well as other microorganisms are usually produced. Aerobic biological methods which are employed for secondary treatment of effluent differ in the transport of oxygen to the microorganisms and the rate organic matter is broken down by these organisms.

High-rate biological reactors possess relatively small volume reactors and highly concentrated microbes when compared with low-rate reactors. Therefore, the pace of growth of new organisms in systems of high rate is quite more due to adequate environmental control. Clarified secondary effluent is produced when there is a separation of the microorganisms from treated wastewater by sedimentation. Secondary clarifiers are the sedimentation tanks that are utilized in secondary effluent treatment. Their mode of operation is similar to that of the primary clarifiers. Solids that are biological in nature which is removed during secondary sedimentation are referred to as either biological sludge or secondary sludge. These are usually combined with primary sludge and are subjected to sludge processing.

(d) **Tertiary and/or advanced treatment (TWWT)**

This is the last stage of wastewater treatment. The TWWT is employed specifically to remove pollutants or contaminants which escaped during secondary treatment. Dissolved solids, heavy metals, refractory organics, suspended solids, phosphorus and nitrogen are necessary to be removed by specific treatment procedures. Advanced treatments are also called tertiary treatments as they are often preceded by high-rate secondary treatment. Tertiary treatment processes, however, are at times coupled with primary or secondary treatments (e.g., the addition of chemicals to primary clarifiers or aeration tanks for phosphorus removal) or used to substitute secondary treatment (e.g., primary effluent overland flow treatment).

Nitrogen content present in influents that is usually in their ammonia form is unaffected as they pass through the first two treatment phases. Near-complete nitrification of the nitrogenous component occurs in the third aerobic zone, converting them to nitrites and nitrates sequentially. The mixed liquor rich in nitrate generated from the aerobic zone is then recycled back to the initial anoxic zone. Nitrates recycled are denitrified anaerobically and nitrogen gas is generated (which completely escapes to the atmosphere) by facultative bacteria, which use influent organics hydrogen donors. Unrecycled nitrates are reduced by endogenous bacteria respiration in the second anoxic zone. Dissolved oxygen levels are raised again in the final re-aeration zone to avert subsequent denitrification that may affect settling in the secondary clarifier. To further prevent denitrification, the level of dissolved oxygen is raised in the final re-aeration zone; this would impair settling in the secondary clarifiers.

(e) **Disinfection**

The disinfectant usually employed for wastewater disinfection is chlorine. It is injected in form of a solution into the wastewater at the chlorine dosage tank. The tank is automated and discharges chlorine at a specific time interval. The dosage of chlorine to be used is dependent on wastewater strength and other factors, but common dosage ranges are between 5 and 20 mg/l. Other disinfectants, though not commonly used are ultraviolet (UV) irradiation and ozone. The chlorine dosage tanks are often automated rectangular channels with designs providing a contact time of about 25–30 min. However, the standard chlorine contact time of 1 h or more may be needed for target irrigation purposes of reclaimed effluents. Other factors which depend on chlorine disinfectants include acidity and alkalinity (pH), impact time, effluent temperature and lastly, organic contents.

8 Other Wastewater Treatment Techniques

Commonly available high rate processes include rotating biological contactors (RBC), activated sludge processes, oxidation ditches and trickling filters otherwise referred to as biofilters. Combining two processes in the serial order for instance; activated sludge subsequent to biofilters is usually utilized to treat community wastewater containing highly concentrated organic matters, which are of industrial origin.

(a) **Activated sludge**

This method typically employs the use of biological floc containing protozoans and bacteria for the treatment of industrial effluents or sewage. In this method, a mixed liquor of microorganism and wastewater suspension is contained in an aeration basin or tank which is the dispersed-growth reactor. The components in the aeration basin are thoroughly mixed by devices that also aerate (i.e. oxygen supplies) the biological suspension. Common devices used for aeration are surface mechanical aerators which input air by shaking the surface of the liquid and submerged diffusers which liberate compressed air. The holding time of hydraulics in aeration basins is between 3 and 8 h. which can be more with wastewaters containing high BODs. Sequel to the aeration phase, separation by sedimentation of microorganisms from liquid occurs and the secondary effluent is obtained as clarified. To have a sustained high level of mixed-liquor suspended solids (MLSS), recycling of aliquots of the biological sludge to the aeration tank is employed. The remaining portion of the sludge is taken off the process and transferred to sludge processing to ensure a consistent microorganism concentration within the system. A different variation of the fundamental activated sludge process, but similar in principle, such as oxidation ditches and extended aeration, are commonly employed.

(b) **Trickling filters**

A trickling filter is a type of treatment system for wastewater. In this system, sewage or wastewater in a downward delivery passes over a filter system consisting of sphagnum peat moss, polyurethane foam, slag, gravel, coke, fixed bed of rocks, ceramic, or plastic media, causing the growth of a layer of microbial slime (biofilm) which covers the media bed. Metabolism of organic components presents in the wastewater occurs as they diffuse into the biofilm. The film is usually supplied by oxygen as unhindered air flows across the media as determined by the wastewater temperature and atmospheric air. On rare occasions, blowers may be needed to supply forced air to the system. An increase in the growth of new organisms consequently leads to an increase in film thickness and at intervals, aliquots of the biofilm slide away from the media. The discarded substances are in the secondary clarifier separated from the liquid and deposited in the sludge processing unit. A portion of the secondary effluent (liquid clarified from the second clarifier) is usually recycled to the biofilter. This is to ensure better wastewater hydraulic spread over the filter.

(c) **Rotating Biological Contactors (RBC)**

This is a biological fixed-film reactor process employed in wastewater treatment. During RBC process, the wastewater is contacted with a biological film so as to remove contaminants in the wastewater. This is done before the remediated wastewater is discharged into water bodies, such as river, lakes and oceans etc. A rotating biological contactor consists of a series of parallel discs that are spaced closely and half-way submerged in the reactor containing flowing wastewater. Biological degradation of the effluents contaminants takes place on the surface of the disc.

It should be noted that 80–90% of the BODs, suspended solids SS and some of the high-density metals initially present in the raw wastewater are been removed during the biological treatment process coupled with primary sedimentation. A little higher quality effluent can be produced from the activated sludge process particularly when disinfection units are coupled with the system. This method gives substantial but incomplete virus and bacteria removal. However, dissolved minerals, non-biodegradable organics, nitrogen or phosphorus are removed in small quantities in the system.

(d) **Effluent storage**

Storage facilities are in most cases links between the irrigation system and the wastewater treatment plant. However, storage is of great need so as:

i. To establish the deviations in daily flow from the plants and to keep excess wastewater when the average flow of wastewater goes beyond irrigation demands (including winter storage).
ii. To keep up with the maximum irrigation demands when influence flow is moderate.
iii. To reduce the effect of interruptions in the irrigation system and treatment plant operations. Storage also provides extra time to address temporary water quality challenges and provides insurance against the chances of unsuitable reclaimed wastewater entering the irrigation system.
iv. To make wastewater available for advanced treatment. Nitrogen, oxygen demands, microorganisms, and suspended solids may reduce during storage.

9 Challenges

In an attempt to solve the problems and challenges of emerging water pollutants, there are needs to focus attention on the following areas:

i. Treating nano-pollutants has remained a big problem to the currently used wastewater treatment technologies.
ii. Water reclamation in form of reuse and recycling should be promoted both in the short and long term.
iii. Basic and fundamental standards of wastewater reclamation should be established; these will include marking/classification of wastewater and the purpose and benefits of re-use.
iv. An established basis for large-scale wastewater reclamation should be put in place when wastewater treatment plants are established.
v. Treatment plants for wastewater in rural areas are both scarce and hardly in existence. This should be given priority attention to prevent water-borne diseases.
vi. In rural areas, facilities for decentralized wastewater treatment should be put in place; this will bring about resource/energy recovery from sludge generated.

vii. Since operational variables are dependent on the type of wastewater generated, modeling and optimization techniques can be used for the simulation of these variables.
viii. If properly harnessed, the evolution of electrocoagulation and hybrid integration coupled with other treatment processes will lead to the sustainable replacement of the conventional treatment systems.
ix. Economic feasibility, parameter optimization and system design should be investigated on wastewater treatment processes.
x. Further studies are needed to develop new low-cost and highly efficient modification technology for activated carbon preparation.
xi. A pragmatic increase in the practical application of locally prepared adsorbents for wastewater treatment, especially in industrial wastewater and municipal wastewater treatment should be encouraged. This will further enhance the use of locally prepared adsorbents in the removal of heavy metals, dyes, emerging contaminants and other pollutants.

10 Conclusions

In conclusion, various water pollution species continue to mount up in environmental components. Although some are beneficial at moderate concentration, however, their presence in high concentration is highly disastrous. Major environmental concern exists especially with the nanopollutants; their various effects on plants, human and the ecosystem are quite enormous. Harnessing the tool of effective wastewater and water treatment may protect the environment and humans hence our recommendations. Focus on friendly water treatment techniques and building the same into an economically advantageous system will be of global benefit.

References and Future Readings

1. Çağatay T, Almet A, Ali RD, Fatih U (2010) Assessment of surface water quality of the Ceyhan River basin, Turkey. Environ Monitor Assess 167:175–184
2. Inyinbor AA, Bello OS, Oluyori AP, Inyinbor HE, Fadiji AE (2019) Waste water conservation and reuse in quality vegetable cultivation: overview, challenges and future prospects. Food Control 98:489–500
3. FAO (2007) Coping with water scarcity challenge of the twenty-first century; Wold water day 22 March, 2007
4. Hasan K, Shahriar A, Jim KU (2019) Water pollution in Bangladesh and its impact on public health. Heliyon 5:e02145
5. Houtman CJ (2010) Emerging contaminants in surface waters and their relevance for the production of drinking water in Europe. J Integr Environ Sci 7(4):271–295
6. Huang Y, Dsikowitzky L, Yang F, Schwarzbauer J (2020) Emerging contaminants in municipal wastewaters and their relevance for the surface water contamination in the tropical coastal city Haikou, China; Estuarine. Coastal Shelf Sci 235

7. Khan AA, Ahmad R, Ahmad I (2020) Density functional theory study of emerging pollutants removal from water by covalent triazine based framework. J Mol Liq 309:113008
8. Schwarzenbach RP, Egli T, Hofstetter TB, Gunten U, Wehrli B (2010) Global water pollution and human health. Ann Rev Environ Resour 35:109–36
9. He R, Yang X, Gassman PW, Wang G, Yu C (2019) Spatiotemporal characterization of nutrient pollution source compositions in the Xiaohong River Basin China. Ecolog Indicator 107:105676
10. Callery O, Healy MG (2019) A novel method to rapidly assess the suitability of water treatment residual and crushed concrete for the mitigation of point and nonpoint source nutrient pollution; Resources, Conservation and Recycling X2:100010FAO, 2007. Coping with Water Scarcity: Challenge of the Twenty-First century. UNWater
11. Farahat A, Abuelgasim A (2019) Role of atmospheric nutrient pollution in stimulating phytoplankton growth in small area and shallow depth water bodies: Arabian Gulf and the sea of Oman. Atmos Environ 219:117045
12. Nwankwegu AS, Li Y, Huang Y, Wei J, Norgbey E, Ji D, Pu Y, Nuamah LA, Yang Z, Jiang Y, Paerl HW (2020) Nitrate repletion during spring bloom intensifies phytoplankton iron demand in Yangtze River tributary China. Environ Pollut 264:114626
13. Zhao CS, Shao NF, Yang ST, Ren H, Ge YR, Zhang ZS, Feng P, Liu WL (2019) Quantitative assessment of the effects of human activities on phytoplankton communities in lakes and reservoirs. Sci Total Environ 665:213–225
14. Jiang Z, Liu J, Li S, Chen Y, Du P, Zhu Y, Chen LQ, Shou L, Yan X, Zeng J, Chen J (2020) Kelp cultivation effectively improves water quality and regulates phytoplankton community in a turbid, highly eutrophic bay. Sci Total Environ 707:135561
15. Kang L, He L, Dai L, He Q, Ai H, Yang G, Liu M, Jiang W, Li H (2019) Interactions between suspended particulate matter and algal cells contributed to the reconstruction of phytoplankton communities in turbulent waters. Water Res 149:251–262
16. Wang J, Hoondert RPJ, Thunnissen NW, Meent D, Hendriks AJ (2020) Chemical fate of persistent organic pollutants in the arctic: evaluation of simple box. Sci Total Environ 720:137579
17. Dai J, Wu S, Wu X, Lv X, Sivakumar B, Wang F, Zhang Y, Yang Q, Gao A, Zhao Y, Yu L, Zhu S (2020) Impacts of a large river-to-lake water diversion project on lacustrine phytoplankton communities. J Hydrol 587:124938
18. Kumwimba MN, Dzakpasu M, Li, X (2020) Potential of invasive watermill foil (Myriophyllum spp.) to remediate eutrophic waterbodies with organic and inorganic pollutants. J Environ Manag 270:110919
19. Guleria A, Kumari G, Lima EC (2020) Cellulose-g-poly-(acrylamide-co-acrylic acid) Polymeric bioadsorbent for the removal of toxic inorganic pollutants from wastewaters. Carbohyd Polym 228:115396
20. Tchounwou PB, Yedjou CG, Patlolla AK, Sutton DJ (2012) Heavy metal toxicity and the environment. In: Luch A (eds) Molecular, clinical and environmental toxicology. Experientia Supplementum, 1 101. Springer, Basel. Supply pages
21. Long H, Zheng Y, Peng Y, Jin G, Deng W, Zhang S, He H (2020) Separation and recovery of arsenic and alkali products during the treatment of antimony smelting residues. Miner En 153:106379
22. Rong Z, Tang X, Wu L, Chen X, Dang W, Wang Y (2020) A novel method to synthesizescorodite using ferrihydrite and its role in removal and immobilization of arsenic. J Mater Res Technol 9(3):5848–5857
23. Nigra AE, Navas-Acien A (2020) Arsenic in US correctional facility drinking water, 2006–2011 Environ Res 188:109768
24. Sodhi KK, Kumar M, Agrawal PK, Singh DK (2019) Perspectives on arsenic toxicity, carcinogenicity and its systemic remediation strategies. Environ Technol Innov 16:
25. Huy DH, Seelen E, Liem-Nguyen V (2020) Removal mechanisms of cadmium and lead ions in contaminated water by stainless steel slag obtained from scrap metal recycling. J Water Process Eng 36:

26. Ali F, Kazi TG, Afridi HI, Baig JA (2018) Exposure of cadmium via smoking and drinking water on zinc levels of biological samples of malnutrition pregnant women: A prospective cohort study. Environ Toxicol Pharmacol 63:48–54
27. World Health Organization (WHO) (1998) Guidelines for Drinking Water Quality, Health Criteria and Other Supporting Information, vol. 2, 2nd ed., World Health Organization, Geneva
28. Environmental Protection Agency, EPA. Edition of the drinking water standards and health advisories, EPA 822-R-02–38, Washington, DC, 2002. http://www.epa.gov/waterscience/drinking/standards.html
29. Redmon JH, Levine KE, Aceituno AM, Litzenberger K, Gibson JM (2020) Lead in drinking water at North Carolina childcare centers: Piloting a citizen science-based testing strategy. Environ Res 183
30. Ravenscroft J, Roy A, Queirolo EI, Mañay N, Martínez G, Peregalli F, Kordas K (2018) Drinking water lead, iron and zinc concentrations as predictors of blood lead levels and urinary lead excretion in school children from Montevideo Uruguay. Chemosphere 212:694–704
31. Meng L, Li Z, Liu L, Chen X, Wu J, Li W, Zhang X, Dong M (2020) Lead removal from water by a newly isolated Geotrichum candidum LG-8 from Tibet kefir milk and its mechanism. Chemosphere 259
32. Gyamfi O, Sorenson PB, Darko G, Ansah E, Bak JL (2020) Human health risk assessment of exposure to indoor mercury vapour in a Ghanaian artisanal small-scale gold mining community. Chemosphere 241
33. Mantey J, Nyarko KB, Owusu-Nimo F, Awua KA, Bempah CK, Amankwah RK, Akatu WE, Appiah-Effah E (2020) Mercury contamination of soil and water media from different illegal artisanal small-scale gold mining operations (galamsey). Heliyon 6
34. Baimenov A, Berillo DA, Moustakas K, Inglezakis VJ (2020) Efficient removal of mercury (II) from water by use of cryogels and comparison to commercial adsorbents under environmentally relevant conditions. J Hazard Mater 399
35. Ottenbros IB, Boerleider RZ, Jubitana B, Roeleveld N, Scheepers PTJ (2019) Knowledge and awareness of health effects related to the use of mercury in artisanal and small-scale gold mining in Suriname. Environ Int 122:142–150
36. Wang L, Hou D, Cao Y, Ok YS, Tack FMG, Rinklebe J, O'Connor D (2020) Remediation of mercury contaminated soil, water, and air: a review of emerging materials and innovative technologies. Environ Int 134
37. Dada AO, Adekola FA, Odebunmi EO (2017) Liquid phase scavenging of Cd (II) and Cu (II) ions onto novel nanoscale zerovalent manganese (nZVMn): Equilibrium, kinetic and thermodynamic studies. Environmental Nanotechnology, Monitoring & Management. 8:63–72. https://doi.org/10.1016/j.enmm.2017.05.001
38. Prasad R (2019) Advanced Research in Nanosciences for Water Technology 341–362. https://doi.org/10.1007/978-3-030-02381-2
39. Dada AO, Adekola FA, Odebunmi EO (2015) Kinetics and equilibrium models for sorption of Cu(II) onto a novel manganese nano-adsorbent. J Dispersion Sci Technol 37(1):119–133
40. Wang Y, Xia Y (2004) Bottom-Up and top-down approaches to the synthesis of mono dispersed spherical colloids of low melting-point metals. Nano Lett 4(10):2047–2050
41. Karn B, Masciangioli T, Zhang W, Colvin V, Alivisatos P (eds) (2005) Nanotechnology and the Environment; Applications and Implications. American Chemical Society, Washington, DC
42. Buzea C, Pacheco II, Robbie K (2007) Nanomaterials and nanoparticles: Sources and toxicity; Biointerphases2, MR17. https://doi.org/10.1116/1.2815690
43. Madl AK, Pinkerton K (2009) Health effects of inhaled engineered and incidental nanoparticles Health effects of inhaled nanoparticles. Crit Rev Toxicol 39(8):629–658
44. Saeed K, Khan I (2013) Carbon nanotubes–properties and applications: a review. Carbon Lett 14(3):131–144
45. Subagia IDGA, Jiang Z, Tijing LD, Kim Y, Kim CS, Lim JK, Shon HK (2014) Hybrid multi-scale basalt fiber-epoxy composite laminate reinforced with Electrospun polyurethane nanofibers containing carbon nanotubes. Fibers Polymers 15:1295–1302. https://doi.org/10.1007/s12221-014-1295-4

46. Thomas S, Harshita BSP, Mishra P, Talegaonkar S (2015) Ceramic nanoparticles: fabrication methods and applications in drug delivery. Curr Pharm Des 21:6165–6188. https://doi.org/10.2174/1381612821666151027153246
47. Malakar A, Snow DD (2020) Nanoparticles as sources of inorganic water pollutants, Inorganic Pollutants in Water. INC. https://doi.org/10.1016/b978-0-12-818965-8.00017-2
48. Rana V, Sharma R (2019) Recent advances in development of nano drug delivery, applications of targeted nano drugs and delivery systems. Elsevier Inc. https://doi.org/10.1016/b978-0-12-814029-1.00005-3
49. Puri A, Loomis K, Smith B, Lee JH, Yavlovich A, Heldman E, Blumenthal R (2009) Lipid-based nanoparticles as pharmaceutical drug carriers: From concepts to clinic. Crit Rev Ther Drug Carrier Syst 26:523–580. https://doi.org/10.1615/CritRevTherDrugCarrierSyst.v26.i6.10
50. Gaiser HF, Popescu R, Gerthsen D, Feldmann C (2020) Ionic-liquid-based synthesis of GaN nanoparticles. Chem Commun 56:2312–2315
51. Ishtiaq M, Al-Rashida M, Alharthy RD, Hameed A (2020) Ionic liquid–based colloidal nanoparticles: applications in organic synthesis, Metal Nanoparticles for Drug Delivery and Diagnostic Applications. Elsevier Inc. https://doi.org/10.1016/b978-0-12-816960-5.00015-x
52. Dahman Y (2017) Nanoshells. Nanotechnol Funct Mater Eng, 175–190. https://doi.org/10.1016/b978-0-323-51256-5.00008-3
53. Nair AK, Mayeen A, Shaji LK, Kala MS, Thomas S, Kalarikkal N (2018) Optical characterization of nanomaterials, Characterization of Nanomaterials: advances and Key Technologies. Elsevier Ltd. https://doi.org/10.1016/B978-0-08-101973-3.00010-9
54. Sharma R, Raghavarao KSMS (2018) Nanoparticle-based Aptasensors for food contaminant detection. Nanomaterials for Food Applications. Elsevier Inc. https://doi.org/10.1016/B978-0-12-814130-4.00006-3
55. Dada AO, Adekola FA, Dada FE, Adelani-Akande AT, Bello MO, Okonkwo CR, Inyinbor AA, Oluyori AP, Olayanju A, Ajanaku KO, Adetunji CO (2019) Silver nanoparticle synthesis by Acalypha wilkesiana extract: phytochemical screening, characterization, influence of operational parameters, and preliminary antibacterial testing. Heliyon. https://doi.org/10.1016/j.heliyon.2019.e02517
56. Dada AO, Adekola FA, Adeyemi OS, Bello OM, Oluwaseun AC, Awakan OJ, Grace F-AA (2018) Exploring the effect of operational factors and characterization imperative to the synthesis of silver nanoparticles. Silver Nanopart Fabr Characte App. https://doi.org/10.5772/intechopen.76947
57. Dada AO, Adekola FA, Odebunmi EO, Dada FE, Bello OS, Ogunlaja A (2020) Bottom-up approach synthesis of core shell nanoscale Zerovalent Iron (CS-nZVI): physicochemical and Spectroscopic Characterization with Cu(II) ions adsorption application. Methods X 100976. https://doi.org/10.1016/j.mex.2020.100976
58. Dreaden EC, Alkilany AM, Huang X, Murphy CJ, El-Sayed MA (2012) The golden age: gold nanoparticles for biomedicine. Chem Soc Rev 41:2740–2779
59. Society T, Industry P (1988) Different types of plastics and their classification. Available on https://www.ryedale.gov.uk/attachments/article/690/Different_plastic_polymer_types.pdf. Accessed 20 July, 2020
60. Uwaegbulam C, Nwannekanma B, Gbonegun V (2018) Producers 'responsibility and plastic pollution crisis, Environment. The Guardian Nigeria News
61. Gangadoo S, Owen S, Rajapaksha P, Plaisted K, Cheeseman S, Haddara H, Truong VK, Ngo ST, Vu VV, Cozzolino D, Elbourne A, Crawford R, Latham K, Chapman J (2020) Nanoplastics and their analytical characterisation and fate in the marine environment: From source to sea. Sci Total Environ 732:138792
62. Wagner S, Reemtsma T (2019) Things we know and don't know about nanoplastics in the environment. Nat Nanotechnol 14:300–301. https://doi.org/10.1038/s41565-019-0424-z
63. Smita S, Gupta SK, Bartonova A, Dusinska M, Gutleb AC, Rahman Q (2012) Nanoparticles in the environment: Assessment using the causal diagram approach. Environ Health 11:S13

64. Bundschuh M, Filser J, Lüderwald S, McKee MS, Metreveli G, Schaumann GE, Schulz R, Wagner S (2018) Nanoparticles in the environment: where do we come from, where do we go to? Environ Sci Europe 30. https://doi.org/10.1186/s12302-018-0132-6
65. Mattsson K, Jocic S, Doverbratt I, Hansson LA (2018) Nanoplastics in the aquatic environment, Microplastic contamination in aquatic environments: an emerging matter of environmental urgency. Elsevier Inc. https://doi.org/10.1016/B978-0-12-813747-5.00013-8
66. Willner MR, Vikesland PJ (2018) Nanomaterial enabled sensors for environmental contaminants Prof Ueli Aebi, Prof Peter Gehr. J Nanobiotechnolo 16:1–16. https://doi.org/10.1186/s12951-018-0419-1
67. Prasad R, Thirugnanasanbandham K (2019) Advances research on nanotechnology for water technology. Springer International Publishing https://www.springer.com/us/book/9783030023805
68. Rizvi SAA, Saleh AM (2018) Applications of nanoparticle systems in drug delivery technology. Saudi Pharmaceutical J 26:64–70. https://doi.org/10.1016/j.jsps.2017.10.012
69. Matsui I (2005) Nanoparticles for electronic device applications: a brief review. J Chem Eng Jpn 38:535–546. https://doi.org/10.1252/jcej.38.535
70. Zhu T, Cloutier SG, Ivanov I, Knappenberger KL, Robel I, Zhang F (2012) Nanocrystals for electronic and optoelectronic applications. J Nanomate 2012:2012–2014. https://doi.org/10.1155/2012/392742
71. Dos Santos MC, Kesler O, Reddy ALM (2012) Nanomaterials for energy conversion and storage. J Nanomater 2012:2012–2014. https://doi.org/10.1155/2012/159249
72. Stark WJ, Stoessel PR, Wohlleben W, Hafner A (2015) Industrial applications of nanoparticles. Chem Soc Rev 44:5793–5805. https://doi.org/10.1039/c4cs00362d
73. Tiquia-Arashiro S, Rodrigues DF (2016) Extremophiles: applications in nanotechnology: biotechnological applications of extremophiles in nanotechnology, Extremophiles: applications in nanotechnology: biotechnological applications of Extremophiles in nanotechnology. https://doi.org/10.1007/978-3-319-45215-9
74. Bergmann M, Gutow L, Klages M, 2015. Marine anthropogenic litter. Marine Anthropogenic Litter, 1–447. https://doi.org/10.1007/978-3-319-16510-3
75. Stephens B, Azimi P, El Orch Z, Ramos T (2013) Ultrafine particle emissions from desktop 3D printers. Atmos Environ 79:334–339. https://doi.org/10.1016/j.atmosenv.2013.06.050
76. Toussaint B, Raffael B, Angers-Loustau A, Gilliland D, Kestens V, Petrillo M, Rio-Echevarria IM, Van den Eede G (2019) Review of micro- and nanoplastic contamination in the food chain. Food Additives & Contaminants.—Part A 36, 639–673. https://doi.org/10.1080/19440049.2019.1583381
77. Rossi G, Barnoud J, Monticelli L (2014) Polystyrene nanoparticles perturb lipid membranes. J Phys Chem Lett 5:241–246. https://doi.org/10.1021/jz402234c
78. Hu G, Cao J (2019) Metal-containing nanoparticles derived from concealed metal deposits: an important source of toxic nanoparticles in aquatic environments. Chemosphere 224:726–733. https://doi.org/10.1016/j.chemosphere.2019.02.183
79. Turan NB, Erkan HS, Engin GO, Bilgili MS (2019) Nanoparticles in the aquatic environment: usage, properties, transformation and toxicity—a review. Process Saf Environ Prot 130:238–249. https://doi.org/10.1016/j.psep.2019.08.014
80. Yen HJ, Horng JL, Yu CH, Fang CY, Yeh YH, Lin LY (2019) Toxic effects of silver and copper nanoparticles on lateral-line hair cells of zebrafish embryos. Aquatic Toxicol 215:105273. https://doi.org/10.1016/j.aquatox.2019.105273
81. Teodosiu C, Gilca A, Barjoveanu G, Fiore S (2018) Emerging pollutants removal through advanced drinking water treatment: A review on processes and environmental performances assessment. J Clean Prod 197:1210–1221
82. Ma J, Hung H, Macdonald RW (2016) The influence of global climate change on the environmental fate of persistent organic pollutants: a review with emphasis on the Northern Hemisphere and the Arctic as a receptor. Global Planet Change 146:89–108
83. El-Salamony RA, Gobara HM, Younis SA (2017) Potential application of MoO_3 loaded SBA-15 photo-catalyst for removal ofmultiple organic pollutants from water environment. J Water Process Eng 18:102–112

84. Wang L, Shen H, Wu Z, Yu Z, Li Y, Su H, Zheng W, Chen J, Xie P (2020b) Warming affects crustacean grazing pressure on phytoplankton by altering the vertical distribution in a stratified lake. Sci Total Environ 734:139195
85. Yanyan L, Kurniawan TA, Zhu M, Ouyang T, Avtar R, Othman MHD, Mohammad BT, Albadarin AB (2018) Removal of acetaminophen from synthetic wastewater in a fixed-bed column adsorption using low-cost coconut shell waste pretreated with NaOH, HNO_3, ozone, and/or chitosan. J Environ Manage 226:365–376
86. Jeevanantham S, Saravanan A, HemavathyR V, Kumar PS, Yaashikaa PR, Yuvaraj D (2019) Removal of toxic pollutants from water environment by phytoremediation: A survey on application and future prospects. Environ Technol Innov 13:264–276
87. Wang Q, Yang Z (2016) Industrila water pollution, water environment treatment and health risks in China. Environ Pollut 218:358–365
88. Manoli K, Morrison LM, Sumarah MW, Nakhla G, Ray AK, Sharma VK (2019) Pharmaceuticals and pesticides in secondary effluent wastewater: Identification and enhanced removal by acid-activated ferrate(VI). Water Res 148:272–280
89. Quesada HB, Baptista ATA, Cusioli LF, Seibert D, Bezerra CO, Bergamasco R (2019) Surface water pollution by pharmaceuticals and an alternative of removal by low-cost adsorbents: a review. Chemosphere 222:766–780
90. Wu P, Huang J, Zheng Y, Yang Y, Zhang Y, He F, Chen H, Quan GYan J, Li T, Gao B (2019) Environmental occurrences, fate, and impacts of microplastics. Ecotoxicol Environ Safety 184:109612
91. Jouhara H, Czajczynska D, Ghazal H, Krzyzynsk R, Anguilano L, Reynolds AJ, Spencer N (2017) Municipal waste management systems for domestic use. Energy 139:485–506
92. Kotowska U, Karpinska J, Kapelewska J, Kowejsza EM, Piotrowska-Niczyporuk A, Piekutin J, Kotowski A (2018) Removal of phthalates and other contaminants from municipal wastewater during cultivation of Wolffiaarrhizal. Process Saf Environ Prot 120:268–277
93. Moora H, Roos I, Kask U, Kask L, Ounapuu K (2017) Determination of biomass content in combusted municipal waste and associated CO_2 emissions in Estonia Energy Procedia 128:222–229
94. Dai Y, Sun Q, Wang W, Lu L, Liu M, Li J, Yang S, Sun Y, Zhang K, Xu J, Zheng W, Hu Z, Yang Y, Gao Y, Chen Y, Zhang X, Gao F, Zhang Y (2018) Utilizations of agricultural waste as adsorbent for the removal of contaminants: a review. Chemosphere 211:235–253
95. Zhou Y, Zhang L, Cheng Z (2015) Removal of organic pollutants from aqueous solution using agricultural wastes: a review. J Mol Liq 212:739–762
96. Mo J, Yanga Q, Zhang N, Zhang W, Zheng Y, Zhang Z (2018) A review of agro-industrial waste (AIW) derived adsorbents for water and waste water treatment. J Environ Manage 227:395–405
97. Lai W (2017) Pesticide use and health outcomes: Evidence from agricultural water pollution in China. J Environ Econ Manag 86:93–120
98. Diendere A, Nguyen G, Corso J-PD, Kephaliacos C (2018) Modeling the relationship between pesticide use and farmer' beliefs about water pollution in Burkina Faso. Ecol Econ 151:114–121
99. Noorhosseini SA, Allahyari MS, Damalas CA, Moghaddam SS (2017) Public Environmental awareness of water pollution from urban growth: the case of Zarjub and Goharrud rivers in Rasht Iran. Sci Total Environ 599–600:2019–2025
100. Qin J, Lin C, Cheruiyot P, Mkpanam S, Duma NG (2017) Potential effects of rainwater-borne hydrogen peroxide on pollutants in stagnant water environments. Chemosphere 174:90–97
101. Pena-Guzman C, Ulloa-Sanchez S, Mora K, Helena-Bustos R, Lopez-Barrera E, Alvarez J, Rodriguez-Pinzon M (2019) Emerging pollutants in the urban water cycle in Latin America: a review of the current literature. J Environ Manage 237:408–423
102. Al-Jandal N, Saeed T, Azad I, Al-Subiai S, Al-Zekri W, Hussain S, Al-Hasan E (2018) Impact of endocrine disrupting compounds in sewage impacted coastal area on seabream. Ecotoxicol Environ Saf 150:280–288

103. Ma Y, Liu Z-H, Xi B-D, He X-S, Li Q-L, Qi Y-J, Jin M-Y, Guo Y (2019) Characteristics of groundwater pollution in a vegetable cultivation area of typical facility agriculture in a developed city. Ecol Ind 105:709–716
104. Iglesias-Gonzalez A, Hardy EM, Appenzeller BMR (2020) Cumulative exposure to organic pollutants of French children assessed by hair analysis. Environ Int 134:
105. Njuguna SM, Makokha VA, Yan X, Gituru RW, Wang Q, Wang J (2019) Health risk assessment by consumption of vegetables irrigated with reclaimed waste water: A case study in Thika (Kenya). J Environ Manage 231:576–581
106. Sun J, Wu Y, Jiang P, Zheng L, Zhang A, Qi H (2019) Concentration, uptake and human dietary intake of novel brominated flame retardants in greenhouse and conventional vegetables. Environ Int 123:436–443
107. Kassotaki E, Pijuan M, Rodriguez-Roda I, Buttiglieri G (2019) Comparative assessment of endocrine disrupting compounds removal in heterotrophic and enriched nitrifying biomass. Chemosphere 217:659–668
108. Crittenden JC, Harza MW (2005) Water treatment: principles and design. J. Wiley publishing. ISBN 0411101839780471110187
109. Liu M, Lü Z, Chen Z, Yu S, Gao C (2011) Comparison of reverse osmosis and nanofiltration membranes in the treatment of biologically treated textile effluent for water reuse. Desalination 281:372–378
110. Nataraj SK, Hosamani KM, Aminabhavi TM (2009) Nano filtration and reverse osmosis thin fi lm composite membrane module for the removal of dye and salts from the simulated mixtures ☆. Desalination 49(1):12–17
111. Sarkar P, Ghosh S, Dutta S, Sen D, Bhattacharjee C (2009) Effect of different operating parameters on the recovery of proteins from casein whey using a rotating disc membrane ultrafiltration cell. Desalination 249(1):5–11. https://doi.org/10.1016/j.desal.2009.06.007
112. Savari S, Sachdeva S, Kumar A (2008) Electrolysis of sodium chloride using composite poly (styrene-co-divinylbenzene) cation exchange membranes 310:246–261
113. Abu H, Moussa H (2004) Removal of heavy metals from wastewater by membrane processes: a comparative study. Desalination 164:105–110
114. Matsuura KCKT, Ethylenediamine EDA (2018) Removal of heavy metals and pollutants by membrane adsorption techniques. Appl Water Sci 8(1):1–30. https://doi.org/10.1007/s13201-018-0661-6
115. Al-bastaki N (2004) Removal of methyl orange dye and Na_2SO_4 salt from synthetic waste water using reverse osmosis. Chem Eng Process 43(12):1561–1567. https://doi.org/10.1016/j.cep.2004.03.001
116. Gonzalez-Vogel A, Rojas OJ (2020) Exploiting electro-convective vortices in electrodialysis with high-frequency asymmetric bipolar pulses for desalination in over limiting current regimes. Desalination 474:114190
117. Abdulgader HA, Kochkodan V, Hilal N (2013) Hybrid ion exchange—Pressure driven membrane processes in water treatment—a review. Sep Purif Technol 116:253–264
118. Jokar M, Mirghaffari N, Soleimani M, Jabbari M (2019) Preparation and characterization of novel bio ion exchanger from medicinal herb waste (chicory) for the removal of Pb^{2+} and Cd^{2+} from aqueous solutions. J Water Process Eng 28:88–99
119. Hassan MM, Carr CM (2018) A critical review on recent advancements of the removal of reactive dyes from dyehouse effluent by ion-exchange adsorbents. Chemosphere 209:10–219
120. Nekouei RK, Pahlevani F, Assefi M, Maroufi S, Sahajwalla V (2019) Selective isolation of heavy metals from spent electronic waste solution by macroporous ion-exchange resins. J Hazard Mater 371:389–396
121. Kefeni KK, Mamba BB (2020) Photocatalytic application of spinel ferrite nanoparticles and nanocomposites in wastewater treatment-a review. Sustain Mater Technol 23, e00140
122. Wang J, Chen H (2020) Catalytic ozonation for water and wastewater treatment: Recent advances and perspective. Sci Total Environ 704:135249
123. Nemati Sani O, Navaei fezabady AA, Yazdani M, Taghavi M (2019) Catalytic ozonation of ciprofloxacin using γ-Al2O3 nanoparticles in synthetic and real waste water. J Water Process Eng 32:100894

124. Kolosov P, Yargeau V (2019) Impact of catalyst load, chemical oxygen demand and nitrite on disinfection and removal of contaminants during catalytic ozonation waste water. Sci Total Environ 651:2139–2147
125. Sushma, Kumari M, Saroha AK (2018) Performance of various catalysts on treatment of refractory pollutants in industrial wastewater by catalytic wet air oxidation: a review. J Environ Manag 228:169–188
126. Wang J, Bai Z (2017) Fe-based catalysts for heterogeneous catalytic ozonation of emerging contaminants in water and wastewater. Chem Eng J 312:79–98
127. Na S-Y, Lee Y (2017) Elimination of trace organic contaminants during enhanced wastewater treatment with horseradish peroxidase/hydrogen peroxide (HRP/H_2O_2) catalytic process. Catal Today 282:86–94
128. Sun Y, Yang Q, Zhang Y, Ge L, Xu Y, Qian G (2019) Synthesis, characterization and catalytic application of ZnPO molecular sieve in wastewater system. J Clean Prod 213:1165–1171
129. Galvan RF, Barranco V, Galvan JC, Batlle, Sebastian FeliuFajardo., García, Piccardo, M, Renzi, M, Terlizzi A, Shen M, Zhang Y, Zhu Y, Song B, Zeng G, Hu D, Wen X, Ren X, Malakar A, Snow DD, Witchel E (2019) Nanoparticles as sources of inorganic water pollutants. Environ Pollut 7(1):337–370
130. Chen D, Wang L, Ma Y, Yang W (2016) Super-adsorbent material based on functional polymer particles with a multilevel porous structure. NPG Asia Materials 8(8):e301–e301. https://doi.org/10.1038/am.2016.117
131. Rashed MN (2012) Adsorption Technique for the Removal of Organic Pollutants from Water and Wastewater. Organic Pollutants—Monitoring, Risk and Treatment—Book Chapter, IntechOpen. https://doi.org/10.5772/54048, i(tourism), 13
132. Inyinbor AA, Adekola FA, Olatunji GA (2016) Kinetic and thermodynamic modeling of liquid phase adsorption of Rhodamine B dye onto Raphia hookerie fruit epicarp. Water Resources and Industry 15:14–27
133. Ghaly M, El-Dars FMSE, Hegazy MM, Abdel Rahman RO (2016) Evaluation of synthetic Birnessite utilization as a sorbent for cobalt and strontium removal from aqueous solution. Chem Eng J 284:1373–1385
134. Atkins PW (1994) Physical Chemistry, 5th edn. Oxford University Press, Oxford, pp 922–926
135. Goodwin HM (1901) Physical chemistry. J Amer Chem Soc 23(7). https://doi.org/10.1021/ja02033a032
136. Patel H (2019) Fixed—bed column adsorption study: a comprehensive review. Appl Water Sci 9(3):1–17. https://doi.org/10.1007/s13201-019-0927-7
137. Cavalcante Jr, CL (2000) Industrial adsorption separation processes: fundamentals, modeling and applications. Latin Amer Appl Res 30:357–364
138. Monash P, Pugazhenthi G (2010) Separation science and technology removal of crystal violet dye from aqueous solution using calcined and uncalcined mixed clay adsorbents removal of crystal violet dye from aqueous solution using calcined and uncalcined mixed clay adsorbents. Sep Sci Technol 45(1):94–104
139. Dada OA, Adekola FA, Odebunmi EO (2016) Kinetics and equilibrium models for sorption of Cu (II) onto a Novel Manganese Nano-adsorbent. J Dispersion Sci Technol 37(1):119–133. https://doi.org/10.1080/01932691.2015.1034361
140. Inyinbor AA, Adekola FA, Olatunji GA (2016) Liquid phase adsorption of Rhodamine B onto acid treated Raphia hookerie epicarp: Kinetics, Isotherm and thermodynamics studies. S Afr J Chem 69:218–226
141. Adekola FA, Jimoh S, Inyinbor AA (2018) P-nitrophenol removal from aqueous solution using raw and modified kaolinite; Orbital: Electron J Chem 10(6):435–445
142. Inyinbor AA, Adekola FA, Dada AO, Oluyori AP, Olatunji GA, Fanawopo OF, Oreofe TA, Abodunrin TO (2019) Novel acid treated biomass: Applications in Cu^{2+} scavenging, Rhodamine B/Cu^{2+} binary solution and real textile effluent treatment. Environ Technol Innov 13:37–47
143. Inyinbor AA, Adekola FA, Olatunji GA (2019) Copper scavenging efficiency of adsorbents prepared from Raphia hookeri fruit waste. Sustain Chem Pharmacy 12:100141

144. Bello OS, Alao OC, Alagbada TC (2019) Biosorption of ibuprofen using functionalized bean husks. Sustain Chem Pharmacy 13: https://doi.org/10.1016/j.scp.2019.100151
145. Rajabi M, Mirza B, Mahanpoor K, Mirjalili M, Najafi F, Moradi O, Sadegh H, Shahryari-ghoshekandi R, Asif M, Tyagi I, Agarwal S, Gupta VK (2016) Adsorption of malachite green from aqueous solution by carboxylate group functionalized multi-walled carbon nanotubes: determination of equilibrium and kinetics parameters. J Ind Eng Chem 34:130–138. https://doi.org/10.1016/j.jiec.2015.11.001
146. Bello OS, Owojuyigbe ES, Babatunde MA, Folaranmi FE (2017) Sustainable conversion of agro-wastes into useful adsorbents. Appl Water Sci 7(7):3561–3571. https://doi.org/10.1007/s13201-016-0494-0
147. Olakunle MO, Inyinbor AA, Dada AO, Bello OS (2017) Combating dye pollution using cocoa pod husks: a sustainable approach. Int J Sustain Eng. https://doi.org/10.1080/19397038.2017.1393023
148. Sadegh H, Ali GAM (2018) Potential Applications of Nanomaterials in Wastewater Treatment. June, 51–61. https://doi.org/10.4018/978-1-5225-5754-8.ch004
149. Adekola FA, Ayodele SB, Inyinbor AA (2019) Efficient Rhodamine B removal using acid and alkaline activated Musa paradisiaca biochar. Polish J Environ Studies 28(5):3063–3070
150. Inyinbor AA, Adekola FA, Olatunji GA (2019) Low cost adsorbent prepared from Vigna subterranean waste: physicochemical, morphological and surface Chemistry data set. Chem Data Collection 24:100294
151. Inyinbor AA, Adekola FA, Olatunji GA (2020) Microwave-assisted Urea Modified Crop residue in Cu^{2+} scavenging. Heliyon 6:e03759
152. Afolabi IC, Popoola SI, Bello OS (2020) Modeling pseudo-second-order kinetics of orange peel-paracetamol adsorption process using artificial neural network. Chemometrics Intell Lab Syst 203:104053
153. Adegoke KA, Agboola OS, Ogunmodede J, Araoye AO, Bello OS (2020) Metal-organic frameworks as adsorbents for sequestering organic pollutants from wastewater. Mater Chem Phys 253:123246

Nano-Porous Graphene as Free-Standing Membranes

Asieh Sadat Kazemi and Mohammad Ali Abdol

Abstract Membrane-based treatments are superior to conventional processes in water purification and desalination in terms of final water quality and smaller footprint. In addition, there are inherent weaknesses of state-of-the-art membranes in reverse osmosis (RO) technology, thus there is pressing constraints for next-generation membranes. Nano-porous graphene membranes as the mother of novel 2D materials have attracted tremendous attention in the last decade in this regard, and many theoretical and experimental attempts have been made towards the realization of these membranes for water desalination. Advancements and challenges facing this approach, research gaps and prospects that shed light on this field are explored in this chapter.

Keywords Graphene · Nanopore · Desalination · Reverse osmosis

1 Introduction

Separation techniques based on membranes have major applications in water desalination [1–4], natural gas purification [5, 6], solvent- and petrochemical-based separations [7, 8], haemodialysis [9] and bioprocessing [10]. Beyond separations, membranes are used in drug delivery [9], bio/chemical sensors [11], fuel cells [12] and energy harvesting [13]. The role of membrane-based treatment technologies in water purification and desalination is undoubtedly unprecedented. These technologies produce water of superior quality and have a much smaller footprint compared with conventional water treatment technologies [2], as they are inherently more energy efficient than thermal approaches [14]. While thermal desalination technologies consume almost five times the energy used in current state-of-the-art reverse

A. S. Kazemi (✉)
Department of Physics, Iran University of Science and Technology, Tehran, Iran
e-mail: asiehsadat_kazemi@iust.ac.ir

M. A. Abdol
Department of Advanced Technologies, Iran University of Science and Technology, Tehran, Iran

R. Das (ed.), *Two-Dimensional (2D) Nanomaterials in Separation Science*,
Springer Series on Polymer and Composite Materials,
https://doi.org/10.1007/978-3-030-72457-3_3

osmosis (RO) technology, the energy required for RO itself, is within a factor of two from the limit of a reversible thermodynamic process [15, 16]. However, there has been not much progress in RO membrane technology based on the same polyamide thin-film composite design in the past three decades [17]. The current membrane design still suffers low permeability even with the most permeable thin-film composite membranes. The membranes become damaged while in contact with chlorine, and fouling tendency is still a major challenge in these designs [18]. RO membrane technology requires new membrane design and fabrication approaches to meet the needs of advanced desalination demands [19]. An ideal membrane should easily allow the flow of the desired species, reject undesired species and exhibit robustness in operation in addition to being economical and manufacturable [20]. Therefore, high permeance, high selectivity, high chemical, mechanical and thermal stability, and low fouling are the critical margins that new membrane technology should focus on. With the emergence of two dimensional (2D) materials in the past decade and beyond, and the enormous amount of attention they captured in the scientific community due to their unique structural, thermal, and electronic properties; the perception of using them as a replacement in RO membrane technology has excelled more than ever. Since the water flux scales inversely proportional to the membrane thickness, it is necessary for new generation of RO membranes to have high aspect ratio, high mechanical stiffness, atomic thickness, nanosized pores and reactivity toward polar and non-polar water pollutants [19, 21, 22]. These extraordinarily properties can be found all together in 2D materials and though still in the early stages of development, they present characteristics that are potentially advantageous for addressing persistent challenges in membrane separations [20]. Monolayer graphene has been one of the first 2D materials isolated in nature and employed in immense variety of applications ever since 2004 [23]. While production of large area graphene is still an ongoing and active research field [24, 25], many attempts have been made to design and fabricate nanopores across individual graphene sheets [26]. Developing 2D nano-porous graphene with tuneable ranges of pore size distribution, pore density and uniformity are critical for water purification and desalination [26–33]. These properties play an essential role in the engineering of the electronic properties [34], surface interactions [35] and chemical reactivity [36] of perforated graphene materials.

We first discuss the state-of-the-art of existing membrane technologies for water purification and desalination, emphasize their inherent weaknesses and determine the pressing constraints for next-generation membranes. Then, we illustrate molecular-level design approaches theoretically towards pristine and functionalized graphene, summarize simulation considerations, explore mechanical aspects, and transport phenomena focusing on monolayer graphene, and in some case on multilayer graphene. Experimentally fabricating highly selective membranes, developments on appropriate support layer and perforation techniques and characterization methods are addressed next. While other nano-porous 2D materials [37–48] and carbon-based materials [49–54] have also attracted extensive attention in separation techniques, they will be discussed in the consecutive chapters. At last, we review computational and experimental gaps and prospects in this field of study.

2 State-of-the-Art Membranes Versus Ideal Membranes

2.1 *State-of-the-Art Membranes, Properties, and Limitations*

The function of RO membranes is to extract fresh water from pressurized saltwater. Membrane technologies for water purification and desal-ination are largely based on size exclusion [15]. Most of Sect. 2.1 is reproduced with permission from Ref. [15], Copyright 2016, *Macmillan Publishers Limited.*

A wide range of membrane materials, both polymeric and inorganic, have been used in RO [55–60]. Polymeric membranes include polysulfone (PSF), polyethylene (PE), polytetrafluoroethylene (PTFE), polypropylene (PP), polyvinylidene fluoride (PVDF) [61], while inorganic membranes including silica, carbon, metal-organic frameworks (MOFs) and various metal oxides [62]. Polymeric membranes are the dominant material of desalination technology today, essentially owing to their high processability and low cost. These membranes are categorized into phase inversion membranes, track etching membranes and thin-film composite (TFC) polyamide membranes [15]. Due to the stochastic nature of the phase inversion process, a wide range of membranes formed by this method, have polydisperse pore size distributions, that adversely influence the selectivity of the resulting active layer [15]. Track-etched membranes, porous systems, consisting of a thin polymer foil with channels from surface to the surface, have a major limitation. Their porosity must be maintained below 5% to prevent pore overlap and consequently, the low porosity results in low water permeability. TFC polyamide membranes consist of a non-porous, highly crosslinked polyamide selective layer and an underlying porous support layer, usually made of PSF. For porous membranes, flow transpires through the active layer and selectivity is inherently linked with active layer pore size [15]. For non-porous and dense membranes, water and solute transports are dominated by the solution–diffusion model [20]. Water and solute molecules separate into the active layer of the membrane, diffuse through the polymer matrix down their chemical potential gradients and desorb into the permeate solution [15].

TFC membranes possess much higher water permeability and salt rejection compared to the first-generation RO membranes (asymmetric cellulose acetate-based membranes) which were developed more than 5 decades ago [63]. Furthermore, TFC membranes proved to be stable over a wide pH range (pH 2–11) in continuous operation [15, 26]. The combined performances of the TFCs has led to their widespread utilization in desalination applications to date, however, commercial membranes still suffer from fairly low permeability and low stability to cleaning chemicals that are inevitable for fouling mitigation [4, 64]. Membrane fouling is the build-up of substances on the membrane surface or within the membrane pores. While inorganic membranes solve most of the challenges involved with polymeric membranes, they endure low scalability and low mechanical strength, which obstructs their large-scale applications [65].

An ideal membrane combines high water permeability and high selectivity. This combination has proven difficult to accomplish and therefore referred as permeability–selectivity trade-off [66, 67]. Investigations on novel active layers for desalination membranes pursue to improve the water permeability and water–solute selectivity [15]. Process modeling for seawater RO desalination shows that increased water permeability above currently achievable levels (2–3 $Lm^{-2}h^{-1}bar^{-1}$); would negligibly decrease energy requirements and capital costs [68, 69], opposing the previous claim of Cohen-Tanugi and Grossman [68]. Such a conflict, irrespective of the membrane permeability, is mainly due to current RO system designs that desalinate seawater using a single membrane stage fed by a high-pressure pump [14, 19]. Note that the use of hydraulic pressure is the main element of the energy consumption by the RO stage. Therefore, instead of realizing higher water permeability, a more effective goal for novel desalination membrane materials research would be to achieve improved selectivity for water over all dissolved solutes [15].

2.2 *Modules for Current RO Membrane Testing*

A typical RO set-up has a simple layout, consisting of feed water input, feed pre-treatment unit, high-pressure pump, modules for the RO membrane and, in some configurations, post-treatment stages [70]. There are four different types of RO membrane modules that are used for desalination processes: plate and frame, tubular, hollow fiber and spiral-wound [1]. Figure 1 shows the schematics of these modules. Plate-and-frame is an old RO module made up of layers of supported plane

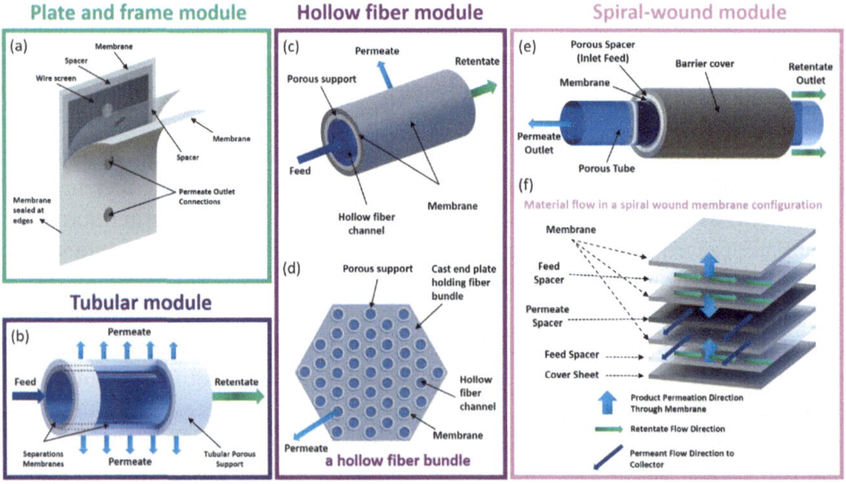

Fig. 1 Schematics of **a** plate and frame, **b** tubular, **c**, **d** hollow fiber and **e**, **f** spiral-wound modules used for water treatments. **d** and **f** are reproduced with permission from [78]. Copyright © 2012 Elsevier Inc

sheet membrane, with a spacer between each support and membrane to drive the feedthrough each membrane, while flowing the permeate outside the module [71–75]. The advantage of this type of module is fouling resistivity, but it has a low packing density which demands a lot of space and makes it expensive. Tubular membrane units, another RO configuration, comprise membrane tubes wrapped around punctured stainless-steel cylinders. The feed runs across the tubes and the permeate flows out from the membrane and support [76]. These two modules share the same advantages and disadvantages and are mainly used for extreme foul feeds [77].

In contrast, hollow-fiber modules have a large packing density, where multiple fine hollow-fiber membranes are placed in a pressure vessel, and the feed runs outside the membranes [72, 73]. While these units have high productivity, they are highly susceptible to fouling. Hollow fiber modules are characteristically 10–20 cm in diameter and 1.0–1.6 m long [78]. One of the most extensively used modules nowadays, is the spiral-wound. A spiral-wound module contains plane layers of membrane separated from each other by spacers and are wrapped around a punctured collection tube. The feed solution passes across the membrane surface and a portion of the feed permeates into the membrane envelope where it spirals toward the center and exits through the collection tube as shown in Fig. 1 [78]. The spiral-wound module has a high packing density in addition to reasonable fouling resistance. Costly wise, it is the cheapest module setup produced from TFC membranes [79, 80]. Advances in materials, manufacturing techniques, feed channels and the size of vessels and spacers in recent years have contributed to better adjustment of the internal fittings between module elements and liquid transport features in this module, thus the multi envelope design of spiral-wound module minimizes the pressure drop encountered by the permeate traveling toward the central pipe and also minimizes fouling [78, 81]. The standard industrial spiral wound module is 20.3 cm in diameter and 101.6 cm long [78]. Four to six spiral wound membrane modules are normally connected in series inside a single pressure vessel. A typical 20.3 cm diameter tube containing six modules has 100–200 m^2 of membrane area [78]. Spiral-wound is superior to the other three modules in terms of lowest energy use, lowest system cost and highest design flexibility but stands in third place in terms of susceptibility to fouling after plate and frame and tubular modules [81].

2.3 Ideal Membranes, Properties, and Prospects

The rise of graphene and other 2D materials in the past 15 years [82], has opened new avenues in membrane technology. The atomic thickness of these materials makes them the thinnest possible barrier, and if combined with their remarkable mechanical strength, chemical robustness, extreme flatness, and ability to sustain selective, nanometre-scale pores [21, 27, 83, 84] at pristine form, become ideal candidates for replacing the current state-of-the-art membranes. Key advantages of nano-porous atomically thin membranes as active layers in RO can be summarized as follows [20]:

(i) their high permeance could enable higher energy efficiency and compact, high-productive membrane separation systems [68, 85, 86].
(ii) their robustness due to high chemical resistance and mechanical strength could prove useful for operation under harsh conditions, e.g., better withstand chlorine, allow more aggressive cleaning procedures [86], overcome challenges of aging, compaction and influence of solvents in organic solvent nanofiltration [7].
(iii) in case of providing universal size-based rejection independent of the solute, it would benefit applications involving a diversity of solutes, e.g., removal of persistent organic pollutants and boron from seawater or in chemical processing [1, 3, 7, 14, 87].
(iv) their flatness could overcome fouling [20].

Built on what is explained so far, an ideal RO membrane based on nano-porous atomically thin materials would be extremely flexible and mechanically stable to fit into the spiral-wound RO modules for the highest efficiency. It would also be extremely flat to mitigate fouling, and it would be decorated with pores of ideal nano-sizes and distributions to break permeability–selectivity trade-off. Besides, it would be extremely thin, to maintain high water permeation as shown schematically in Fig. 2.

Another important aspect of the RO membrane is the substrate underneath the active layer or the so-called 'support layer'. The principal role of the substrate is to provide mechanical support for the active layer via bearing most of the hydraulic load and at the same time, distributing the pressure from the water onto patches of the active layer. However, the substrate has a minor direct role in the salt rejection process, while water simply percolates through its pore network after permeating through the active layer [88–90]. In conventional TFC membranes, the active layer (typically composed of polyamide) bears the salt rejection role and extends 100–200 nm in thickness [91].

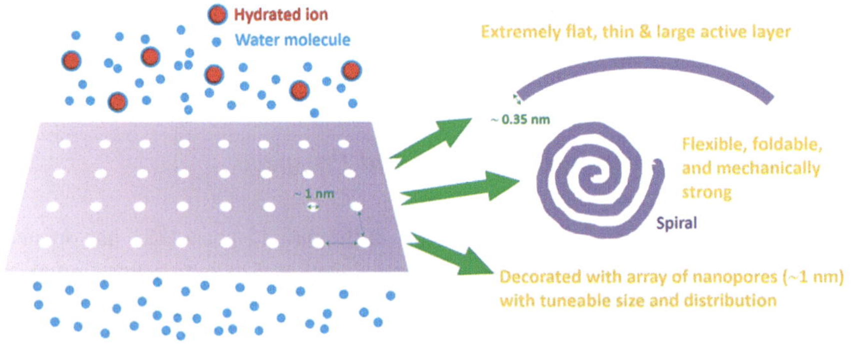

Fig. 2 Ideal RO membrane based on supreme characteristics of nano-porous atomically thin materials

The support layer (usually made of PSF) is significantly more porous and thicker than the active layer with cavities around 0.1−0.5 μm and an overall layer thickness >100 μm [20, 21, 88, 89]. Ultrafiltration membranes with pores as narrow as 10 nm have also been utilized as support layers for RO membranes [92]. Like polyamide active layers, nano-porous atomically thin membranes must also lie on top of a porous substrate for the reasons explained earlier. Figure 3 schematically compares an ideal nano-porous atomically thin membrane structure and its transport mechanism with state-of-the-art-membranes. Note that for simplicity, the active layer in the latter is depicted as non-porous while, there are porous versions employed in RO technology as well. Furthermore, the structure of the support layer in this image is inspired by an asymmetric ultrafiltration membrane with the finger-like macro-voids cast from PSF in [93].

In an ideal situation, the atomically flat membrane as the active layer, with the least thickness (~0.35 nm) would sit on top of a porous support layer. This layer has wide pores and a much lower thickness than its counterpart in polymeric membranes but still provides sufficient mechanical support for the active layer. Through any of the mechanisms of perforation techniques (see Sect. 4.3), the active layer would become nano-porous with ideal size and distribution of nanopores across the membrane. Therefore, the transport mechanism of salts and water across this membrane would obey the size exclusion process (Fig. 3). In contrast, the polymeric membrane with an average thickness of 200 nm, with high surface roughness, sits on top of a thick, porous support layer with narrow pores. The dominant transport mechanism in this structure is based on the solution-diffusion process particularly if the active layer is

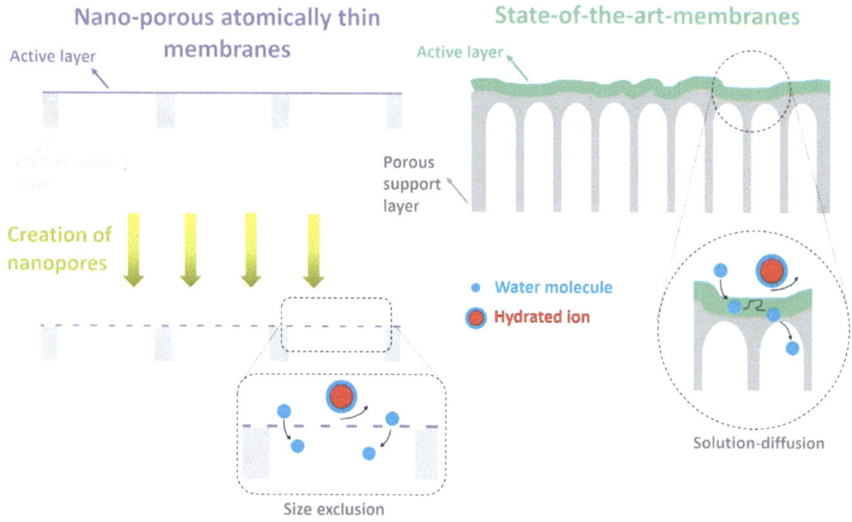

Fig. 3 Cross-section schematic illustration of ideal nano-porous atomically thin membrane (on the left) versus state-of-the-art membranes structure (on the right) and a comparison between ion rejection mechanisms in the two

non-porous. Even for a porous active layer, regarding the thickness of the polyamide layer, the solution diffusion process would accompany the size exclusion. Now let us compare the performance of the two structures when pressurized salty water is applied to the membranes. Figure 4 illustrates schematically the situation and compares what would happen to the two structures over time. If identical water pressure with similar salinity is applied to the two membranes, the hydrated salt ions (which are typically larger than the ions) would be completely rejected from the ideal nano-porous membrane but some would leak across the polymeric membrane.

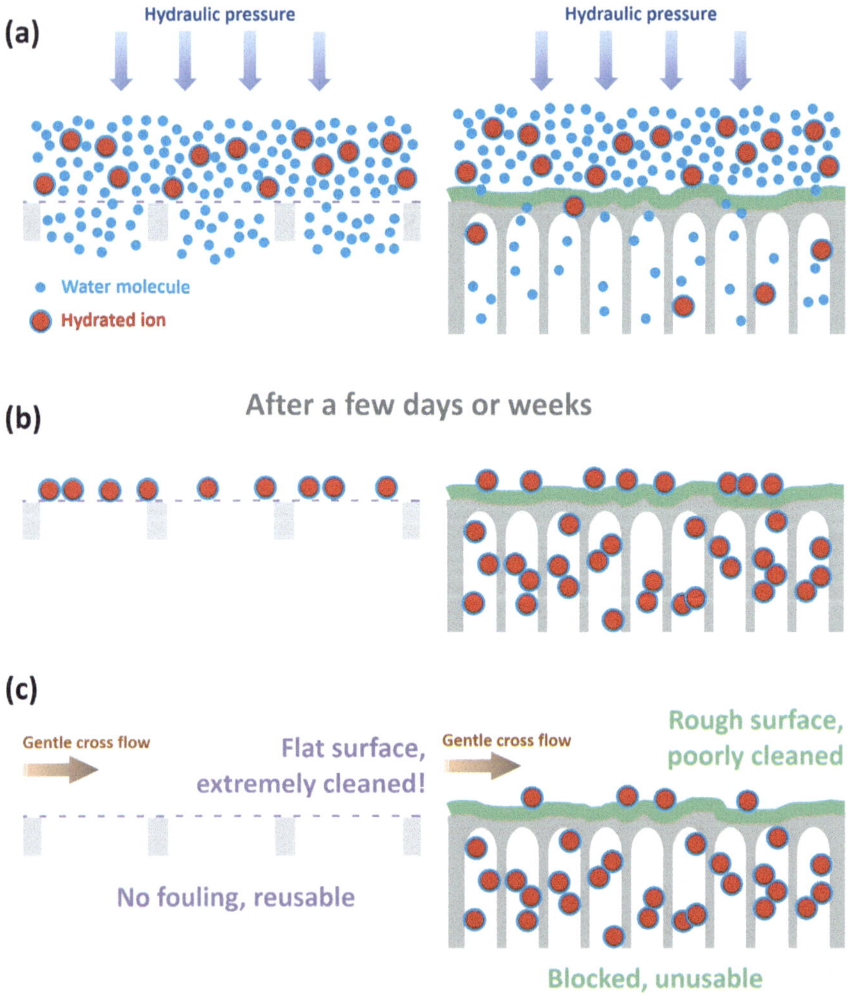

Fig. 4 Schematic comparison between the relative performance of an atomically thin nano-porous membrane (on the left) and a conventional polyamide membrane (on the right) under applied pressurized salty water over time

Water molecules would pass through the nanopores of the ideal membrane much faster than across the polymeric membrane due to the different transport mechanisms, active layer thickness and support layer thickness in addition to support layer porosity (Fig. 4a). Therefore, the performance of the ideal membrane is already better in terms of salt rejection and water permeation. After a few days or weeks (Fig. 4b), there would be an accumulation of salt ions on the surface of both membranes. However, there would also be an accumulation of salt ions in the pores of the support layer of polymeric membranes due to the leakage (however low) in time. This would not happen for the ideal membrane if its thickness and porosity are carefully designed. A gentle cross-flow of water or other chemicals (Fig. 4c) would clean the flat surface of the ideal membrane while the rough surface of the polymeric membrane would not be cleaned entirely because of the surface roughness that traps ions. The support layer with its complicated porous structure is not easily cleaned either, therefore the polymeric membrane may be fouled and not useable anymore. That is if the membrane is not damaged by chlorine during the cleaning process. In contrast, the ideal membrane is completely clean and can be reused further in the RO process. It is worthy to note that in polyamide membranes it is hard to synthesize films thinner than 100 nm because the film thickness is governed by the mass transfer of the diamine to the organic phase during interfacial polymerization [94]. It is also uncertain that a significantly thinner polyamide film would exhibit high enough salt rejection at any rate [43].

With this introduction on the advantages and prospects of nano-porous atomically thin membranes over polymeric membranes for RO technology, it is inevitable to explore nano-porous graphene as the mother of the 2D materials. Graphene-based desalination membranes have been envisaged to serve as selective layers and operated as molecular sieves with the size-based exclusion of undesired solutes in two forms: nano-porous graphene and graphene-based frameworks [15]. Graphene-based framework membranes comprise a multi-layered stack of finite-sized graphene sheets [95–97], formed as laminates through vacuum filtration or layer-by-layer deposition of graphene oxide and differ in morphology and water transport mechanism from nano-porous graphene [15] which are comprehensively explored in the literature and are not the scope of this chapter.

3 Computer Simulations on Nano-Porous Graphene Membranes; Water Desalination Perspective

In this section, we will review the recent computational findings on pristine and functionalized nano-porous monolayer graphene membranes, inspect and summarize simulation considerations and explore the latest results on the mechanical stability of these membranes in a RO system. Further in this section, we will discuss simulation achievements in the bilayer and multilayer nano-porous graphene and the raised concerns in adding additional layers to permeability and selectivity. Finally, we will discuss the transport of water and ions and selectivity issues that were the center of attention in the corresponding literature. Some of the materials in this section are reproduced with permission from Ref. [98], *Copyright © 2012, American Chemical Society* and Ref. [99] *Copyright © 2016, American Chemical Society.*

(a) **Pristine and functionalized membranes**

A graphene monolayer can be considered as the thinnest membrane, as its thickness is only one carbon atom. Because of the superior strength of graphene [100], porous graphene is a potential membrane for molecular sieving or water filtration. Nanopores of various diameters can be realized in graphene via mono vacancies or multi vacancies [101]. One of the first attempts in functionalization of nanopores in graphene monolayers showed that they could serve as ionic sieves of high selectivity and transparency [44]. The chemically modified graphene nanopores were terminated by negatively charged nitrogen and fluorine, favoring the passage of cations. These were compared with nanopores terminated by positively charged hydrogens, favoring the passage of anions [44].

Transport of water through porous graphene membrane was studied by Suk and Aluru [102] using molecular dynamics (MD) simulations where various diameters of nanopores ranging from 0.75 to 2.75 nm were tested. Their simulation set-up with graphene membrane constituted 6 nm water baths on each side of the membrane. The pressure-driven flow was simulated by applying a 100 MPa pressure drop across the pore and water flux was calculated by counting the net amount of water molecules transported through the pore. For larger pores, the flow rate of water was higher because of the reduced energy barrier at the entrance of the graphene sheet pore. The dipole orientation of water molecules that formed the single file chain in the membrane was also calculated. Dipole orientation is defined as the angle between the water dipole vector and the tube axis and is averaged over all the water molecules in the membrane [102]. They found that the dipole orientation flipped frequently in the graphene membrane which indicated frequent breaking of the hydrogen-bonding. They also investigated water transport through graphene by computing the energetics of water permeation. The energy barrier at the entrance was estimated to be a relatively low value of 0.32 k_BT by computing the Potential of Mean Force (PMF) of water. Since the energy barrier arises from reduced interaction energies, directly connected water baths may have been the source of low energy barrier by creating a more bulk-like environment.

Cohen-Tanugi and Grossman [98] found that both the size and chemical functionalization of graphene pores play an important role in blocking salt ions while allowing water to flow through the graphene membranes. Their MD examination of the structure of water in the pore vicinity revealed that the hydrophobic character of hydrogenated pores reduces the water flow by imposing additional conformational order on the system, even as the limited hydrogen bonding allows for greater salt rejection relative to hydroxylated pores. Additionally, the water flux through pristine and OH-functionalized pores membranes was predicted to be 2–3 times faster than that typical of the current state-of-the-art desalination technology at an equal pressure drop [98]. Note that these studies with the aim of reducing computational time employed pressures that are several orders of magnitude higher (1000–2000 bars) than those in real RO systems (10–100 bars). Moreover, they have kept the atoms in the nano-porous graphene membrane frozen to decouple the desalination performance from mechanical effects. However, graphene membranes synthesized experimentally are flexible and the functional groups at the edge of each nanopore are susceptible to deformation under the influence of water molecules and salt ions. This in turn may affect the permeability and salt rejection of the membrane [103]. This was addressed in [103] where the permeability of nano-porous graphene remained approximately constant down to very low pressures. it was demonstrated by MD simulations that nano-porous graphene maintained its desalination performance even when the flexibility of the membrane atoms was considered [103].

MD simulations were employed by Konatham et al. [104] to comprehensively study the transport of water and ions through pores created on the basal plane of one graphene sheet. They quantified the effect of functionalizing pores with carboxyl anion COO^-, amine cation $NH3^+$, and hydroxyl OH^- groups on the pores' ability to reject NaCl. The pores that were considered had diameters of 14.5, 10.5, and 7.5 Å. The ease of Na^+ and Cl^- ions and water translocation across the pores was monitored by calculating the PMF along the direction perpendicular to the graphene sheet pore. The results indicated that effective ion exclusion can be achieved only using pristine pores of diameter ~7.5 Å, whereas the ions can easily penetrate pristine pores of diameters ~10.5 and 14.5 Å. It was shown that not only the ion size (the diameters of Na^+ and Cl^- are 2.58 and 4.40 Å, respectively) but also the ion hydration structure might affect the free-energy profile. It was demonstrated that carboxyl functional groups enhance ion exclusion for all pores considered, but the effect becomes less pronounced as both the ion concentration and the pore diameter increase. The results in [104] suggest that narrow graphene pores functionalized with hydroxyl groups remain effective at excluding Cl^- ions even at moderate solution ionic strength.

In another study [105], graphene nanopores with diameters of 3.8–8.2 Å, with different geometries were terminated by hydrogen and hydroxyl functional groups and studied with MD under 10–200 MPa pressures. Their results demonstrated that water flux permeating the membranes scaled linearly with external pressure and pore diameter. Both Na^+ and Cl^- ions permeated through the membrane with the largest pore, and the selectivity of the permeated ions exhibited a significant correlation with the functional group [105]. Water desalination across Si-passivated nanopores graphene was investigated for systems with fixed and unfixed membranes [106]. The results indicated that membrane curvature reduces water flux as well as salt passage through the pore rim. In contrast with fixed membranes, salt rejection for systems with unfixed membranes increases with increasing the applied pressure. It was shown that the symmetrical shape of the density surface at low pressure increases the salt rejection [106].

Methyl, ethyl and a combination of fluorine and hydrogen molecules were distributed around the graphene nanopores [107]. The different number of functional molecules was employed to find an optimum distribution of the groups at hand. The results showed that an appropriate distribution of alkyl groups properly stops the salt ions from passing through the graphene, where the flow rate of water molecules is sufficient. Moreover, the investigation of graphene membrane's performance in different salt concentrations revealed that with an increase in salt concentration of feed water, the water flow rate falls and the number of salt ions on the other side of the membrane increases [107]. The desalination performances simulated for monolayer graphene nanopores with pyridinic N doped functionalization under various N-doping levels in [108] demonstrated that all N-doped graphene membranes exhibit higher water flux with several orders of magnitude higher than polymeric RO membranes and high salt rejections. Further simulations of the PMF profiles showed a moderate free energy barrier for water and large free energy barrier for ions. The interfacial properties of water and ions, as well as their free energy landscapes in passing through the graphene nanopores, were simulated to explore the desalination mechanism. Density Functional Theory (DFT) simulations revealed that the salt rejection mechanism of the functionalized graphene was the pore size exclusion of hydrated ions and the charged repulsion from pore surfaces was not the main factor [108].

(b) **Simulation considerations and summary**

Figure 5 shows a snapshot of a typical MD simulation box in various reports. The model graphene sheet membrane (L_x Å × L_y Å), with a pore in its center, is placed parallel to the xy plane in the center of the simulation box. The pore diameter (R) is a few angstroms and is obtained by removing carbon atoms, as necessary. The membrane is surrounded by L_{z1} and L_{z2} Å of water along the z-direction on both sides. Different functional groups are schematically added adjacent to the pore.

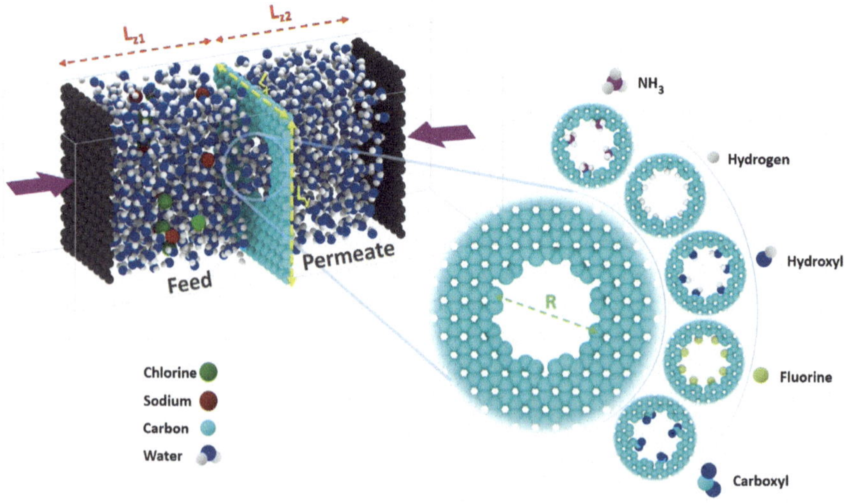

Fig. 5 Snapshot of a typical MD simulation box in various reports. L_x and L_y are the dimensions of the graphene sheet, with a pore in its center with a diameter R. The sheet is placed parallel to the xy plane and is surrounded by L_{z1} and L_{z2} Å of water along the z-direction on both sides. Different functional groups are schematically added adjacent to the graphene pore

Table 1 summarizes the configurations and parameters of MD results on pristine and functionalized nano-porous graphene for water desalination.

MD simulations are ideal to study nano-porous graphene system for water desalination since they compute the evolution of a system of atoms from an original configuration and under a set of constraints [111]. In these simulations, the forces between atoms at each time step are calculated, then, the positions of all the atoms at the following time step is updated using Newton's equations of motion [99]. Depending on the choice of force field, the performance of relatively large molecular systems (10^2-10^9 atoms) can be explored over physically meaningful time scales (typically between 1 ns and 1 μs) [99]. With this introduction, MD simulations allow us to probe the kinetics and thermodynamics of desalination while accounting for the physics of water, ions, and graphene layers with high accuracy [99].

MD simulation is usually performed with the temperature maintained at 300 K by using a Nose-Hoover thermostat. However, there are a few reports that studied water desalination parameters of functionalized nano-porous graphene membrane at a range of temperatures 275–343 K [17]. Periodic boundary conditions are employed to maintain a continuous 2D membrane. In some simulations, the functional groups were usually pointed toward the center of the pore due to steric constraints [104]. Thus, the diameter of the pore changed depending on the size of the functional groups grafted to it. The diameter of the functionalized pore was calculated as the distance between the ends of two opposite functional groups. In each functionalized pore, a specific number of functional groups were spread uniformly around the rim of the pore [17, 104]. In other works, the size of each pore was measured by plotting atoms as van der Waals spheres and calculating the amount of contiguous area not obstructed by any atomic representations [98]. Pore diameters were obtained from the open pore area measurements by the straightforward formula $d = 2\sqrt{A/\pi}$ [98], which resulted in nominally smaller diameters than the center-to-center measurements. In some simulations [104], when charged functional groups were used, an equivalent number of counterions were placed far from the pore and were fixed during the simulations to maintain the electroneutrality of the system. This prevented the ions from accumulating near the functional groups, affecting the results.

The salt ions were usually allowed to move freely within the simulation box. The number of water molecules was maintained constant in different studies. Carbon atoms in the graphene sheet were held stationary and modeled as Lennard-Jones spheres using the parameters proposed by Cheng and Steele in 1990 [112, 113]. Na^+ and Cl^- ions and carbon-water interactions were modeled using the Lennard-Jones [114] parameters. The functional groups on the pores were rigid. They were modeled using the flexible Optimized Potentials for Liquid Simulations (OPLS) force field parameters proposed by Jorgensen et al. [115].

There is a debate on the significance of polarization effects in monovalent salt solutions. Therefore, water is modeled differently based on different simulation aims and perspectives. Modeling the water using TIP4P potential [116] allows for water polarization arising at the intermolecular level via orientational rearrangement [117]. However, intramolecular contributions to water polarizability (due to bond

Table 1 Summary of the configurations and parameters of MD results on pristine and functionalized nano-porous graphene for water desalination used in the literature

References	$L_x \times L_y$	L_{z1}/L_{z2}	Pressure	Pore diameter	Water diameter	Ion ddiameter	Number of water molecules	Salt molarity	Functional groups
[104]	56.6 Å × 55.3 Å	20 Å/20 Å	–	7.5 Å 10.5 Å 14.5 Å	–	Na^+ in bulk ~5.9 ± 0.6 Å Cl^- in bulk ~7.2 ± 0.9 Å	4100 water molecules	0.025 M (NaCl) 0.25 M (NaCl) 0.6 M (NaCl)	COO^-, $NH3^+$, OH^-
[102]	40 Å × 40 Å 60 Å × 60 Å	40 Å/40 Å 60 Å/60 Å	100 MPa 100 MPa	7.5 Å 27.5 Å	–	–	6479–14974	–	–
[98]	30 Å × 30 Å	15 Å/60 Å	100–200 MPa	1.5 $Å^2$ 62 $Å^2$	–	–	825 water molecules	16 Na^+, 16 Cl^-	OH, H
[103]	30 Å × 30 Å	15 Å/60 Å	2.9 MPa 14. 6 MPa 146 MPa	23 $Å^2$			955 water molecules	2 Na^+, 2 Cl^-	H
[109]	–	10 Å/10 Å	–	5 $Å^2$	–	–	–		N, F, H
[110]	2.9 Å × 2.9 Å	40 Å/40 Å	–	3.7 Å 17.6 Å	–	–	–	0.5 M NaCl	F, N
[106]	30 Å × 30 Å	–	–	2 Å 6.9 Å 11 Å	–	–	1240 water molecules	8 Na^+, 8 Cl^-	Si

(continued)

Table 1 (continued)

References	$L_x \times L_y$	L_{z1}/L_{z2}	Pressure	Pore diameter	Water diameter	Ion ddiameter	Number of water molecules	Salt molarity	Functional groups
[105]	29.8 Å × 29.5 Å	30 Å/30 Å	10–200 MPa	3.2 Å 4.6 Å 5.2 Å 5.6 Å 6.6 Å 8.2 Å	5–6.5 Å	6.4 Å (Na^+) 7.7 Å (Cl^+)	–	0.5 M (NaCl)	H, OH
[107]	30 Å × 30 Å	–	100 MPa	15 Å	4.7 Å	4.14 Å (Na^+) 3.9 Å (Cl^+)	978 water molecules	0.6, 1.2, 1.8 M (NaCl)	Methyl, Ethyl, H, F
[108]	34.09 Å × 34.44 Å	–	50–530 MPa	7.5 Å 8.9 Å	–	9.09 (Na^+) 9.94 (Cl^-)	–	0.6 M NaCl	H, NH_3, N, NOH, NH

and angle deformations as well as changes in electronic structure) are not explicitly included in this model [98]. Cohen-Tanugi et al. [98], performed an auxiliary set of simulations that added the component of polarizability to the orientational component. This allowed for enhanced water polarizability within individual water molecules by permitting flexible bonds and angles, using the SPC/F (flexible) force field. The results of these flexible water simulations indicated that the rigid TIP4P model provided similar dynamics at a lower computational cost. They concluded that intramolecular vibrations and polarizability play a negligible role in the dynamics of saltwater transport and desalination performance. Other studies, modeled water molecules using the nonpolarizable, rigid point charge extended (SPC/E) model [102, 104, 118] because of its simplicity, reliability in reproducing water structure and dynamics, and the availability of ion—water potentials specifically parametrized for SPC/E water [17, 113, 119, 120].

Since the main components of seawater are Na^+ and Cl^- ions [121], the salinity of seawater is often expressed as 0.6 M NaCl or 0.6 mol·L^{-1} which is ~35 g/L. The focus of most MD simulations is on separating water and NaCl and usually, they consider a higher salinity than seawater in order to increase the occurrence of ion—pore interactions and obtain more precise results for given system size and simulation time [98]. Golchoobi et al. [122] have investigated the effect of feed water salinity on permeability and salt rejection of 7.4 Å diameter hydroxylated pores in graphene membranes. Concentration polarization graphs and density map plots for water molecules near the nanopores clearly displayed the orientation of water molecules as a function of time and coordinate. Their results proved that higher salinity causes lower congestion of water molecules near the pore as a result of hydration effects or decreased water permeability. The effect of hydrated ions on water transport and salt rejection revealed the importance of fine manipulation of the feed composition.

The pressures employed in MD simulations are considerably greater than the pressures applied in RO plants (<8 MPa). High simulated pressures on the order of ~100 MPa have the advantage of obtaining more precise data for water flux and salt rejection in a finite simulation time (i.e., order of 10 ns) [123]. This approach is reasonable since water flux scales linearly with the net driving pressure, i.e., the results obtained at hundreds of MPa can be extrapolated to calculate the water flux that would result from lower net driving pressures in an RO system [103].

(c) **Mechanical stability**

Recent developments in the fabrication of nano-porous graphene are promising for the future of water supply by RO desalination. Although previous studies have highlighted the potential of nano-porous graphene membranes, there are few reports exploring their strength and their mechanical integrity under the high hydraulic pressures essential in the RO desalination process. Graphene exhibits exceptional mechanical properties in its pristine, defect-free state; however, the structural strength of nano-porous realistic defected graphene has not been examined in the specific context of water desalination. It is well-known that pores tend to weaken graphene by reducing its fracture strength and its elastic modulus [124–126]. Like polyamide active layers, nano-porous graphene must also lie on top of a porous substrate. It has been shown that the maximum pressure in supported sheets of graphene decreases as a function of nanopore size and substrate pore size [98]. Moreover, the continuous presence of water may further impact the fracture toughness of graphene as a RO membrane [127]. RO process relies on the molecular-level separation of salt ions from water molecules, thus, mechanical failures at the small scales may undermine the entire system [123]. Considering d, as the thickness of the membrane, the stress in a thin membrane under pressure scales with $d^{-2/3}$ [128], and realizing that graphene is ~1,000 times thinner than the polyamide layers used in conventional TFC membranes, the important question arises: *Is nano-porous graphene strong enough to withstand similar pressures without fracturing?*

The combined role of applied pressure, membrane morphology, elastic properties, fracture stress and the effect of water have been systematically studied in [128]. Therein, using MD simulations and continuum fracture mechanics, it is demonstrated that the maximum pressure that nano-porous graphene can withstand depends on (1) the size and spacing of the nanopores and (2) the radius of the pores in the substrate material. Therefore, a nano-porous graphene membrane can maintain its mechanical integrity in RO but the choice of substrate is crucial to this execution. An appropriate substrate with pores smaller than 1 μm would allow nano-porous graphene to withstand pressures exceeding 57 MPa (570 bar). Furthermore, it is revealed that nano-porous graphene membranes with greater porosity may help to withstand even higher pressures [128].

While experimental tuning of the shape of nanopores at the atomic level remains a challenge, Qin et al. [129], performed MD simulations on rectangular nanopores with different areas, aspect ratios (length/width ratios) and orientations to explore their effect on the tensile strength of defective graphene. It was shown that defective graphene with a larger area of rectangular nanopore shows a bigger drop in tensile strength. Furthermore, changes in aspect ratio and orientation of rectangular nanopore can either decrease or increase the tensile strength of defective graphene, dependent on the tensile direction [129]. Simulations of nano-porous graphene were performed by distributing vacancy defects, randomly and uniformly, across the lattice by Anastasi et al. [130]. It was concluded that the fracture stress decreases substantially with increasing defect density and nano-porous graphene sheets with uniformly distributed defects can withstand higher loads when compared to their counterparts with random defects. They found that the fracture stress decreased substantially with increasing defect density. The elastic modulus was found to be constant up to around 5% of vacancy defects and decreased for higher defect densities [130].

Structural response and fracture characteristics of nano-porous graphene membranes subjected to shear loading were studied by Fang et al. [131]. Although shear loading or strength is not directly a problem in RO membranes, considering nano-porous graphene over a support substrate, examining these effects may be helpful in design parameters. A shear load is a force that tends to generate a sliding failure on a material along a plane that is parallel to the direction of the force. When a paper is cut with scissors, the paper fails in shear. The effects of porosity, temperature, and shear velocity on the mechanical responses of nano-porous graphene membranes were examined in [131]. Their results showed that the wrinkling of the membrane became more obvious with increasing strain. Fracture stress in the nano-porous graphene membrane decreased with increasing porosity. In addition, the effect of shear velocity on the shear modulus decreased with increasing porosity [131].

(d) **Multilayer graphene**

Single-layer nano-porous graphene has shown potential as a RO desalination membrane, and multiple efforts have targeted the synthesis of large-scale graphene films [132]. However, producing perfect monolayer graphene over large areas remains challenging. The primary method for synthesizing graphene sheets, chemical vapor deposition (CVD) results in substantial multilayer coverage, defects, tears and wrinkles [28]. However, multilayer graphene membranes can be synthesized more economically than the monolayer material and have higher mechanical stability. Some of the physical implications of these multiple graphene layers on the membrane's performance have been studied recently [99, 133]. The major concerns are *whether these multilayers will maintain high water permeability? what is the interplay between nanopores on different layers? How would they influence the membrane's water permeation and salt rejection ability? Do we have the tool to create arrays of pores with diameters ~1 nm, with suitable distribution and deep enough into all layers across large-scale structure?*

MD simulation was employed to investigate the effects of layering graphene on RO performance. This was done by taking a bilayer nano-porous membrane as a model system [99]. The model assumed that the alteration from a monolayer to two layers, demonstrated the most important effects of adding layers and that subsequent layers have a qualitatively similar impact. It was shown that multilayer nano-porous graphene membranes display similar promising desalination properties to monolayer membranes. However, manipulating layer separation and pore alignment is essential to improve desalination performance [99]. Figure 6 shows a snapshot of a typical bilayer simulation with effective parameters on the desalination performance. The key parameters that were explored in multilayer nano-porous graphene sheet efficiency for water desalination studies were the feed pressure P, the offset O between nanopores in the upstream and downstream graphene layer, and the spacing H between graphene layers. The interlayer spacing can be as fine as H = 3.35 Å ($\propto$ layer spacing in graphite) or can be much larger. Some physically relevant spacing are 10 and 14 Å that correspond to two and three water layers, between the graphene sheets [99].

Cohen-Tanugi et al. [99] demonstrated that for completely aligned pores in bilayer graphene, the flow rate decays nonlinearly with increasing layer separation. Free energy barrier studies reveal that a small separation between the pores results in a system possessing a single barrier for transport. This height of this barrier is larger than a monolayer nano-porous graphene membrane partially due to the stronger surface adsorption of water on the bilayer nano-porous graphene surface. Conversely, for maximally offset pores, the interspace region does not allow any water passage

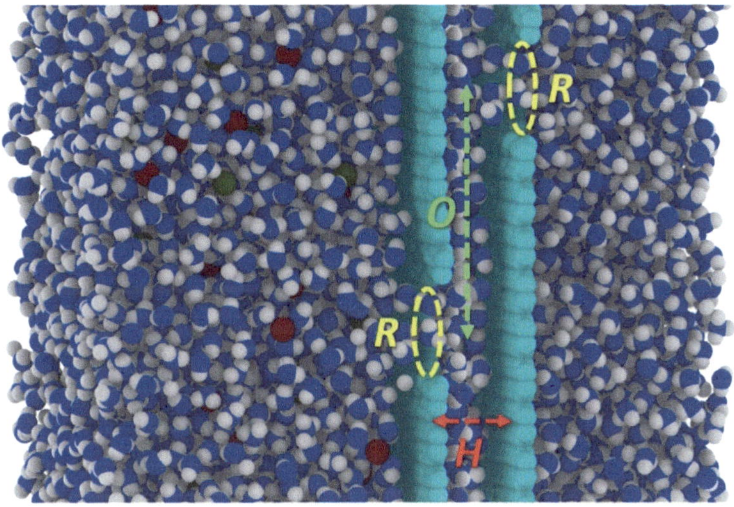

Fig. 6 Snapshot of typical bilayer simulation with effective parameters on the desalination performance

[99]. They also found that the bilayer nano-porous graphene membrane is capable of rejecting salt for sufficiently small nanopores (<4 Å) where salt rejection decreases for greater pore alignment, larger layer spacing, or higher pressure. Pore offset and layer separation considerably influences the membrane's ability to reject salt. In the limit of large H, it is expected that the two layers act as independent membranes. For N-layer membranes, the permeability can be predicted using a resistance-in-series model while increasing the number of layers results in greater salt rejection due to a higher diffusion resistance for ions. Although nano-porous graphene membrane's permeability will drop linearly as the number of graphene layers increases, the resulting permeability loss may still represent an acceptable trade-off [99].

Zhang et al. in [134] also investigated design parameters in multilayer nano-porous graphene systems for water desalination. In addition to the offset O (12.6 Å) between the nanopores and the interlayer spacing H (7–12 Å) between graphene layers, they explored the number of pores and layers, considered more realistic pore diameters (8–12 Å), optimized the parameters and examined salt rejection for NaCl, $MgCl_2$ and $CaCl_2$ solutions. Similar to [99], they found that salt rejection is influenced substantially by the interlayer spacing distance. This was due to the large free energy between ions and graphene sheets and the relatively large size of the hydration layer around the ions. With the increase in the number of graphene sheets, the water flow rate was slowed down. Alternatively, the salt rejection rate increased notably with the number of graphene sheets in agreement with the report by Mooney et al. [135]. Water flux per area per time analyses revealed the dependence of the desalination performance on the number of pores. With the fluctuation of salt rejection around 100% owing to the trapping phenomenon of ions in small interlayer spacing, water flux increased with the increase in the pore numbers.

Cohen-Tanugi et al. [99] summarized the most efficient scenarios for multilayer experimental situation based on their simulation findings in four points:

(1) If both pore alignment and layer separation could be precisely controlled, a multilayer nano-porous graphene membrane with the smallest possible layer separation and fully aligned pores would represent the most promising choice.
(2) If only layer separation could be precisely controlled, sufficiently large layer separation (i.e., 8 Å) would be preferable, to avoid fully obstructing water passage across misaligned pores.
(3) If only pore offset could be controlled, a membrane with fully aligned pores is desirable.
(4) If neither pore neither alignment nor layer separation could be controlled, having a nano-porous graphene membrane with the greatest possible pore density is recommended to enhance the likelihood of having aligned pores.

In an innovative MD study, Abdol et al. [136] created aligned pores in multilayer graphene using focused ion beam irradiation and strengthened the stability of the whole structure simultaneously. They proposed that incident ion beams, besides creating pores, can reinforce the lateral strength of multilayer graphene by creating some permanent covalent bonds around the pores between graphene layers. They also showed that by changing the kinetic energy of the incident beams the density of the covalent bonds can be tuned. Graphene layers in a multilayer structure have a great tendency to be delaminated or swelled in an aquatic environment. Considering that, the proposed technique could be a promising approach to overcome the weak points of multilayer graphene structures and make them suitable candidates for use as water purification membranes. However, these ideas are yet to be explored in the experiment.

(e) **Ion transport and its implications**

Membranes functioning over different length scales, achieve selective transport through a variety of mechanisms [20]. At the smallest scale, dense polymeric non-porous RO or gas separation membranes, such as RO membranes for water desalination, operate by a solution-diffusion mechanism [5, 61]. In these systems, differences in solubilities and diffusivities of the species in the membrane material trigger selectivity. While solubility relates to the molecular structure, porosity and chemical affinity of the membranes; diffusivity is regulated by thermally activated rearrangements of the polymer chains [20]. More permeable materials typically provide less selectivity when selectivity is only governed by diffusion, and result in a trade-off between permeability and selectivity. The incorporation of additional mechanisms, such as chemical affinity or molecular sieving assists in overcoming this trade-off [13].

The atomic thickness and composition in 2D membranes induce several discrepancies with solid-state nanopores. Sahu and Zwolak [137] have comprehensively discussed ionic phenomena in nanoscale pores through 2D materials. For pores in 2D membranes with diameters above ~2 nm, 'access resistance' is dominant over 'pore resistance'. When the pore radius is below 1–2 nm, the 'pore resistance' plays the dominant role. The resistance for ions to transfer from one end of the pore to the other is defined as the 'pore resistance'. On the contrary, the resistance for ions to converge from the bulk electrolyte away from the membrane to the mouth of the pore is the 'access resistance' and occurs on both sides of the membrane. While both resistances influence ion transport, pores in 2D membranes, balance these two contributions differently than other membranes [137].

As earlier mentioned, pore functionalization can drastically change ionic transport, especially if the pore is smaller than the size of the hydrated ion [20]. Charged or partially charged functional groups can reduce the energy barrier for ions of opposite charge and increase the barrier for ions of like charge [104] along the pore edge. This will lead to cation/anion selectivity [109]. Astonishingly, the flexibility of the membrane can also impact the 'pore resistance'. In contrast to solid-state pores, graphene pores are more flexible, and their dynamic area can be larger than the static area [138].

In a less sophisticated view to ion transport, Thomas et al. [139] proposed six main mechanisms for salt rejection by nano-porous (sub-nanometer) monolayer graphene membranes: size exclusion, dehydration effects (steric exclusion of the hydration shell), charge repulsion, interactions with the pore, interactions of solutes with specific chemical structures of the pore and entropic differences. The strong electric field around dissolved ions forces the nearby water molecules to orient into hydration layers. The first hydration layer is strongly bound to the ion with an energy range from ~1 eV in monovalent ions to ~10 eV in bivalent ions. This hydration layer tends to move along with the ion. The second layer is only partially oriented, and the third hydration layer is diffuse and only weakly defined [140]. As the diameters of the first hydration shell for Na^+, K^+, Ca^{2+}, Mg^{2+} and Cl^- are 0.72 nm, 0.66 nm, 0.82 nm, 0.86 nm and 0.66 nm, respectively; which are larger than the effective size of a water molecule (0.26 nm), attributing the size exclusion and dehydration effects are the most important salt rejection mechanisms [17, 20, 141]. Neutrally charged pores smaller than the ion hydration size, therefore, present a blockade to ion transport. The transport of water is influenced by hydrogen bonding and structuring of water molecules when the pore size is below ~2 nm [20]. Again, 'pore resistance issues should be carefully taken into account. The smallest pores in graphene that allow water to pass through, can accommodate only a single water molecule in their cross-section and therefore exhibit single-file movement of water molecules [104]. In pores with the diameter <1.5 nm, water molecules adopt certain preferential configurations as they pass through the pores [104, 142].

While much effort is focused on the molecular understanding of microscopic mechanisms driving water and ion transport through nano-porous graphene, only few works [130, 143] have connected interfacial properties, such as surface tension and friction to water and ion transport through graphitic monolayers. The experimental determination of the contact angle depends on the impurities and defects on the surface, which may lead to the scattering of contact angle values. Because the material surface is critical for compatibility with the surrounding environment, several molecular simulations have been performed to describe the interfacial region at the atomistic scale. Some of these atomistic simulations have been used to predict the contact angle of water on different surfaces [144, 145]. The contact angle value depends on the chemical nature of the surface and other parameters, such as roughness and chemical heterogeneity, which have been much less investigated from this theoretical approach. Another way for estimating solid—liquid interactions is to compute the solid—liquid interfacial tension. Thermodynamic definitions of the interfacial tension were used to compute the solid—liquid interfacial tension for graphene—water systems [146]. Water transport mechanisms including velocity, viscosity and slip length have also been studied in nano-porous atomically graphene membranes. Since water structure, is confined in the radial direction and layered in the axial direction of the pore, water viscosity and slip length increase with a decrease in the pore radius [127].

4 Experimental Work on Nano-porous Graphene Membranes

Practical attainment of nano-porous atomically thin membranes in RO systems involves the following: (1) synthesis of a continuous large layer of the material, (2) engineer of nanopores in terms of size and distribution and (3) the ability to handle the materials using suitable porous supports. The latest achievements at the experimental level in these fields are discussed in the following subsections.

4.1 Fabrication of Large-Area Graphene

CVD has emerged as a tunable and versatile method for producing continuous layers of 2D materials over large areas [24, 147] and holds promise in RO membrane fabrication. The CVD method for graphene synthesis was first reported in 2006 on a Ni substrate [148]. Later in 2009, Li et al. [149] demonstrated large-area CVD growth of monolayer graphene on copper, followed by roll-to-roll transfer of 0.76 m sheets of CVD graphene by Samsung in 2010 [150]. Kobayashi et al. [151] from SONY exhibited roll-to-roll synthesis and transfer of 100 m long graphene-coated films in 2013. Other techniques developed in recent years for the fabrication of graphene sheets including micromechanical cleavage, liquid-phase exfoliation, and epitaxial growth on SiC, result in high quality, small flakes, and low yield. Other methods like self-assembly, electrochemical exfoliation, unzipping of carbon nanotubes (CNTs) and reduction of graphene oxide either produce very small flakes or flakes of low quality. Overall, apart from CVD and epitaxial growth, all the other methods lack appropriate control over the number of layers [20, 27, 147, 152–155]. Figure 7 summarizes the main fabrication methods of graphene in terms of sheet area and cost.

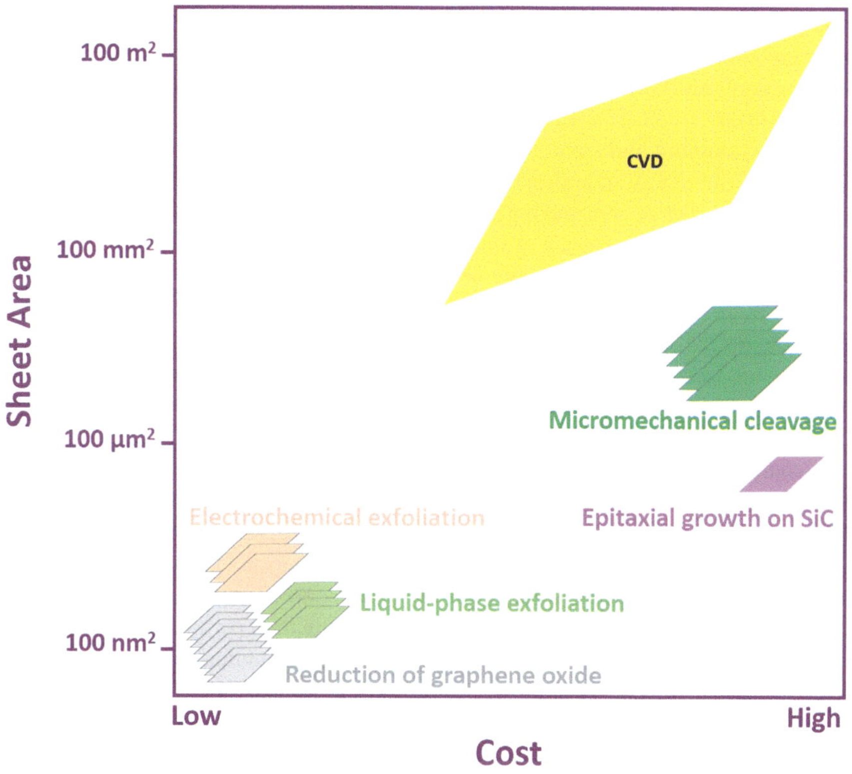

Fig. 7 Sheet area per cost of different graphene fabrication methods

Experimental work on nano-porous atomically thin membranes for RO has largely focused on CVD graphene. The advantages and problems regarding this method is technically advanced and tackled [20].

4.2 *Appropriate Porous Support Layer for Free-Standing Graphene*

Although there is the possibility of creating nanopores onto the large area material prior to its transfer over the appropriate support layer [156], but the inverse process is usually practiced [31, 33, 157–159]. Since CVD graphene is used for membrane fabrication, a transfer of the graphene layer is necessary from the underlying growth support to the substrate of interest [160]. The transfer usually involves chemical etching of the metal catalyst like Ni or Cu on which graphene has been grown. A successful etching process and survival of graphene requires a very thin polymer being spin-coated on the sheet beforehand. Once the metal is completely etched from the backside of graphene and the polymer layer/graphene is fished out of the water onto the substrate of interest, the polymer layer is removed by a combination of cleaning methods (i.e., chemical and heating). However, this polymeric layer is never fully removed [31, 33, 158, 159] and leaves behind some residues that undermine the mechanical stability and continuity of the graphene sheet. For RO applications, the substrate of interest is an appropriate porous support layer that allows graphene to free-stand over a large area and together with the properties of graphene, accomplish high selectivity in the presence of imperfections and defects [161].

Ideal support or coating for the active layer will add a transport resistance in series to both selective pores and defects, thus limiting leakage without adversely impacting the permeance [161]. This resistive layer must be thinner than the spacing between non-selective defects in the atomically thin layer [6, 161]. More critically, the support must have high surface porosity for efficient use of the surface area of the active layer, have high mechanical strength, higher permeance than the active layer, good chemical resistance, provide stable adhesion and facilitate transfer or coating of the active layer [161]. Most experimental studies have used specialized supports for RO applications. Polycarbonate track-etched [33], poly(1-methyl silyl-1-propyne) [162], polypropylene [163] and polyvinyl idene difluoride [163] membranes and microfabricated Si supports [29, 31, 158, 159] are some examples. Microfabricated Si supports have low surface porosity, which results in high permeance with respect to the graphene area but low permeance based on total membrane area. They are also expensive and require several fabrication steps. Additionally, they are not flexible to fit a spiral-wound module for practical large-scale separations [159]. Even the polymeric substrates are suboptimum [20, 33], thus there is a need for the design and fabrication of a new generation of supporting layers; layers that meet all the mentioned requirements to match nano-porous atomically thin membranes for pressure-driven separations.

4.3 Fabrication of Nanopores on the Active Layer

Nano-porous graphene membranes as the active layer in the RO process can be fabricated by creating pores in an initially non-porous atomically thin material, or by synthesis of an intrinsically porous material. The former attains large area nano-porous graphene sheets useful for scale-up in industry and the latter, gives lab-scale nano-porous sheets. An active layer of RO membrane requires nanopores large enough that enable water passage and ion blockage when size-exclusion separation technique is employed. Although nanopores could be developed from localized defective sites, precise control of the size and density distributions of the pores involves specific perforation strategies [26, 164]. Most of Sect. 4.3 is reproduced with permission from Ref. [26], Copyright 2019, *Royal Society of Chemistry.*

The generation of pores across pristine graphene or other graphitic materials can be achieved by various perforation routes usually classified into four categories: stochastic etching, guided-etching, guided growth techniques [165] and self-assembly. Among them, irradiation etching, and guided growth are ideal for RO membranes.

(i) Etching routes are conducted via removing a few carbon clusters from the graphitic basal plane of graphene sheets by providing the minimum ablating energy to break the C = C sp^2 bonds and generate a pore behind in the graphene structure [26]. This activation energy can be delivered by several excitation sources, such as irradiation [166, 167], thermal [168] and chemical/plasma [27, 31, 158, 169–171] processes. Irradiation itself can be via gamma [172], electron [173, 174], ion [33, 166, 173, 175], electric pulses [176, 177] and ultraviolet [27] routes. Etching based on irradiation techniques displayed a pore size range of 0.5–1.2 nm and pore densities across graphitic structures between 10^3 and 10^6 pore/cm^2 based on treatment conditions [178, 179]. Control of power and time of oxygen plasma which also induces nanometre-scale pores in suspended single-layer graphene; has resulted in 1–10 nm wide pores at a high density [31, 158, 171, 180, 181]. Etching based on chemical etchants and thermal activations were dependent on the development of intrinsic defective sites at different etchant concentrations, operating pH and temperatures [182–184]. These techniques exhibited a wide range of pore sizes between 10–100 nm and a limited surface density <10^2, with kinetic rates varying from a few minutes to several hours [27, 168]. While these etching methods produce a distribution of pore sizes and possibly work better in single-layer materials, Focused Ion Beams (FIB) can precisely machine a large number of individual pores down to ~8 nm in diameter [29] and electron beams focused below 1 nm can generate precise sub-nanometer to few-nanometer pores [185, 186]. One of the early works that showed that nanopores can be introduced onto 2D suspended graphene was reported by Fischbein and Drndic in 2008, who used focused electron to drill nanometer sized holes in multilayer graphene [187]. However, machining using FIB is less scalable and appropriate for smaller membranes, whereas electrochemical methods and machining using atomic force microscopy (AFM) or tightly focused electron beams are suitable for the creation of a few pores for microscale membranes [170].

(ii) Guided-etching perforation routes involve a temporary porous template via lithography [188] and pore mediators [189]. Several templates were used that may include continuous porous structures in addition to nanospheres and nanoparticles [189, 190] for localizing the oxidative energies during the perforation process. This method offered a pore size distribution of 20–200 nm and pore density distribution between 10^4 and 10^1 pores/μm^2. However, functionalization of such pore mediators for patterning into ordered arrays via linkers [191] may increase the porosity level and provide better distribution and anchoring across graphitic surfaces.

(iii) The guided-growth perforation method can produce pores either during or after the nucleation process of graphitic nanostructures [26]. During the CVD growth, the graphitic nucleation can be terminated for specific locations on the substrate to control gaps between graphene islands and to define pores between them [190, 192–194]. In this method, pores could potentially be created by controlling the ratio of the growth precursor and etchant (for example, CH_4 and H_2). It has been shown that guided-growth perforation routes can generate pores with dimensions ranging from a few Å up to sub-nanometres and densities of 10^3–10^5 pores/cm^2 [26].

(iv) In contrast to the creation of pores in initially non-porous material, recent advances in self-assembly techniques have the potential for creating a high density of atomically precise pores, directly in covalently bonded single-layer materials [41, 195–200]. For example, surface-assisted synthesis of nano-porous graphene by aryl–aryl coupling of polyphenylene-based precursors or cyclodehydrogenation has been proposed [195]. Furthermore, graphdiyne, a new two-dimensional (2D) carbon allotrope, has been created at the millimeter scale by cross-linking of hexaethynylbenzene on copper under nitrogen atmosphere [201]. Nevertheless, these approaches face key challenges in the synthesis of continuous layers over sufficiently large areas for membrane applications [202].

4.4 Characterization Tools of Nano-Porous Graphene Membranes

Once mono or few layer graphene is fabricated via the CVD method, transferred onto an appropriate porous support layer and perforated through any of the methods discussed in the previous part, and prior to RO measurements; it is necessary to characterize the surface of the membranes and inspect various surface features and properties.

For assessment of continuity of free-standing surface over large scales, optical microscopy, scanning electron microscopy (SEM) and field emission scanning electron microscopy (FE-SEM) would be safe options. These methods allow large area inspection, without damaging the membrane. FESEM is carried out on conductive surfaces or on non-conductive surfaces coated with few nm of a metal. Although pristine graphene is conductive, polymeric residues left from the transfer process on CVD graphene are non-conductive and resolved images could not be captured with the electron beam and often, a deposition of Au is necessary. AFM inspection for continuity or number of layers is not recommended with contact mode, again due to the numerous polymeric residues left on the surface but if careful enough with tapping or non-contact mode, one may be lucky to scan across free-standing membranes without having the tip ripping the surface. Yet, evaluating the number of layers is highly recommended with Raman spectroscopy with 2D peak intensity indications [31, 158, 203].

Nanopore assessment is possible through several techniques. First, careful imaging with high-resolution transmission electron microscopy (HRTEM) is the most non-invasive method for this purpose. However, it requires a transfer of the nano-porous graphene membrane over TEM grids which can be done following the protocols in [158, 159]. In this method, not only the pore size and density can be evaluated across the surface by multiple imaging, but the nanopore shape and form can be assessed. Low temperature (LT) and ultra-high vacuum (UHV) scanning tunneling microscopy (STM) allows for inspection of membrane flatness, defects, nanopore size and terminations, topographically and electronically with atomic resolution. However, for resolving the nanopores, it is recommended that the membrane be transferred onto an atomically flat and clean conductor like Au surface. A more precise technique for nanopore examinations at the atomic scale is advanced non-contact (NC-) AFM operating in UHV and LT. This technique provides similar topography information as STM but is superior to STM for two reasons: first, it allows force spectroscopy on the surface providing mechanical information of the membrane. Second, it does not require a conductive sample so the membrane can be transferred onto any arbitrary atomically flat surface. If the nanopores are created using oxygen plasma, investigating the evolution of Raman D peak and G peak of graphene with plasma exposure time is very useful. Correlating the Raman results with resolved topographic results on the nanopores obtained from HRTEM, STM or NC-AFM gives insight into the distribution of the nanopores across large-area graphene membranes [31, 158, 159].

If the nano-porous graphene membrane is functionalized with different functional agents, a combination of X-ray photoelectron spectroscopy (XPS), Fourier transform infrared spectroscopy (FTIR) and energy-dispersive X-ray spectroscopy (EDX) gives information on the percentage of these additives, defect interplay with the surface, type of the agents and the bonds they have made with the carbonaceous membrane in bulk. Again, investigating the evolution of Raman D peak and G peak of graphene at different functionalization conditions gives insight into alterations induced on the surface by functional groups [50, 51]. However, to demonstrate the termination of the nanopores and what the electronic and mechanical impacts of the functional groups are on the surface, atomic resolution imaging such as STM and NC-AFM in UHV and LT are vital.

4.5 Realization of Nano-Porous Graphene Membranes for Water Desalination

Early experimentally realized suspended atomically thin membranes were produced by a combination of photolithography and mechanical exfoliation of graphene [27, 83]. The SiO_2 support layer consisted array of circles with diameters of 5 mm and 7 mm defined by photolithography. Reactive ion etching (RIE) was then used to etch the circles into cylindrical cavities with a depth of 250–500 nm, leaving a series of wells on the wafer. Mechanical exfoliation of graphite was then used to deposit suspended graphene sheets over the wells and pores were introduced by ultraviolet-induced oxidative etching in only 5 μm wide standing graphene membranes. The membranes were tested up to 200 kPa and successfully used as molecular sieves [27]. Later, O'Hern et al. [28] fabricated a millimeter-scale (25 mm^2) monolayer CVD graphene membrane with intrinsic pores instead of extrinsically generated ones and transferred it onto a polycarbonate membrane (PCTE). Natural defects were formed during the fabrication process with 1–15 nm in size. The rejection percentage for potassium chloride (KCl), tetramethylammonium chloride, Allura red, and thiuram disulfide were 46, 71, 23, and 17%, respectively. Two years later, the same group reported ionic selectivity in macroscale monolayer graphene membranes [166]. This time, they introduced sub-nanometre pores in graphene by ion bombardment followed by chemical etching. Although there was significant leakage through defects, the mem-branes displayed some K^+/Cl^- selectivity. Further etching caused the membrane to permit selective transport of KCl over a larger organic molecule (~1.0 nm size). In the same year, Celebi et al. [29] stacked two layers of micrometer-sized graphene on silicon support and formed an almost impermeable layer. Ordered arrays of pores were machined by FIB onto bilayer graphene with diameters of 8–1000 nm. Water transport measured across the membrane under pressures up to 200 kPa exhibited water flow rates five to seven folds greater than commercial ultra-filtration mem-branes. It is noteworthy that there are no reports of creating aligned sub-nanometre pores in multilayer materials [20]. In an attempt to overcome leakage across large intrinsic tears generated in the transferred graphene sheets, O'Hern et al. [33], plugged the gaps with inorganic materials in 2015. For this, they activated the graphene sheet in a KOH solution and deposited a 3.5 nm layer of Hafnia (HfO_2) on the sheet to fill the intrinsic defects, followed by intrusion of nylon-6,6 plugs within the pores by interfacial polymerization. Sub-nanometer pores with an average size of 0.5 nm and density of 3.8×10^{13} pore/cm^2 were then created by ion irradiation and etch-ing on the graphene sheet. The designed membranes demonstrated water permeance of ~1.41 L m^{-2} h^{-1} bar^{-1}, in the same order as RO commercial membranes. However, the membranes exhibited different rejections under osmotically driven flow at 70, 90 and 83% for $MgSO_4$ (0.86 nm size), Allura red (~1 nm size), dextran (~3.7 nm size), respectively. Additionally, the membrane exhibited a negative rejection ratio toward NaCl (0.716 nm). Although the membrane this team produced was comparatively large-sized (cm-scale), it was only tested using an osmotic pressure-driven configuration. Higher pressures were avoided since the sealed defects were still prone to failure in RO-driven configuration [204]. Later in 2017, Qin et al. [205] used an innovative liquid-casting method to transfer 63 cm^2 graphene nano-porous membrane onto a porous polymer substrate with the minimal defect. Nanopores were generated on the graphene surface using similar method by O'Hern et al. [33]. The resulting membrane was not able to reject salt ions (0% NaCl rejection) but showed flux a few times higher than typical polymeric membranes.

Back in 2015, Surwade et al. [31] demonstrated water/ion selectivity by nanoporous monolayer CVD transferred graphene onto 5 μm apertures of SiN in gravity-driven configuration up to a pressure of 17 kPa. Sub-nanometre pores were etched on the graphene membrane by O_2 plasma treatment which led to a pore size of 0.5–1 nm and a density of around 10^{12} pore/cm^2. Over 200 samples were fabricated in this work, with 20% success, meaning 80% of the samples were ruptured with either large or small tears and ruled out of the measurements. The tearless samples gained 93–100% rejection of 6 mM KCl. Their study also provided a reproducible protocol for pore generation by tuning the power and time of O_2 plasma evidenced in Raman spectroscopy spectra and analyses of D and G peaks intensities. Water permeability in Surwade work was exceedingly high, several orders of magnitude higher than current RO membranes, when gravity-driven configuration was used. Extremely high permeability was later observed by Kazemi et al., in 2019 [158, 159] when a similar configuration was used. It is likely that the water transport mechanism in the configuration of water-filled feed-empty permeate is significantly different from the configuration of water-filled feed-filled permeate [204]. Another explanation is the role of the SiN/graphene interface and the small aperture size of the SiN contribution to achieving such high permeance [26, 158, 159]. Note that the water flux from osmotic diffusion dropped to much lower values due to ions binding with unsaturated edges around pores that preceded the nanopores blockage [31].

In contrast to previous studies, graphene exfoliated by ribonucleic acid was used as a raw material by Park et al. [206]. Nanopores with a 5 nm nominal size was formed with the assistance of Cu nanoparticles as a template during a guided HNO_3 etching procedure. At a feed pressure of 83 kPa, 1.662×10^4 L m^{-2} h^{-1} bar^{-1} water permeance was achieved. The salt rejection was measured at ~25% for Mn^{2+} and ~20% for Fe^{3+} ions by electron paramagnetic resonance. In 2017, FIB etching was employed in a few layers of graphene to generate pores by Jang et al. [32]. This was followed by O_2 plasma treatment to modify the surface chemistry of the porous structure and further extend the dimension of the pore which generated a wide range of pore-size from 0.18 nm up to 15 nm with and density of 1.67×10^{13} pore/cm^2 in 50 s of etching. Interfacial polymerization was used to mitigate large intrinsic tears in the graphene layer surface. About 98% rejection was obtained for $MgSO_4$ with water permeance comparable to other nanofiltration membranes. Faster water transport rate was achieved for the membranes etched by the combination of ion bombardment and plasma treatment.

On a different approach, Boutilier et al. [30] focused on designing an appropriate porous support layer to minimize defects side effects. Using a resistance model, they chose supports with similar resistance as the selective pores in graphene, which was proved to be effective in isolating small defects and limiting leakages through large defects [207]. They successfully fabricated a cm-scale membrane which was tested up to 200 kPa. A year later, Yang et al. [208] employed the interwoven network of single-walled carbon nanotubes (SWCNT) as a novel support layer to mechanically reinforce nano-porous graphene membranes. This method assisted in avoiding solute leakage during water desalination. A SWCNT layer was transferred over graphene, and then covered by SiO_2, as a mesoporous template, prior to O_2 plasma treatment. The resulting hybrid membranes exhibited pore sizes of 0.32 nm, 0.55 nm and 1.14 nm when treated for <5 s to 20 s of O_2 plasma. Using crossflow nanofiltration, water desalination performance was assessed with these membranes. They showed salt rejections from 85.2 to 93.4% towards Na_2SO_4, $MgCl_2$, NaCl and KCl while experiencing a range of pressure. 2–4 MPa of pressure for non-supported stainless-steel mesh (30 mm) and 8–10 MPa for supported polycarbonate track etch membrane (0.2 mm) were sustained, respectively. Further advancements in applying efficient support layers for nano-porous graphene membranes were conducted by Kazemi et al., [158] in 2019, using various TEM grids as porous primary mechanical supports to maintain the mechanical integrity of the membranes. 1.27×10^6 μm^2 to 7.92×10^5 μm^2 spanned area monolayer CVD graphene, were transferred onto hole-through Ni TEM grids with aperture sizes of 6.5–70 μm. O_2 plasma etching was used to create nanopores with sizes from 1 nm to 8 nm in every 100×100 nm^2 unit area. TEM analysis showed that the distribution of 1 and 2 nm wide pores were several times higher than 5–8 nm wide pores. The team used gravity driven configuration to examine water desalination parameters at 10 kPa pressure. NaCl and KCl rejections varied from 7 to 46% from the largest TEM aperture size to the lowest. To improve the salt rejections and minimize the leakage from the defected graphene surface, they used a secondary support provided by porous SiN wafer with an overall smaller aperture than the entire TEM mesh above. This reduced the graphene effective area to 2.77×10^4 μm^2 but increased the NaCl rejection to ~76% while maintaining a very high water permeability. In a more in-depth study, the same team investigated large area mobile superimposed nano-porous monolayer graphene membrane/grid on to different SiN/Si holes with distinct size (6.5, 10.5, 30 and 75 μm) and spacing (21.4, 60, 200 and 238 μm), experimentally and by modeling [159]. Their studies showed NaCl rejection of 58–100% in different configurations of graphene/grid/holes with very high water permeation ~5×10^7 L m^{-2} h^{-1} bar^{-1}.

By synthesizing bilayer graphene membranes containing pores with diameters from ~6 to 1000 nm, Buchheim et al. [133] investigated liquid permeation over a wide range of viscosities and pressures. With the goal of assessing the thickness—permeation paradigm, they revealed that nano-porous membranes with thicknesses up to 90 nm demonstrate dominance of the entrance resistance for aspect ratios up to one. Furthermore, they showed that liquid permeation across these atomically thin pores was limited by viscous dissipation at the pore entrance independent of thickness and universal for porous materials.

There are some recent advances in the literature that report large area mono or bilayer graphene transfer over porous supports with high mechanical durability. Although they have not directly used these membranes for water desalination but exploiting new ideas from these reports is recommended for more efficient fabrication of RO membranes. Huang et al. [209] reported a novel nano-porous carbon-assisted transfer technique, yielding the transfer of relatively large area (1 mm^2) monolayer CVD graphene onto macro-porous support (with 5 μm pores) with an ultra-low density of the intrinsic defects (porosity of 0.025%) without generating cracks. Wang et al. [210] directly investigated the ability of cm-scale monolayer CVD graphene withstanding high pressure when placed on porous polycarbonate track-etched membranes with diameters in the range of 30 nm to 3 μm using electron microscopy, AFM, and gas flow measurements. Surprisingly, non-wrinkled areas withstood pressures exceeding 100 bar at which many kinds of membrane suffer from compaction [210]. They also showed that if wrinkles on single-layer graphene membranes were isolated using supports with small pores, the sheet could sustain ultrahigh pressure. Schmidt et al. [211] fabricated nanomesh by patterning large-area monolayer graphene with nanometer-scale pores with 10 nm pitch and <4 nm pore diameter by the direct helium ion beam milling requiring no post-patterning process. Akbari et al. [25] fabricated ultra-clean single and bilayer CVD graphene membranes with diameters up to 500 μm and 750 μm, respectively, using inverted floating method (IFM) followed by thermal annealing in vacuum. By measuring the dynamic mechanical properties of the membranes, they observed a reduction of the effective intrinsic stress in the graphene membranes with annealing.

5 Research Gaps

Although a lot of research has been conducted on the physics and chemistry of nano-porous graphene membranes in the last decade, the incorporation of these materials into RO systems at an industrial scale is still far from reality. There are aspects that have not been fully understood or limitations that have not been tackled and analyses that are yet to be done both theoretically and experimentally to overcome all the existing challenges in this field. Being part of a much larger family of 2D materials, any advancement in understanding the fundamental science at the atomic level across the nano-porous graphene membranes or progress in fabrication techniques will develop similar characteristics in other 2D materials. The research gaps in the study of nano-porous graphene membranes can be categorized into two main parts: Computational and experimental.

Computational gaps: In general, the computation suffers from limitations on simulations due to computational cost, regarding both spatial cell size and time scales. MD simulations as the main computational tool in capturing the overall essence of water and ion dynamics across the membranes have the following constraints [137]: uncertainties in the applied force fields, quasi-long range convergence of the bulk to the pore, resistance to "normal" bulk flow and Newtonian mechanics that govern dynamics of the species. While quantum mechanics needs to be incorporated in the simulations to grasp a more legitimate picture of the interactions between atomic features in these studies, larger spatial cell size and lower time scales are necessary to allow an envision of the bulk of the system. Due to these limitations, many realistic aspects of transport phenomena are ignored or simplified that hinders a genuine understanding and envisage of these materials. Another important constraint due to computational costs is considering the pristine form of the membranes as the basis of computational studies, while a large area of free-standing CVD graphene with all its intrinsic defects, cannot be simulated realistically. Even with a very small cell size of pristine graphene and well-defined pore shape, the applied pressure and salinity of the solution in the simulations are assumed much higher than conventional RO systems to avoid additional computational costs. Fouling, as the major obstacle for the operation of the current RO membranes, is almost ignored in computational studies. Only recently, Kohler et al. [212] employed MD simulations to explore the flocculation as a prominent factor in desalination through nano-porous graphene and MoS_2 membranes. Although there are some insights now on the ideal size and distribution of nanopores on the free-standing membranes regarding mechanical robustness, there is no precise protocol with a clear definition of size and distribution of the pores available to experimentalists.

Experimental gaps: Unlike simulations, experimental efforts towards realizing nano-porous graphene membranes for RO technology, have confronted us with the reality of this approach. With all the fascinating properties of 2D materials and indeed graphene, there are some crucial experimental challenges we need to focus on. Fabrication of industrial-scale continuous monolayer graphene without intrinsic or extrinsic defects is at the forefront. With all the advances in perforation techniques, we are still lacking a technology that creates well-defined pores ~1 nm in diameter. Pore density is also an important factor for controlling the water permeability of membranes based. Is the predicted pore density 10^{12} pores/cm^2 an ideal number? Likewise, the chemistry and geometry of pores can significantly impact the water flux [26]. What about the support layer? There are few studies focused on the impact of the support layer but there is no conclusion on the ideal type, thickness, porosity, chemical and physical properties of this layer. No comprehensive experimental work is yet reported on fouling challenges of nano-porous graphene membranes while fouling mitigation is the most fundamental reason that motivates replacement of the current RO membranes with nano-porous atomically thin membranes. Excessive fouling deteriorates membrane performance and demands chemical cleaning which results in short membrane life and increases energy consumption and consequently the operating costs [63, 213]. Production at appropriate scale and cost, packaging into modules that minimize concentration, and demonstration of long-term performance under realistic conditions is essential [20]. These require further investigation to achieve enhanced performance, toward a reduction of operating costs for desalination plants.

6 Conclusions

Unique properties of pristine graphene such as atomic thickness, extreme flatness and mechanical robustness, have motivated theoreticians and experimentalists to exploit it for a new generation of membrane-based materials in water purification and desalination technologies. Latest findings at the atomic level by MD simulations, and experimental advancements in the fabrication of large-area free-standing nano-porous membrane was reviewed, the physical interplay of nanopores across graphene membranes with ion transport was investigated and the challenges, prospects and gaps were highlighted. Nano-porous graphene membranes represent tremendous potential for advancing membrane technology and are likely to endure into the future.

References and Future Readings

1. Greenlee LF, Lawler DF, Freeman BD, Marrot B, Moulin P (2009) Reverse osmosis desalination: water sources, technology, and today's challenges. Water Res 43:2317–2348. https://doi.org/10.1016/j.watres.2009.03.010
2. Shannon MA, Bohn PW, Elimelech M, Georgiadis JG, Marinas BJ, Mayes AM (2010) Science and technology for water purification in the coming decades. World Scientific 337–346. https://doi.org/10.1142/9789814287005_0035
3. Malaeb L, Ayoub GM (2011) Reverse osmosis technology for water treatment: state of the art review. Desalination 267:1–8. https://doi.org/10.1016/j.desal.2010.09.001
4. Pendergast MM, Hoek EM (2011) A review of water treatment membrane nanotechnologies. Energy Environ Sci 4:1946–1971. https://doi.org/10.1039/c0ee00541j
5. Baker RW, Low BT (2014) Gas separation membrane materials: a perspective. Macromolecules 47:6999–7013. https://doi.org/10.1021/ma501488s
6. Baker RW (2002) Future directions of membrane gas separation technology. Ind Eng Chem Res 41:1393–1411. https://doi.org/10.1016/s0958-2118(01)80332-3
7. Marchetti P, Jimenez Solomon MF, Szekely G, Livingston AG (2014) Molecular separation with organic solvent nanofiltration: a critical review. Chem Rev 114:10735–10806. https://doi.org/10.1021/cr500006j
8. Ravanchi MT, Kaghazchi T, Kargari A (2009) Application of membrane separation processes in petrochemical industry: a review. Desalination 235:199–244. https://doi.org/10.1016/j.desal.2007.10.042
9. Stamatialis DF, Papenburg BJ, Girones M, Saiful S, Bettahalli SN, Schmitmeier S, Wessling M (2008) Medical applications of membranes: drug delivery, artificial organs and tissue engineering. J Membr Sci 308:1–34. https://doi.org/10.1016/j.memsci.2007.09.059
10. van Reis R, Zydney A (2007) Bioprocess membrane technology. J Membr Sci 297:16–50. https://doi.org/10.1016/j.memsci.2007.02.045
11. De Marco R, Clarke G, Pejcic B (2007) Ion-selective electrode potentiometry in environmental analysis. Electroanal Int J Devoted Fund Pract Aspects Electroa 19:1987–2001. https://doi.org/10.1002/elan.200703916
12. Wang Y, Chen KS, Mishler J, Cho SC, Adroher XC (2011) A review of polymer electrolyte membrane fuel cells: Technology, applications, and needs on fundamental research. Appl Energy 88:981–1007. https://doi.org/10.1016/j.apenergy.2010.09.030
13. Geise GM, Lee HS, Miller DJ, Freeman BD, McGrath JE, Paul DR (2010) Water purification by membranes: the role of polymer science. J Polym Sci, Part B: Polym Phys 48:1685–1718. https://doi.org/10.1002/polb.22037
14. Elimelech M, Phillip WA (2011) The future of seawater desalination: energy, technology, and the environment. Science 333:712–717. https://doi.org/10.1126/science.1200488
15. Werber JR, Osuji CO, Elimelech M (2016) Materials for next-generation desalination and water purification membranes. Nature Rev Mater 1:1–15. https://doi.org/10.1038/natrevmats.2016.18
16. Semiat R (2008) Energy issues in desalination processes. Environ Sci Technol 42:8193–8201. https://doi.org/10.1021/es801330u
17. Cohen-Tanugi D, Grossman JC (2015) Nanoporous graphene as a reverse osmosis membrane: recent insights from theory and simulation. Desalination 366:59–70. https://doi.org/10.1016/j.desal.2014.12.046
18. Lee KS (2011) Three ways of linking laboratory endeavours to the realm of policies. Eur J History of Econ Thought 18:755–776. https://doi.org/10.1080/09672567.2011.616593
19. Homaeigohar S, Elbahri M (2017) Graphene membranes for water desalination. NPG Asia Materials 9: https://doi.org/10.1038/am.2017.135
20. Wang L, Boutilier MS, Kidambi PR, Jang D, Hadjiconstantinou NG, Karnik R (2017) Fundamental transport mechanisms, fabrication and potential applications of nanoporous atomically thin membranes. Nat Nanotechnol 12:509. https://doi.org/10.1038/nnano.2017.72

21. Lee C, Wei X, Kysar JW, Hone J (2008) Measurement of the elastic properties and intrinsic strength of monolayer graphene. Science 321:385–388. https://doi.org/10.1126/science.1157996
22. Ni Z, Wang H, Kasim J, Fan H, Yu T, Wu YH, Feng Y, Shen Z (2007) Graphene thickness determination using reflection and contrast spectroscopy. Nano Lett 7:2758–2763. https://doi.org/10.1021/nl071254m
23. Akinwande D, Huyghebaert C, Wang C-H, Serna MI, Goossens S, Li L-J, Wong H-SP, Koppens FH (2019) Graphene and two-dimensional materials for silicon technology. Nature 573:507–518. https://doi.org/10.1038/s41586-019-1573-9
24. Yang X, Zhang G, Prakash J, Chen Z, Gauthier M, Sun S (2019) Chemical vapour deposition of graphene: layer control, the transfer process, characterisation, and related applications. Int Rev Phys Chem 38:149–199. https://doi.org/10.1080/0144235x.2019.1634319
25. Akbari SA, Ghafarinia V, Larsen T, Parmar MM, Villanueva LG (2020) Large suspended monolayer and Bilayer Graphene membranes with diameter up to 750 μm. Scientific Reports 10:1–8. https://doi.org/10.1038/s41598-020-63562-y
26. Guirguis A, Maina JW, Zhang X, Henderson LC, Kong L, Shon H, Dumée LF (2020) Applications of nano-porous graphene materials–critical review on performance and challenges. Mater Horizons 7:1218–1245. https://doi.org/10.1039/c9mh01570a
27. Koenig SP, Wang L, Pellegrino J, Bunch JS (2012) Selective molecular sieving through porous graphene. Nat Nanotechnol 7:728–732. https://doi.org/10.1038/nnano.2012.162
28. O'Hern SC, Stewart CA, Boutilier MS, Idrobo J-C, Bhaviripudi S, Das SK, Kong J, Laoui T, Atieh M, Karnik R (2012) Selective molecular transport through intrinsic defects in a single layer of CVD graphene. ACS Nano 6:10130–10138. https://doi.org/10.1021/nn303869m
29. Celebi K, Buchheim J, Wyss RM, Droudian A, Gasser P, Shorubalko I, Kye J-I, Lee C, Park HG (2014) Ultimate permeation across atomically thin porous graphene. Science 344:289–292. https://doi.org/10.1126/science.1249097
30. Boutilier MS, Jang D, Idrobo J-C, Kidambi PR, Hadjiconstantinou NG, Karnik R (2017) Molecular sieving across centimeter-scale single-layer nanoporous graphene membranes. ACS Nano 11:5726–5736. https://doi.org/10.1021/acsnano.7b01231
31. Surwade SP, Smirnov SN, Vlassiouk IV, Unocic RR, Veith GM, Dai S, Mahurin SM (2015) Water desalination using nanoporous single-layer graphene. Nat Nanotechnol 10:459–464. https://doi.org/10.1038/nnano.2015.37
32. Jang D, Idrobo J-C, Laoui T, Karnik R (2017) Water and solute transport governed by tunable pore size distributions in nanoporous graphene membranes. ACS Nano 11:10042–10052. https://doi.org/10.1021/acsnano.7b04299
33. O'Hern SC, Jang D, Bose S, Idrobo J-C, Song Y, Laoui T, Kong J, Karnik R (2015) Nanofiltration across defect-sealed nanoporous monolayer graphene. Nano Lett 15:3254–3260. https://doi.org/10.1021/acs.nanolett.5b00456
34. Jiao L, Zhang L, Wang X, Diankov G, Dai H (2009) Narrow graphene nanoribbons from carbon nanotubes. Nature 458:877–880. https://doi.org/10.1038/nature07919
35. Niu F, Liu J-M, Tao L-M, Wang W, Song W-G (2013) Nitrogen and silica co-doped graphene nanosheets for NO_2 gas sensing. J Mater Chem A 1:6130–6133. https://doi.org/10.1039/c3ta11070b
36. Fan X, Zhang G, Zhang F (2015) Multiple roles of graphene in heterogeneous catalysis. Chem Soc Rev 44:3023–3035. https://doi.org/10.1039/c5cs00094g
37. Shen J-W, Li J, Liu F, Zhang L, Liang L, Wang H, Wu J-Y (2020) A molecular dynamics study on water desalination using single-layer MoSe2 nanopore. J Membr Sci 595: https://doi.org/10.1016/j.memsci.2019.117611
38. Li H, Ko T-J, Lee M, Chung H-S, Han SS, Oh KH, Sadmani A, Kang H, Jung Y (2019) Experimental realization of few layer two-dimensional MoS_2 membranes of near atomic thickness for high efficiency water desalination. Nano Lett 19:5194–5204. https://doi.org/10.1021/acs.nanolett.9b01577
39. Qiu H, Xue M, Shen C, Zhang Z, Guo W (2019) Graphynes for water desalination and gas separation. Adv Mater 31:1803772. https://doi.org/10.1002/adma.201803772

40. Li W, Yang Y, Weber JK, Zhang G, Zhou R (2016) Tunable, strain-controlled nanoporous MoS_2 filter for water desalination. ACS Nano 10:1829–1835. https://doi.org/10.1021/acsnano.5b05250
41. Lin L-C, Choi J, Grossman JC (2015) Two-dimensional covalent triazine framework as an ultrathin-film nanoporous membrane for desalination. Chem Commun 51:14921–14924. https://doi.org/10.1039/c5cc05969k
42. Heiranian M, Farimani AB, Aluru NR (2015) Water desalination with a single-layer MoS_2 nanopore. Nature communications 6. https://doi.org/10.1038/ncomms9616
43. Seo DH, Pineda S, Woo YC, Xie M, Murdock AT, Ang EY, Jiao Y, Park MJ, Lim SI, Lawn M (2018) Anti-fouling graphene-based membranes for effective water desalination. Nature Commun 9:1–12. https://doi.org/10.1038/s41467-018-02871-3
44. Kang Y, Xia Y, Wang H, Zhang X (2019) 2D laminar membranes for selective water and ion transport. Adv Func Mater 29:1902014. https://doi.org/10.1002/adfm.201902014
45. Das R, Vecitis CD, Schulze A, Cao B, Ismail AF, Lu X, Chen J, Ramakrishna S (2017) Recent advances in nanomaterials for water protection and monitoring. Chem Soc Rev 46:6946–7020. https://doi.org/10.1039/c6cs00921b
46. Xu G-R, Xu J-M, Su H-C, Liu X-Y, Zhao H-L, Feng H-J, Das R (2017) Two-dimensional (2D) nanoporous membranes with sub-nanopores in reverse osmosis desalination: latest developments and future directions. Desalination https://doi.org/10.1016/j.desal.2017.09.024
47. Das R, Khayet M (2019) Nanotechnology based platforms for efficient water desalination. Desalination 451:1. https://doi.org/10.1016/j.desal.2018.11.011
48. Xu G-R, An Z-H, Xu K, Liu Q, Das R, Zhao H-L Metal organic framework (MOF)-based micro/nanoscaled materials for heavy metal ions removal: The cutting-edge study on designs, synthesis, and applications. Coordination Chem Rev 427:213554. https://doi.org/10.1016/j.ccr.2020.213554
49. Banerjee P, Das R, Das P, Mukhopadhyay A (2018) Membrane Technology. Springer, 127–150. https://doi.org/10.1007/978-3-319-95603-9_3
50. Kazemi AS, Nataj ZE, Abdi Y (2020) Sidewall hydrogenation impact on the structure and wettability of spaghetti MWCNTs. Appl Phys A 126:1–10. https://doi.org/10.1007/s00339-020-03885-9
51. Kazemi AS, Nataj ZE, Abdi Y, Abdol MA (2021) Tuning wettability and surface order of MWCNTs by functionalization for water desalination. Desalination 508:115049. https://doi.org/10.1016/j.desal.2021.115049
52. Nataj ZE, Kazemi AS, Abdi Y Contact angle hysteresis in fluorinated MWCNTs. J Fluorine Chem Peer Rev
53. Kazemi AS, Noroozi AA, Khamsavi A, Mazaheri A, Hosseini SM, Abdi Y (2019) engineering water and solute dynamics and maximal use of CNT surface area for efficient water desalination. ACS omega 4:6826–6847. https://doi.org/10.1021/acsomega.9b00188
54. Eslami J, Abdi Y, Khamsavi A, EbrahimNataj Z, Kazemi AS (2018) Effect of surface area of carbon nanotubes on membrane performance for effective water desalination. Appl Phys A 124:791. https://doi.org/10.1007/s00339-018-2214-0
55. Yang Z, Zhou Y, Feng Z, Rui X, Zhang T, Zhang Z (2019) A review on reverse osmosis and nanofiltration membranes for water purification. Polymers 11:1252. https://doi.org/10.3390/polym11081252
56. Yin J, Deng B (2015) Polymer-matrix nanocomposite membranes for water treatment. J Membr Sci 479:256–275. https://doi.org/10.1016/j.memsci.2014.11.019
57. Bassyouni M, Abdel-Aziz M, Zoromba MS, Abdel-Hamid S, Drioli E (2019) A review of polymeric nanocomposite membranes for water purification. J Ind Eng Chem 73:19–46. https://doi.org/10.1016/j.jiec.2019.01.045
58. Das R (2019) Polymeric materials for clean water. Springer. https://doi.org/10.1007/978-3-030-00743-0
59. Liu Q, Xu G-R, Das R (2019) Inorganic scaling in reverse osmosis (RO) desalination: mechanisms, monitoring, and inhibition strategies. Desalination 468:114065. https://doi.org/10.1016/j.desal.2019.07.005

60. Das R (2017) Advanced membrane materials for desalination: carbon nanotube and graphene. Inorganic Pollutants in Wastewater: Methods Anal Removal Treatment 16:322. https://doi.org/10.21741/9781945291357-10
61. Yampolskii Y (2012) Polymeric gas separation membranes. Macromolecules 45:3298–3311. https://doi.org/10.1021/ma300213b
62. Cot L, Ayral A, Durand J, Guizard C, Hovnanian N, Julbe A, Larbot A (2000) Inorganic membranes and solid state sciences. Solid State Sci 2:313–334. https://doi.org/10.1016/s1293-2558(00)00141-2
63. Baker RW (2012) Membrane technology and applications. Wiley. https://doi.org/10.1002/9781118359686
64. Qasim M, Badrelzaman M, Darwish NN, Darwish NA, Hilal N (2019) Reverse osmosis desalination: a state-of-the-art review. Desalination 459:59–104. https://doi.org/10.1016/j.desal.2019.02.008
65. Shannon M, Bohn P, Elimelech M, Georgiadis J, Marinas B, Mayes A (2010) Science and technology for water purification in the coming decades. Nanosc Technol, 337–346. https://doi.org/10.1142/9789814287005_0035
66. An X, Hu Y, Wang N, Wang T, Liu Z (2019) Breaking the permeability–selectivity trade-off in thin-film composite polyamide membranes with a PEG-b-PSF-b-PEG block copolymer ultrafiltration membrane support through post-annealing treatment. NPG Asia Mater 11:1–12. https://doi.org/10.1038/s41427-019-0114-1
67. Freeman BD (1999) Basis of permeability/selectivity tradeoff relations in polymeric gas separation membranes. Macromolecules 32:375–380. https://doi.org/10.1021/ma9814548
68. Cohen-Tanugi D, McGovern RK, Dave SH, Lienhard JH, Grossman JC (2014) Quantifying the potential of ultra-permeable membranes for water desalination. Energy Environ Sci 7:1134–1141. https://doi.org/10.1039/c3ee43221a
69. Deshmukh A, Yip NY, Lin S, Elimelech M (2015) Desalination by forward osmosis: Identifying performance limiting parameters through module-scale modeling. J Membr Sci 491:159–167. https://doi.org/10.1016/j.memsci.2015.03.080
70. Sanza MA, Bonnélyea V, Cremerb G (2007) Fujairah reverse osmosis plant: 2 years of operation. Desalination 203:91–99. https://doi.org/10.1016/j.desal.2006.03.526
71. Parekh BS (1988) Reverse osmosis technology: applications for high-purity-water production. M. Dekker. ISBN: 0824779851
72. Baker RW (1991) Membrane separation system: recent developments and future directions. Elsevier Science ISBN: 9780815518273
73. Porter MC (1989) Handbook of industrial membrane technology. William Andrew. ISBN: 9780815517559
74. Basile A, Figoli A, Khayet M (2015) Pervaporation, vapour permeation and membrane distillation: principles and applications. Elsevier. https://doi.org/10.1016/c2013-0-16500-2
75. Basile A, Cassano A, Rastogi NK (2015) Advances in membrane technologies for water treatment: materials, processes and applications. Elsevier. https://doi.org/10.1016/c2013-0-16469-0
76. Bhattacharyya D, Williams M, Ray R, McCray S (1992) Membrane Handbook. Van Nostrand Reinhold. https://doi.org/10.1007/978-1-4615-3548-5_21
77. Xu W, Chen Q, Ge Q (2017) Recent advances in forward osmosis (FO) membrane: Chemical modifications on membranes for FO processes. Desalination 419:101–116. https://doi.org/10.1016/j.desal.2017.06.007
78. McKeen LW (2012) Permeability properties of plastics and elastomers. William Andrew. https://doi.org/10.1016/C2010-0-66502-3
79. Pearce G (2007) Water and wastewater filtration: Membrane module format. Filtr Sep 44:31–33. https://doi.org/10.1016/s0015-1882(07)70117-9
80. Polasek V, Talo S, Sharif T (2003) Conversion from hollow fiber to spiral technology in large seawater RO systems—process design and economics. Desalination 156:239–247. https://doi.org/10.1016/s0011-9164(03)00346-1

81. ElMekawy A, Hegab HM, Pant D (2014) A Review on the Near-Future Integration of Microbial Desalination Cell with Reverse Osmosis Technology. Energy Environ Sci 7:3921–3933. https://doi.org/10.1039/c4ee02208d
82. Geim AK, Novoselov KS (2010) The rise of graphene. World Scientific 11–19. https://doi.org/10.1142/9789814287005_0002
83. Bunch JS, Verbridge SS, Alden JS, Van Der Zande AM, Parpia JM, Craighead HG, McEuen PL (2008) Impermeable atomic membranes from graphene sheets. Nano Lett 8:2458–2462. https://doi.org/10.1021/nl801457b
84. Chen Y, Zou J, Campbell SJ, Le Caer G (2004) Boron nitride nanotubes: pronounced resistance to oxidation. Appl Phys Lett 84:2430–2432. https://doi.org/10.1063/1.1667278
85. Mohammad AW, Teow Y, Ang W, Chung Y, Oatley-Radcliffe D, Hilal N (2015) Nanofiltration membranes review: recent advances and future prospects. Desalination 356:226–254. https://doi.org/10.1016/j.desal.2014.10.043
86. Zhao D, Yu S (2015) A review of recent advance in fouling mitigation of NF/RO membranes in water treatment: pretreatment, membrane modification, and chemical cleaning. Desalination Water Treatment 55:870–891. https://doi.org/10.1080/19443994.2014.928804
87. Buonomenna MG (2013) Membrane processes for a sustainable industrial growth. RSC advances 3:5694–5740. https://doi.org/10.1039/c2ra22580h
88. Ghosh AK, Hoek EM (2009) Impacts of support membrane structure and chemistry on polyamide–polysulfone interfacial composite membranes. J Membr Sci 336:140–148. https://doi.org/10.1016/j.memsci.2009.03.024
89. Singh PS, Joshi S, Trivedi J, Devmurari C, Rao AP, Ghosh P (2006) Probing the structural variations of thin film composite RO membranes obtained by coating polyamide over polysulfone membranes of different pore dimensions. J Membr Sci 278:19–25. https://doi.org/10.1016/j.memsci.2005.10.039
90. McCutcheon JR, Elimelech M (2008) Influence of membrane support layer hydrophobicity on water flux in osmotically driven membrane processes. J Membr Sci 318:458–466. https://doi.org/10.1016/j.memsci.2008.03.021
91. Baker R (2004) Ion exchange membrane processes—electrodialysis. Membrane technology and applications, 2nd edn. Wiley, New York. doi 10:9781118359686. https://doi.org/10.1002/0470020393.ch10
92. Kim HI, Kim SS (2006) Plasma treatment of polypropylene and polysulfone supports for thin film composite reverse osmosis membrane. J Membr Sci 286:193–201. https://doi.org/10.1016/j.memsci.2006.09.037
93. Tiraferri A, Yip NY, Phillip WA, Schiffman JD, Elimelech M (2011) Relating performance of thin-film composite forward osmosis membranes to support layer formation and structure. J Membr Sci 367:340–352. https://doi.org/10.1016/j.memsci.2010.11.014
94. Fugallo G, Cepellotti A, Paulatto L, Lazzeri M, Marzari N, Mauri F (2014) Thermal conductivity of graphene and graphite: collective excitations and mean free paths. Nano Lett 14:6109–6114. https://doi.org/10.1021/nl502059f
95. Joshi R, Carbone P, Wang F-C, Kravets VG, Su Y, Grigorieva IV, Wu H, Geim AK, Nair RR (2014) Precise and ultrafast molecular sieving through graphene oxide membranes. Science 343:752–754. https://doi.org/10.1126/science.1245711
96. Nair R, Wu H, Jayaram P, Grigorieva I, Geim A (2012) Unimpeded permeation of water through helium-leak–tight graphene-based membranes. Science 335:442–444. https://doi.org/10.1126/science.1211694
97. Mi B (2014) Graphene oxide membranes for ionic and molecular sieving. Science 343:740–742. https://doi.org/10.1126/science.1250247
98. Cohen-Tanugi D, Grossman JC (2012) Water desalination across nanoporous graphene. Nano Lett 12:3602–3608. https://doi.org/10.1021/nl3012853
99. Cohen-Tanugi D, Lin L-C, Grossman JC (2016) Multilayer nanoporous graphene membranes for water desalination. Nano Lett 16:1027–1033. https://doi.org/10.1021/acs.nanolett.5b04089

100. Booth TJ, Blake P, Nair RR, Jiang D, Hill EW, Bangert U, Bleloch A, Gass M, Novoselov KS, Katsnelson MI (2008) Macroscopic graphene membranes and their extraordinary stiffness. Nano Lett 8:2442–2446. https://doi.org/10.1021/nl801412y
101. Girit ÇÖ, Meyer JC, Erni R, Rossell MD, Kisielowski C, Yang L, Park C-H, Crommie M, Cohen ML, Louie SG (2009) Graphene at the edge: stability and dynamics. Science 323:1705–1708. https://doi.org/10.1126/science.1166999
102. Suk ME, Aluru N (2010) Water transport through ultrathin graphene. J Phys Chem Lett 1:1590–1594. https://doi.org/10.1021/jz100240r
103. Cohen-Tanugi D, Grossman JC (2014) Water permeability of nanoporous graphene at realistic pressures for reverse osmosis desalination. J Chem Phys 141: https://doi.org/10.1063/1.4892638
104. Konatham D, Yu J, Ho TA, Striolo A (2013) Simulation insights for graphene-based water desalination membranes. Langmuir 29:11884–11897. https://doi.org/10.1021/la4018695
105. Wang Y, He Z, Gupta KM, Shi Q, Lu R (2017) Molecular dynamics study on water desalination through functionalized nanoporous graphene. Carbon 116:120–127. https://doi.org/10.1016/j.carbon.2017.01.099
106. Ebrahimi S (2016) Influence of curvature on water desalination through the graphene membrane with Si-passivated nanopore. Comput Mater Sci 124:160–165. https://doi.org/10.1016/j.commatsci.2016.07.036
107. Chogani A, Moosavi A, Sarvestani AB, Shariat M (2020) The effect of chemical functional groups and salt concentration on performance of single-layer graphene membrane in water desalination process: A molecular dynamics simulation study. J Mol Liq 301: https://doi.org/10.1016/j.molliq.2020.112478
108. Chen Q, Yang X (2015) Pyridinic nitrogen doped nanoporous graphene as desalination membrane: molecular simulation study. J Membr Sci 496:108–117. https://doi.org/10.1016/j.memsci.2015.08.052
109. Sint K, Wang B, Král P (2008) Selective ion passage through functionalized graphene nanopores. J Am Chem Soc 130:16448–16449. https://doi.org/10.1021/ja903655u
110. Gai J-G, Gong X-L, Wang W-W, Zhang X, Kang W-L (2014) An ultrafast water transport forward osmosis membrane: porous graphene. J Mater Chem A 2:4023–4028. https://doi.org/10.1039/C3TA14256F
111. Xu M, Tabarraei A, Paci JT, Oswald J, Belytschko T (2012) A coupled quantum/continuum mechanics study of graphene fracture. Int J Fract 173:163–173. https://doi.org/10.1007/s10704-011-9675-x
112. Cheng A, Steele W (1990) Computer simulation of ammonia on graphite. I. Low temperature structure of monolayer and bilayer films. J Chem Phys 92:3858–3866. https://doi.org/10.1063/1.458562
113. Rajan AG, Silmore KS, Swett J, Robertson AW, Warner JH, Blankschtein D, Strano MS (2019) Addressing the isomer cataloguing problem for nanopores in two-dimensional materials. Nat Mater 18:129–135. https://doi.org/10.1038/s41563-018-0258-3
114. Joung IS, Cheatham TE III (2008) Determination of alkali and halide monovalent ion parameters for use in explicitly solvated biomolecular simulations. J Phys Chem B 112:9020–9041. https://doi.org/10.1021/jp8001614
115. Jorgensen WL, Maxwell DS, Tirado-Rives J (1996) Development and testing of the OPLS all-atom force field on conformational energetics and properties of organic liquids. J Am Chem Soc 118:11225–11236. https://doi.org/10.1021/ja9621760
116. Jorgensen WL, Chandrasekhar J, Madura JD, Impey RW, Klein ML (1983) Comparison of simple potential functions for simulating liquid water. J Chem Phys 79:926–935. https://doi.org/10.1063/1.445869
117. Yu H, Van Gunsteren WF (2005) Accounting for polarization in molecular simulation. Comput Phys Commun 172:69–85. https://doi.org/10.1016/j.cpc.2005.01.022
118. Berendsen H, Grigera J, Straatsma T (1987) The missing term in effective pair potentials. J Phys Chem 91:6269–6271. https://doi.org/10.1021/j100308a038

119. Palmer BJ, Pfund DM, Fulton JL (1996) Direct modeling of EXAFS spectra from molecular dynamics simulations. J Phys Chem 100:13393–13398. https://doi.org/10.1021/jp960160q
120. Zhang Z, Huang L, Wang Y, Yang K, Du Y, Wang Y, Kipper MJ, Belfiore LA, Tang J (2020) Theory and simulation developments of confined mass transport through graphene-based separation membranes. Phys Chem Chem Phys 22:6032–6057. https://doi.org/10.1039/c9cp05551g
121. Millero FJ, Feistel R, Wright DG, McDougall TJ (2008) The composition of Standard Seawater and the definition of the Reference-Composition Salinity Scale. Deep Sea Res Part I 55:50–72. https://doi.org/10.1016/j.dsr.2007.10.001
122. Dervin S, Dionysiou DD, Pillai SC (2016) 2D nanostructures for water purification: graphene and beyond. Nanoscale 8:15115–15131. https://doi.org/10.1039/c6nr04508a
123. Cohen-Tanugi D, Grossman JC (2014) Mechanical strength of nanoporous graphene as a desalination membrane. Nano Lett 14:6171–6178. https://doi.org/10.1021/nl502399y
124. Mysels KJ (1960) Structure and Properties of Thin Films (Neugebauer, CA; Newkirk, JB; Vermilyea, DA; eds.). ACS Publications. https://doi.org/10.1021/ed037pa822
125. Kovacs A, Kovács Á, Pogány M, Mescheder U (2007) Mechanical investigation of perforated and porous membranes for micro-and nanofilter applications. Sens Actuat B: Chem 127:120–125. https://doi.org/10.1016/j.snb.2007.07.044
126. Han T, Jiang T, Wang X, Li P, Qiao L, Zhang X (2019) Tuning the mechanical properties of nanoporous graphene: a molecular dynamics study. Mater Res Express 6: https://doi.org/10.1088/2053-1591/ab3331
127. Suk M, Aluru N (2013) Molecular and continuum hydrodynamics in graphene nanopores. RSC Adv 3:9365–9372. https://doi.org/10.1039/c3ra40661j
128. Kim JY, Lee J-H, Grossman JC (2012) Thermal transport in functionalized graphene. ACS Nano 6:9050–9057. https://doi.org/10.1021/nn3031595
129. Qin X, Yan W, Guo X, Gao T (2018) Effects of area, aspect ratio and orientation of rectangular nanohole on the tensile strength of defective graphene–a molecular dynamics study. RSC Adv 8:17034–17043. https://doi.org/10.1039/c8ra02415d
130. Agius Anastasi A, Ritos K, Cassar G, Borg MK (2016) Mechanical properties of pristine and nanoporous graphene. Mol Simul 42:1502–1511. https://doi.org/10.1080/08927022.2016.1209753
131. Fang T-H, Lee Z-W, Chang W-J (2017) Molecular dynamics study of the shear strength and fracture behavior of nanoporous graphene membranes. Curr Appl Phys 17:1323–1328. https://doi.org/10.1016/j.cap.2017.07.003
132. Reina A, Jia X, Ho J, Nezich D, Son H, Bulovic V, Dresselhaus MS, Kong J (2009) Large area, few-layer graphene films on arbitrary substrates by chemical vapor deposition. Nano Lett 9:30–35. https://doi.org/10.1021/nl801827v
133. Buchheim J, Schlichting K-P, Wyss RM, Park HG (2018) Assessing the thickness-permeation paradigm in nanoporous membranes. ACS Nano 13:134–142. https://doi.org/10.1021/acsnano.8b04875
134. Zhang J, Chen C, Pan J, Zhang L, Liang L, Kong Z, Wang X, Zhang W, Shen J-W (2020) Atomistic insights into the separation mechanism of multilayer graphene membranes for water desalination. Phys Chem Chem Phys 22:7224–7233. https://doi.org/10.1039/d0cp00071j
135. Mooney D, Müller-Plathe F, Kremer K (1998) Simulation studies for liquid phenol: properties evaluated and tested over a range of temperatures. Chem Phys Lett 294:135–142. https://doi.org/10.1016/s0009-2614(98)00860-4
136. Abdol MA, Sadeghzadeh S, Jalaly M, Khatibi MM (2019) Constructing a three-dimensional graphene structure via bonding layers by ion beam irradiation. Scientific Reports 9. https://doi.org/10.1038/s41598-019-44697-z
137. Sahu S, Zwolak M (2019) Colloquium: Ionic phenomena in nanoscale pores through 2D materials. Rev Mod Phys 91: https://doi.org/10.1103/revmodphys.91.021004
138. Sahu S, Zwolak M (2018) Maxwell-Hall access resistance in graphene nanopores. Phys Chem Chem Phys 20:4646–4651. https://doi.org/10.1039/c7cp07924a

139. Thomas M, Corry B, Hilder TA (2014) What have we learnt about the mechanisms of rapid water transport, ion rejection and selectivity in nanopores from molecular simulation? Small 10:1453–1465. https://doi.org/10.1002/smll.201302968
140. Sahu S, Di Ventra M, Zwolak M (2017) Dehydration as a universal mechanism for ion selectivity in graphene and other atomically thin pores. Nano Lett 17:4719–4724. https://doi.org/10.1021/acs.nanolett.7b01399
141. Xu K, Feng B, Zhou C, Huang A (2016) Synthesis of highly stable graphene oxide membranes on polydopamine functionalized supports for seawater desalination. Chem Eng Sci 146:159–165. https://doi.org/10.1016/j.ces.2016.03.003
142. Zhu C, Li H, Meng S (2014) Transport behavior of water molecules through two-dimensional nanopores. J Chem Phys 141:18C528. https://doi.org/10.1063/1.4898075
143. Tocci G, Joly L, Michaelides A (2014) Friction of water on graphene and hexagonal boron nitride from ab initio methods: very different slippage despite very similar interface structures. Nano Lett 14:6872–6877. https://doi.org/10.1021/nl502837d
144. Werder T, Walther JH, Jaffe R, Halicioglu T, Koumoutsakos P (2003) On the water—carbon interaction for use in molecular dynamics simulations of graphite and carbon nanotubes. J Phys Chem B 107:1345–1352. https://doi.org/10.1021/jp8083106
145. Li H, Zeng XC (2012) Wetting and interfacial properties of water nanodroplets in contact with graphene and monolayer boron–nitride sheets. ACS Nano 6:2401–2409. https://doi.org/10.1021/nn204661d
146. Garnier L, Szymczyk A, Malfreyt P, Ghoufi A (2016) Physics behind water transport through nanoporous boron nitride and graphene. J Phys Chem Lett 7:3371–3376. https://doi.org/10.1021/acs.jpclett.6b01365
147. Ferrari AC, Bonaccorso F, Fal'Ko V, Novoselov KS, Roche S, Bøggild P, Borini S, Koppens FH, Palermo V, Pugno N (2015) Science and technology roadmap for graphene, related two-dimensional crystals, and hybrid systems. Nanoscale 7:4598–4810. https://doi.org/10.1039/C4NR01600A
148. Worsley MA, Pauzauskie PJ, Olson TY, Biener J, Satcher Jr JH, Baumann TF (2010) Synthesis of graphene aerogel with high electrical conductivity. J Amer Chem Soc 132:14067–14069. https://doi.org/10.1021/ja1072299
149. Li X, Cai W, An J, Kim S, Nah J, Yang D, Piner R, Velamakanni A, Jung I, Tutuc E (2009) Large-area synthesis of high-quality and uniform graphene films on copper foils. Science 324:1312–1314. https://doi.org/10.1126/science.1171245
150. Bae S, Kim H, Lee Y, Xu X, Park J-S, Zheng Y, Balakrishnan J, Lei T, Kim HR, Song YI (2010) Roll-to-roll production of 30-inch graphene films for transparent electrodes. Nat Nanotechnol 5:574. https://doi.org/10.1038/nnano.2010.132
151. Kobayashi T, Bando M, Kimura N, Shimizu K, Kadono K, Umezu N, Miyahara K, Hayazaki S, Nagai S, Mizuguchi Y (2013) Production of a 100-m-long high-quality graphene transparent conductive film by roll-to-roll chemical vapor deposition and transfer process. Appl Phys Lett 102: https://doi.org/10.1063/1.4776707
152. Yan Z, Yao W, Hu L, Liu D, Wang C, Lee C-S (2015) Progress in the preparation and application of three-dimensional graphene-based porous nanocomposites. Nanoscale 7:5563–5577. https://doi.org/10.1039/c5nr00030k
153. Chen W, Yan L (2011) In situ self-assembly of mild chemical reduction graphene for three-dimensional architectures. Nanoscale 3:3132–3137. https://doi.org/10.1039/c1nr10355e
154. Moon G-h, Shin Y, Choi D, Arey BW, Exarhos GJ, Wang C, Choi W, Liu J (2013) Catalytic templating approaches for three-dimensional hollow carbon/graphene oxide nano-architectures. Nanoscale 5:6291–6296. https://doi.org/10.1039/c3nr01387a
155. Wang Y, Wang F, He J (2013) Controlled fabrication and photocatalytic properties of a three-dimensional ZnO nanowire/reduced graphene oxide/CdS heterostructure on carbon cloth. Nanoscale 5:11291–11297. https://doi.org/10.1039/c3nr03969b
156. Waduge P, Bilgin I, Larkin J, Henley RY, Goodfellow K, Graham AC, Bell DC, Vamivakas N, Kar S, Wanunu M (2015) Direct and scalable deposition of atomically thin low-noise MoS_2 membranes on apertures. ACS Nano 9:7352–7359. https://doi.org/10.1021/acsnano.5b02369

157. Alemán B, Regan W, Aloni S, Altoe V, Alem N, Cl G, Geng B, Maserati L, Crommie M, Wang F (2010) Transfer-free batch fabrication of large-area suspended graphene membranes. ACS Nano 4:4762–4768. https://doi.org/10.1021/nn100459u
158. Kazemi AS, Hosseini SM, Abdi Y (2019) Large total area membrane of suspended single layer graphene for water desalination. Desalination 451:160–171. https://doi.org/10.1016/j.desal.2017.12.050
159. Kazemi AS, Abdi Y, Eslami J, Das R (2019) Support based novel single layer nanoporous graphene membrane for efficacious water desalination. Desalination 451:148–159. https://doi.org/10.1016/j.desal.2018.03.003
160. Zaretski AV, Lipomi DJ (2015) Processes for non-destructive transfer of graphene: widening the bottleneck for industrial scale production. Nanoscale 7:9963–9969. https://doi.org/10.1039/c5nr01777g
161. Boutilier MS, Sun C, O'Hern SC, Au H, Hadjiconstantinou NG, Karnik R (2014) Implications of permeation through intrinsic defects in graphene on the design of defect-tolerant membranes for gas separation. ACS Nano 8:841–849. https://doi.org/10.1021/nn405537u
162. Kim HW, Yoon HW, Yoon S-M, Yoo BM, Ahn BK, Cho YH, Shin HJ, Yang H, Paik U, Kwon S (2013) Selective gas transport through few-layered graphene and graphene oxide membranes. Science 342:91–95. https://doi.org/10.1126/science.1236098
163. Kafiah FM, Khan Z, Ibrahim A, Karnik R, Atieh M, Laoui T (2016) Monolayer graphene transfer onto polypropylene and polyvinylidenedifluoride microfiltration membranes for water desalination. Desalination 388:29–37. https://doi.org/10.1016/j.desal.2016.02.027
164. Yasaei P, Kumar B, Hantehzadeh R, Kayyalha M, Baskin A, Repnin N, Wang C, Klie RF, Chen YP, Král P (2014) Chemical sensing with switchable transport channels in graphene grain boundaries. Nature communications 5:1–8. https://doi.org/10.1038/ncomms5911
165. Guirguis A, Maina JW, Kong L, Henderson LC, Rana A, Li LH, Majumder M, Dumée LF (2019) Perforation routes towards practical nano-porous graphene and analogous materials engineering. Carbon 155:660–673. https://doi.org/10.1016/j.carbon.2019.09.028
166. O'Hern SC, Boutilier MS, Idrobo J-C, Song Y, Kong J, Laoui T, Atieh M, Karnik R (2014) Selective ionic transport through tunable subnanometer pores in single-layer graphene membranes. Nano Lett 14:1234–1241. https://doi.org/10.1021/nl404118f
167. Lehtinen O, Dumur E, Kotakoski J, Krasheninnikov A, Nordlund K, Keinonen J (2011) Production of defects in hexagonal boron nitride monolayer under ion irradiation. Nucl Instrum Methods Phys Res, Sect B 269:1327–1331. https://doi.org/10.1016/j.nimb.2010.11.027
168. Lin Y, Han X, Campbell CJ, Kim JW, Zhao B, Luo W, Dai J, Hu L, Connell JW (2015) Holey graphene nanomanufacturing: structure, composition, and electrochemical properties. Adv Func Mater 25:2920–2927. https://doi.org/10.1002/adfm.201500321
169. Wang X, Jiao L, Sheng K, Li C, Dai L, Shi G (2013) Solution-processable graphene nanomeshes with controlled pore structures. Scientific reports 3:1996. https://doi.org/10.1038/srep01996
170. Wang L, Drahushuk LW, Cantley L, Koenig SP, Liu X, Pellegrino J, Strano MS, Bunch JS (2015) Molecular valves for controlling gas phase transport made from discrete ångström-sized pores in graphene. Nat Nanotechnol 10:785–790. https://doi.org/10.1038/nnano.2015.158
171. Rozada R, Solís-Fernández P, Paredes J, Martínez-Alonso A, Ago H, Tascón J (2014) Controlled generation of atomic vacancies in chemical vapor deposited graphene by microwave oxygen plasma. Carbon 79:664–669. https://doi.org/10.1016/j.carbon.2014.08.015
172. Dumée LF, Feng C, He L, Yi Z, She F, Peng Z, Gao W, Banos C, Davies JB, Huynh C (2014) Single step preparation of meso-porous and reduced graphene oxide by gamma-ray irradiation in gaseous phase. Carbon 70:313–318. https://doi.org/10.1016/j.carbon.2013.12.094
173. Russo CJ, Golovchenko JA (2012) Atom-by-atom nucleation and growth of graphene nanopores. Proc Natl Acad Sci 109:5953–5957. https://doi.org/10.1073/pnas.1119827109

174. Garaj S, Hubbard W, Reina A, Kong J, Branton D, Golovchenko J (2010) Graphene as a subnanometre trans-electrode membrane. Nature 467:190–193. https://doi.org/10.1038/nature09379
175. Lehtinen O, Kotakoski J, Krasheninnikov A, Tolvanen A, Nordlund K, Keinonen J (2010) Effects of ion bombardment on a two-dimensional target: atomistic simulations of graphene irradiation. Physical review B 81: https://doi.org/10.1103/physrevb.81.153401
176. Kuan AT, Lu B, Xie P, Szalay T, Golovchenko JA (2015) Electrical pulse fabrication of graphene nanopores in electrolyte solution. Appl Phys Lett 106: https://doi.org/10.1063/1.4921620
177. Feng J, Liu K, Graf M, Lihter M, Bulushev RD, Dumcenco D, Alexander DT, Krasnozhon D, Vuletic T, Kis A (2015) Electrochemical reaction in single layer MoS_2: nanopores opened atom by atom. Nano Lett 15:3431–3438. https://doi.org/10.1021/acs.nanolett.5b00768
178. Standop S, Lehtinen O, Herbig C, Lewes-Malandrakis G, Craes F, Kotakoski J, Michely T, Krasheninnikov AV, Busse C (2013) Ion impacts on graphene/Ir (111): interface channeling, vacancy funnels, and a nanomesh. Nano Lett 13:1948–1955. https://doi.org/10.1021/nl304659n
179. Moreno C, Vilas-Varela M, Kretz B, Garcia-Lekue A, Costache MV, Paradinas M, Panighel M, Ceballos G, Valenzuela SO, Peña D (2018) Bottom-up synthesis of multifunctional nanoporous graphene. Science 360:199–203. https://doi.org/10.1126/science.aar2009
180. Xie G, Yang R, Chen P, Zhang J, Tian X, Wu S, Zhao J, Cheng M, Yang W, Wang D (2014) A general route towards defect and pore engineering in graphene. Small 10:2280–2284. https://doi.org/10.1002/smll.201303671
181. Zandiatashbar A, Lee G-H, An SJ, Lee S, Mathew N, Terrones M, Hayashi T, Picu CR, Hone J, Koratkar N (2014) Effect of defects on the intrinsic strength and stiffness of graphene. Nature communications 5:1–9. https://doi.org/10.1038/ncomms4186
182. Bagri A, Grantab R, Medhekar N, Shenoy V (2010) Stability and formation mechanisms of carbonyl-and hydroxyl-decorated holes in graphene oxide. J Phys Chem C 114:12053–12061. https://doi.org/10.1021/jp908801c
183. Tracz A, Wegner G, Rabe JP (2003) Scanning tunneling microscopy study of graphite oxidation in ozone – air mixtures. Langmuir 19:6807–6812. https://doi.org/10.1021/la034103h
184. Liu L, Ryu S, Tomasik MR, Stolyarova E, Jung N, Hybertsen MS, Steigerwald ML, Brus LE, Flynn GW (2008) Graphene oxidation: thickness-dependent etching and strong chemical doping. Nano Lett 8:1965–1970. https://doi.org/10.1021/nl0808684
185. Feng J, Liu K, Bulushev RD, Khlybov S, Dumcenco D, Kis A, Radenovic A (2015) Identification of single nucleotides in MoS_2 nanopores. Nat Nanotechnol 10:1070–1076. https://doi.org/10.1038/nnano.2015.219
186. He K, Robertson AW, Gong C, Allen CS, Xu Q, Zandbergen H, Grossman JC, Kirkland AI, Warner JH (2015) Controlled formation of closed-edge nanopores in graphene. Nanoscale 7:11602–11610. https://doi.org/10.1039/c5nr02277k
187. Fischbein MD, Drndić M (2008) Electron beam nanosculpting of suspended graphene sheets. Appl Phys Lett 93: https://doi.org/10.1063/1.2980518
188. Safron NS, Brewer AS, Arnold MS (2011) Semiconducting two-dimensional graphene nanoconstriction arrays. Small 7:492–498. https://doi.org/10.1002/smll.201001193
189. Zhang J, Song H, Zeng D, Wang H, Qin Z, Xu K, Pang A, Xie C (2016) Facile synthesis of diverse graphene nanomeshes based on simultaneous regulation of pore size and surface structure. Scientific Reports 6:1–9. https://doi.org/10.1038/srep32310
190. Cun H, Iannuzzi M, Hemmi A, Osterwalder Jr, Greber T (2014) Two-nanometer voids in single-layer hexagonal boron nitride: formation via the "can-opener" effect and annihilation by self-healing. ACS Nano 8:7423–7431. https://doi.org/10.1021/nn502645w
191. Liu X, Feng H, Zhang J, Zhao R, Liu X, Wong DK (2012) Hydrogen peroxide detection at a horseradish peroxidase biosensor with a Au nanoparticle–dotted titanate nanotubel hydrophobic ionic liquid scaffold. Biosens Bioelectron 32:188–194. https://doi.org/10.1016/j.bios.2011.12.002

192. Safron NS, Kim M, Gopalan P, Arnold MS (2012) Barrier-guided growth of micro-and nano-structured graphene. Adv Mater 24:1041–1045. https://doi.org/10.1002/adma.201104195
193. Kidambi PR, Ducati C, Dlubak B, Gardiner D, Weatherup RS, Martin M-B, Seneor P, Coles H, Hofmann S (2012) The parameter space of graphene chemical vapor deposition on polycrystalline Cu. J Phys Chem C 116:22492–22501. https://doi.org/10.1021/jp303597m
194. Wei D, Liu Y, Wang Y, Zhang H, Huang L, Yu G (2009) Synthesis of N-doped graphene by chemical vapor deposition and its electrical properties. Nano Lett 9:1752–1758. https://doi.org/10.1021/nl803279t
195. Blankenburg S, Bieri M, Fasel R, Müllen K, Pignedoli CA, Passerone D (2010) Porous graphene as an atmospheric nanofilter. Small 6:2266–2271. https://doi.org/10.1002/smll.201001126
196. Zhang H, He X, Zhao M, Zhang M, Zhao L, Feng X, Luo Y (2012) Tunable hydrogen separation in sp–sp2 hybridized carbon membranes: a first-principles prediction. J Phys Chem C 116:16634–16638. https://doi.org/10.1021/jp304908p
197. Xue M, Qiu H, Guo W (2013) Exceptionally fast water desalination at complete salt rejection by pristine graphyne monolayers. Nanotechnology 24: https://doi.org/10.1088/0957-4484/24/50/505720
198. Lin S, Buehler MJ (2013) Mechanics and molecular filtration performance of graphyne nanoweb membranes for selective water purification. Nanoscale 5:11801–11807. https://doi.org/10.1039/c3nr03241h
199. Kou J, Zhou X, Chen Y, Lu H, Wu F, Fan J (2013) Water permeation through single-layer graphyne membrane. J Chem Phys 139: https://doi.org/10.1063/1.4817596
200. Peng Q, Dearden AK, Crean J, Han L, Liu S, Wen X, De S (2014) New materials graphyne, graphdiyne, graphone, and graphane: review of properties, synthesis, and application in nanotechnology. Nanotechnol Sci Appl 7:1. https://doi.org/10.2147/nsa.s40324
201. Li G, Li Y, Liu H, Guo Y, Li Y, Zhu D (2010) Architecture of graphdiyne nanoscale films. Chem Commun 46:3256–3258. https://doi.org/10.1039/b922733d
202. Cai SL, Zhang WG, Zuckermann RN, Li ZT, Zhao X, Liu Y (2015) The organic flatland—recent advances in synthetic 2D organic layers. Adv Mater 27:5762–5770. https://doi.org/10.1002/adma.201500124
203. Yuan Z, Benck JD, Eatmon Y, Blankschtein D, Strano MS (2018) Stable, temperature-dependent gas mixture permeation and separation through suspended nanoporous single-layer graphene membranes. Nano Lett 18:5057–5069. https://doi.org/10.1021/acs.nanolett.8b01866.s001
204. Ang EY, Toh W, Yeo J, Lin R, Liu Z, Geethalakshmi K, Ng TY (2020) A review on low dimensional carbon desalination and gas separation membrane designs. J Membr Sci 598: https://doi.org/10.1016/j.memsci.2019.117785
205. Qin Y, Hu Y, Koehler S, Cai L, Wen J, Tan X, Xu WL, Sheng Q, Hou X, Xue J (2017) Ultrafast nanofiltration through large-area single-layered graphene membranes. ACS Appl Mater Interfaces 9:9239–9244. https://doi.org/10.1021/acsami.7b00504
206. Park J, Bazylewski P, Fanchini G (2016) Porous graphene-based membranes for water purification from metal ions at low differential pressures. Nanoscale 8:9563–9571. https://doi.org/10.1039/c5nr09278g
207. Tincu B, Avram M, Avram A, Tutunaru O, Tucureanu V, Matei A, Burinaru T, Comanescu F, Demetrescu I (2020) Progress and control in development of single layer graphene membranes. Vacuum 175: https://doi.org/10.1016/j.vacuum.2020.109269
208. Yang Y, Yang X, Liang L, Gao Y, Cheng H, Li X, Zou M, Ma R, Yuan Q, Duan X (2019) Large-area graphene-nanomesh/carbon-nanotube hybrid membranes for ionic and molecular nanofiltration. Science 364:1057–1062. https://doi.org/10.1126/science.aau5321
209. Huang S, Dakhchoune M, Luo W, Oveisi E, He G, Rezaei M, Zhao J, Alexander DT, Züttel A, Strano MS (2018) Single-layer graphene membranes by crack-free transfer for gas mixture separation. Nature Commun 9:1–11. https://doi.org/10.1038/s41467-018-04904-3
210. Wang L, Williams CM, Boutilier MS, Kidambi PR, Karnik R (2017) Single-layer graphene membranes withstand ultrahigh applied pressure. Nano Lett 17:3081–3088. https://doi.org/10.1021/acs.nanolett.7b00442

211. Schmidt ME, Iwasaki T, Muruganathan M, Haque M, Van Ngoc H, Ogawa S, Mizuta H (2018) Structurally controlled large-area 10 nm pitch graphene nanomesh by focused helium ion beam milling. ACS Appl Mater Interfaces 10:10362–10368. https://doi.org/10.1021/acsami.8b00427
212. Köhler MH, Bordin JR, Barbosa MC (2019) Ion flocculation in water: from bulk to nanoporous membrane desalination. J Mol Liq 277:516–521. https://doi.org/10.1016/j.molliq.2018.12.077
213. Belfort G, Davis RH, Zydney AL (1994) The behavior of suspensions and macromolecular solutions in crossflow microfiltration. J Membr Sci 96:1–58. https://doi.org/10.1016/0376-7388(94)00119-7

Three-Dimensional and Lamellar Graphene Oxide Membranes for Water Purification

Mateus H. Köhler, Mayara B. Leão, José Rafael Bordin, and Carolina F. de Matos

Abstract Graphene oxide (GO) is one of the most prominent nanoscaled membranes for water purification. Thanks to a combination of massive theoretical and experimental efforts, large-scale production of both three-dimensional (3D) and lamellar GO membranes is at hand. Countless methods to synthesize, functionalize, and characterize GO membranes are available, which inspire tremendous excitement about the possibilities of increasing the efficiency of current reverse osmosis (RO) desalination plants. Here, we reveal some of the main physical–chemical insights as well as manufacturing techniques of GO, reduced GO, and related material-based separation techniques.

Keywords Graphene oxide · Lamellar membrane · Solute separation · Water purification

1 Introduction

The misuse of freshwater supplies in the last century has led humanity to face considerable challenges to prevent water scarcity. According to the United Nations World Water Development Report [1], nearly half of the world population suffers from water stress, a problem that affects people even in developed countries. The amount of available clean water is vanishing at an unprecedented rate, leading to environmental, health, geopolitical, and humanitarian issues. Sadly, the projections are not optimistic—the freshwater crisis will get worse in the next decades due to factors such

M. H. Köhler (✉)
Department of Physics, Federal University of Santa Maria, Santa Maria 97105-900, Brazil
e-mail: mateus.kohler@ufsm.br

M. B. Leão · C. F. de Matos
Campus Caçapava do Sul, Universidade Federal do Pampa, Caçapava do Sul 96570-000, Brazil

J. R. Bordin
Department of Physics, Institute of Physics and Mathematics, Federal University of Pelotas, Capão do Leão 96050-500, Brazil

R. Das (ed.), *Two-Dimensional (2D) Nanomaterials in Separation Science*, Springer Series on Polymer and Composite Materials,
https://doi.org/10.1007/978-3-030-72457-3_4

as population growth, climate change, political decisions, and our current lifestyle [2]. About 80% of current residuals' water, from industry to city sewers, returns to water bodies without the proper treatment [3]. A consequence is the increasing contamination and degradation of aqueous environments. Not to mention that health issues related to contaminated water tend to weigh even more on the low-income population.

This is a humanitarian challenge that needs urgent technological solutions. In this direction, scientists are making efforts to find cheaper, greener materials to be used in water purification plants. A wide variety of methods have been employed, such as screening, filtration, centrifugation, crystallization, sedimentation and separation by gravity, flotation, precipitation, coagulation, oxidation, solvent extraction, evaporation, distillation, reverse osmosis (RO), ion exchange, electrodialysis, electrolysis, and adsorption. These distinct methods employ a large variety of materials. However, many of these materials cannot remove some pollutants, especially at natural water pH or in high trace level concentration [4].

In this sense, one of the most significant tasks of modern material science is to develop a material able to adsorb a broad spectrum of water pollutants. Among the several materials proposed in recent years, graphene and other carbon-based nanostructures arise as the most prominent for water purification [4, 5]. It all started in the early 50s when Radushkevich obtained carbon nanotubes (CNTs) by decomposition of carbon monoxide (CO) in iron [6]. Oberlinv and co-workers also produced CNTs from benzene decomposition back in the 70s [7]. However, the scientific and technological possibilities of carbon-based nanostructures only became notorious after Ijima's work on the fabrication of CNTs in the early 90s [8]. A decade after, the hypothesis of a hierarchical self-assembly of carbon foams from nanostructured graphite was first discussed by Umemoto and co-workers [9], giving rise to the idea of two-dimensional (2D) carbon-based nanostructures. In 2004, a seminal work by Novoselov et al. [10] revolutionized materials science. They were able to strip graphene sheets out of graphite with a Scotch Tape, isolating this 1-carbon atom-thick nanostructure. The technique was awarded the 2010 Nobel Prize in Physics and represents one of the greatest revolutions in the chemical, physical, and engineering sciences of our century. The extraordinary graphene properties, as mechanical flexibility, chemical and thermal stability, and, especially, its high surface area [11], make this material one of the most promising for new water purification technologies.

Pure graphene sheets, or pristine graphene, will interact with contaminants mainly by van der Waals (vdW) forces, since their carbon atoms are highly stabilized in an sp^2 configuration. This property is useful, for instance, to separate heavy metals from water. These ions are naturally observed in aqueous environments. However, we are facing an increase in this kind of pollutant in water supplies due to human-made phenomena, as mining and mineral processing [12] and the excessive use of pesticides [13]. Graphene can be very effective in separating ions from water, but for some pollutants, the energetic penalty for leaving bulk solution and be absorbed in graphene's surface is too high. A solution to adsorb polar pollutants is to decorate pristine graphene with specific functional groups. This increases the number of possible applications of graphene-based materials for water treatment and purification, including the removal of dyes, organic solvents, pesticides, heavy metals, and inorganic pollutants.

Another way to include polar groups is by using graphene oxide (GO). This material differs from the standard pristine graphene once it has oxidized sites. A GO compound can be obtained through chemical oxidation of carbon sources, such as graphite, using oxidizing agents (e.g., $KMnO_4$, H_2SO_4, etc.) to produce graphite oxide. Then, distinct exfoliation methods can be used to yield GO nanosheets. The possibility of large-scale production of GO makes it inexpensive in comparison with other materials [14]. GO sheets have hexagonal structure as pristine graphene with oxygen-based functionalized sites. It shows high electronic mobility, thermal conductivity, and remarkable mechanical strength, with the advantage of being highly stable in water. As well, its surface can be easily functionalized by organic biomolecules, strong π–π stacking, and vdW interactions.

Spontaneous self-assembly is one of the most prominent and efficient strategies to build 2D or 3D macroscopic structures from nanosized chemical building blocks [15]. It allows exploring the unique properties of nanomaterials, like GO, in macroscopic devices. One of the main quests in self-assembly is to control the resulting morphology. In the case of GO, the control of the oxidation degree can be obtained through reducing agents. This control over the polar-oxygenated functional sites allows for GO manipulation in water and distinct 2D or 3D structures' self-assembly [14]. Reduced GO (rGO) sheet with a controlled number of functional groups is itself a strategy to explore the separation properties of GO-based materials. As well, distinct reducing agents can be employed to reduce and create new and unique binding sites for pollutants.

A huge variety of structures are obtained by the assembly of GO/rGO nanosheets. In suspension, these layers can be manipulated to obtain either lamellar or 3D nanostructures, as shown in Fig. 1. 2D lamellar multilayered structures of GO and rGO have shown great potential for separation. These 2D materials have nanochannels that can be tuned in size by distinct reducing agents [16, 17] to allow the flow of specific atoms and molecules, controlling both water permeability and pollutant removal. As well, the lamellar structures can be functionalized by distinct materials to achieve specific adsorption and filtration properties [18, 19]. On the other hand, 3D-GO-based macrostructures have shown high adsorption capacity and recyclability due to their unique superficial area and porosity [20, 21]. Briefly stated, 3D-GO multifunctional structures are obtained by employing methods that prevent the GO/rGO sheets to stack in the suspended solution. This 3D configuration keeps the GO/rGO sheet properties, and increase applications based in characteristics that single sheets do not have, as porosity. In this way, several studies have been dedicated to synthesize 3D-GO/rGO structures, as foam, sponges, and aerogels, with distinct morphology, structure, and properties [20]. Both lamellar and 3D structures have physical–chemical properties that diverge from their chemical building blocks (i.e., the GO/rGO nanosheets and the functionalization molecules), making them interesting materials for new water treatment technologies.

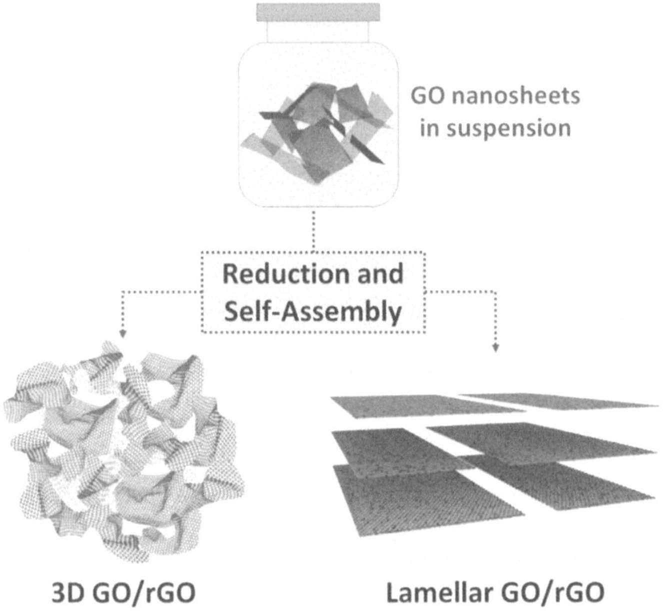

Fig. 1 Distinct self-assembled morphologies can be obtained through different physical–chemical routes to create lamellar or 3D GO/rGO structures

Experiments are not the only tool to propose new materials. Molecular modeling at different levels—from quantum simulations to coarse-grained approaches—is a powerful method to understand phenomena at the nanoscale. The power of computational modeling has helped not only to predict and understand the applications of graphene-based membranes but also allowed the prediction of new carbon allotropes. This is the case of the graphyne (GY) family [22]—carbon allotropes featuring assembled sp- and sp^2-hybridized layers. It was predicted in 1987 [23] as a lattice of benzene rings (sp^2 hybridized) linked together with acetylenic bonds ($-C{\equiv}C-$, sp hybridized). This particular geometry is known as γ-graphyne. Since this seminal work, a large number of GY nanostructures were proposed [24]. Countless members of the GY family can be obtained by controlling the number n of acetylenic linkages within the material. As computer simulations have indicated, GY-based membranes stand a great potential for water desalination [22, 24]. However, to date, only one case was experimentally synthesized: the $n = 2$ case or graphdiyne (GDY) [25]. So far, the experiments have corroborated the computational findings, showing that GDY-based membranes can effectively remove pollutants (e.g., oil [26] and heavy metals [27]) from water, even though the most prominent member is expected to be the γ-graphyne-3—not yet experimentally synthesized.

In this chapter, we review recent advances in 2D lamellar and 3D porous GO/rGO membranes for water purification. First, we present in Sect. 2 the primary methods used to produce GO and rGO sheets. Next, in Sect. 3, we discuss the experimental works on lamellar and 3D membranes, while the computational studies are shown in Sect. 4. Current research gaps, perspectives, and conclusions end the chapter in Sect. 5.

2 Preparation Methods of Graphene Oxide and Reduced Graphene Oxide

Graphene is a 2D material formed by sp^2-hybridized carbon atoms densely packed in a regular atomic-scale chicken wire or honeycomb (hexagonal) pattern. Thanks to this structure, graphene shows a significant number of desirable properties such as high electrical and thermal conductivities, mechanical strength, molecular barrier capability, high specific surface area, and many others. Due to all these amazing features, it is natural to think of graphene as an excellent candidate for water purification membranes. However, the use of crystalline graphene in aqueous environments has been challenging because of its low solubility and the vdW interactions that lead to aggregation. The problematic scheduling of bottom-up synthesis methods does not help either. Alternatively, compounds structurally similar to graphene can be chemically obtained from graphite or even other carbon sources. These graphene-based species present many of the advantages of pure graphene, while also having other properties thanks to the presence of oxygen groups functionalizing their surfaces.

The chemical oxidation of graphite leads to the formation of graphite oxide, which consists of several stacked layers of GO. In the current literature, most of the methods used in the oxidation of graphite are based on the first preparation by Brodie in 1859 and the subsequently developed Staudenmaier and Hummers methods, all of which depending on the homogeneous reaction of graphite with strong mixed oxidants [28, 29]. For example, routes based on the Hummers method—undoubtedly the most employed—usually involve a combination of oxidizing agents such as $KMnO_4$ and H_2SO_4, leading to the formation of a super oxidizing agent Mn_2O_7:

$$KMnO_{4(s)} + 6H^+_{(aq)} + 3SO^{2-}_{4(aq)} \rightarrow K^+_{(aq)} + MnO^+_{3(aq)} + H_3O^+ + 3HSO^-_{4(aq)}$$
$$MnO^+_{3(aq)} + MnO^-_{4(aq)} \rightarrow Mn_2O_{7(aq)}$$

New approaches to obtain GO involve different methods, many of them seeking to decrease the use of strong oxidizers, the risk of explosion, environmental pollution (due to synthesis-generated residues), and shorten the long reaction time. Recently, a group from the Chinese Academy of Sciences at Shenyang described a scalable, green synthesis of GO from graphite in a timescale of seconds [30]. They used a water electrolytic oxidation method, where the oxidation degree of GO sheets can be modulated by the concentration of H_2SO_4 in solution, and the number of sheets and lateral size can be controlled by the sonication time.

The GO can be isolated through exfoliation (e.g., by mechanical stirring or sonication) of graphite oxide. It exhibits a hexagonal carbon structure similar to graphene.

Still, it also contains functional groups such as hydroxyl (−OH), alkoxy (C−O−C), carbonyl (C−O), the carboxylic acid (−COOH), and others with oxygen-based functions. At the basal level, GO is essentially hydrophobic, formed by a network of unoxidized aromatic polycyclic benzene ring islands. Given these characteristics, GO is seen as an amphiphilic molecule with a basal plane mostly hydrophobic and hydrophilic edges [31].

The maximum dispersibility of GO varies from 1 to 4 mg per mL of water, an essential factor for processing and subsequent application. This value depends on both the solvent and the degree of surface functionalization during oxidation. Due to these amphiphilic characteristics, GO has also been applied as an alternative to surfactant materials.

The presence of oxygenated groups also allows GO to act as a platform for various chemical reactions. From the functionalization—with the most different organic, inorganic, and biomolecule groups—to the reduction of GO, the particular conditions result in graphene materials with modified properties. Some of these properties are fundamental in the construction of membranes for water purification, increasing their permeability and selectivity.

2.1 Reduced Graphene Oxide

The reduction of GO, that is, the removal of oxygenated groups, results in the formation of rGO. GO reduction can be accomplished through chemical (with reducing agents such as hydrazine, metal hydrides, NaOH, hydroiodic and ascorbic acids), thermal, or electrochemical processes [32]. All of them lead to products that resemble, in different degrees, the pristine graphene obtained by peeling the graphite, particularly in terms of electrical, thermal, and mechanical properties as well as its surface morphology.

An important characteristic of graphene obtained by graphite oxidation methods is the presence of a high number of structural defects. These defects consist mainly of (i) the organization of carbon atoms in the form of pentagons and heptagons, known as stone-walled structures, forming perfect graphene "islands" (carbons hexagonally linked to other sp^2-hybridized carbons) surrounded by defective regions (carbons linked to other carbons with hybridizations other than sp^2) and (ii) in the remaining oxygenated groups after reduction.

Topological defects, as well as vacancies, impurities, and functionalization, can introduce important changes in the properties of the material. This, in turn, can influence the membrane's behavior [33, 34].

The form of the reduction is also essential to define the material's structure. For example, some thermochemical reductions lead to the formation of porous three-dimensional macrostructures with fascinating characteristics for environmental applications in the purification of water and gas separation [35]. These materials are discussed further in Sect. 3.2.2.

The low cost and the massive scalability are some of the advantages associated with graphene's reduction. Furthermore, the starting material is pure graphite, and the technique can easily be adapted for the production of "chemically modified graphenes". An additional advantage is that chemical oxidation of graphite is currently considered the only viable method of synthesis for obtaining industrial-scale membranes for water purification based on graphene [36].

3 Experimental Works

3.1 GO/rGO Lamellar Membranes

The last decade witnessed significant advances in the experimental synthesis and characterization of GO and rGO. There are different stacking possibilities for these structures according to the preparation method we choose to use. In order to get their maximum potential, a smart approach could be to organize them in a lamellar architecture. Lamellar GO and rGO are used as building blocks in advanced membrane devices. Compared with others, these structures come up with many advantages such as high energy efficiency and price-performance ratio, reduced dimensions, and easy operation [37].

In addition, lamellar GO/rGO membranes present a highly active chemical surface—favoring adsorptive processes and selectivity—and very stable porous structures—with internal spaces of high permeability and the possibility of multi-functionality. Nevertheless, to obtain all these advantages, it is necessary to guarantee an efficient, low-cost assembly methodology.

The most common preparation methods of graphene-based separation membranes include vacuum-assisted/pressure-assisted, self-assembly, casting/coating, and layer-by-layer (LBL). The vacuum filtration method is a straightforward process based on the filtration of GO/rGO dispersions. It allows for almost parallel deposition of GO/rGO sheets on top of each other over a porous base membrane (e.g., filtration or ultrafiltration membranes), as depicted in Fig. 2a. In this process, the membrane's thickness can be easily controlled by parameters such as the dispersion concentration and volume, and the degree of functionalization of the graphene species [38]. After drying, the deposited GO/rGO lamellar membrane can be peeled off using tweezers or, in exceptional cases, immersing the base membrane containing the GO/rGO in a bath where the two membranes have different affinities with the solvent, in such a way that they separate from each other (in a process called phase inversion). The organization of the sheets along the membrane depends on the filtration process: while the pressure-assisted approach results in a highly ordered membrane, only a partially organized membrane can be produced by vacuum-assisted filtration. The self-assembly method, with solvent evaporation, leads to a very heterogeneous membrane [39].

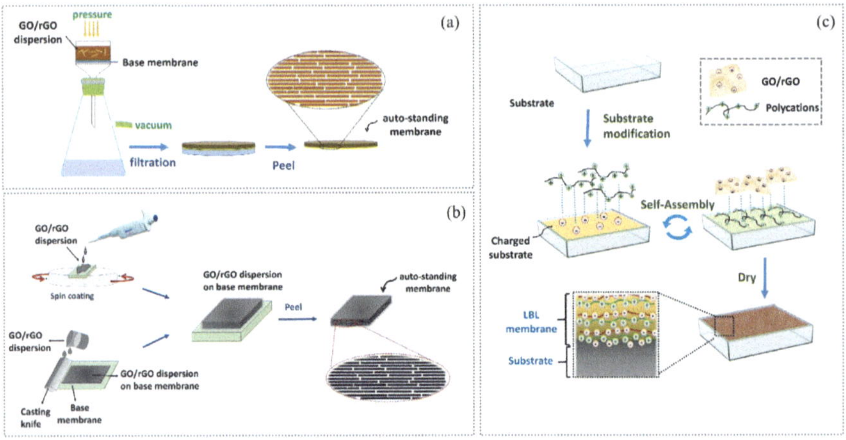

Fig. 2 Preparation of GO/rGO-based membranes by different methods: **a** vacuum-assisted/pressure-assisted, **b** casting/coating, and **c** LBL self-assembly methods

The casting/coating is also a widespread method to obtain lamellar GO/rGO membranes, including dip-coating, drop-casting, spin-coating, and spray-coating [40]. It usually consists of dripping off GO/rGO nanosheet dispersions onto a substrate or immersing the substrate in the dispersion, repeatedly. Then, the sample can be heated to remove the solvent or rotated in a high-speed spin coating system, where a centrifugal force will be responsible for the formation of a thin layer. Another deposition approach involves shaping a glass plate using a casting knife, Fig. 2b. Then, the GO/rGO membrane can be peeled off from the sample at the base membrane, similar to the vacuum filtration. The membranes formed by this technique generally are created by a relatively heterogeneous GO/rGO deposition caused by electrostatic repulsion between nanosheet edges [41].

The LBL self-assembly method is an ideal method for mounting GO/rGO membranes [42]. Conventional approaches to assemble LBL membranes involve polycations and polyanions. They can be deposited on the substrate (with a previously charged surface), producing thin films that are molecularly charged by the auto-assembly of oppositely charged electrolytes, Fig. 2c. These polyions are uniformly inserted to functionalize or intercalate GO/rGO membranes, being the main responsible for the distance between layers. GO derivatives present laminar structure and oxygenated functional groups, such characteristics make these species suitable for LBL self-assembly technology [43]. New methodologies have been sought, where the main challenge is to obtain membranes that are simultaneously scalable, selective, and cost-effective.

GO/rGO lamellar membrane assembling methods, properties, and applications have been extensively investigated in recent years. Most of these studies look for materials that are both ecologically and economically sustainable, with multifunctional structures for aquatic environmental pollution remediation. In this scenario, the membrane must stand at the same time high rates of permeability and selectivity.

Although the difference between permeability and separation performance remains a challenge, there have been some huge advancements. For instance, a recent work by Thebo et al. [44] evaluated a class of rGO membranes, where the distance between the lamellae was controlled using the amino acid theanine and tannic acid as a reducing and cross-linking agent. They found a water permeability above 10,000 $L \cdot m^{-2} \cdot h^{-1} \cdot bar^{-1}$, which is up to three orders of magnitude times higher than the rates described for similar materials and commercial membranes. The material also showed high selectivity, retaining 100% of some dyes, such as rhodamine B and methylene blue, with no sign of damage or delamination in water, acid, and basic media, even after several months. It strikes that we can make use of small organic molecules in order to design enhanced 2D nanofluidic channels based on GO lamellar membranes.

Adjusting GO's degree of reduction can also be a viable strategy to improve permeability. Membranes that are prepared using different types of GO, with varying degrees of reduction or originating from different graphite sources, have very different purification performances. These differences are related to the oxygenated functional groups—hydroxyl (−OH), epoxy (−COC−), and carboxyl (−COOH). However, the role of these different functional groups is not yet fully understood.

Yu et al. [45] prepared and systematically studied three types of GO membranes. It included dominant oxygenated functional groups: dominant carboxyl, dominant hydroxyl GO, and dominant epoxy GO. They noted that the functional groups containing oxygen with different hydrophilicity and electricity controlled the spacing between membrane layers, resulting in different water permeability and ion retention rates. A combination of hydrophilic and electrostatic interactions was then appointed as the main ingredient affecting those properties.

The ionic permeability rate can be controlled by adapting the nanochannel size within a sub-nanometer range. This can be achieved by controlling the reduction degree of the GO nanosheets exposing them to hydroiodic acid vapor, based on time variation. As reduction increases, the size of the nanometric channels decreases, enhancing ionic retention in the GO membrane [46].

GO processing also plays a crucial role in the membrane's main characteristics. For example, the sonication time during preparation affects the size of the GO nanosheet, the amount and distribution of defects, the surface morphology, the degree of oxidation, and consequently, the spacing between lamellae. These characteristics directly influence the permeability and the ability of the membrane to retain molecules and ions of interest [47].

The path length that water must travel along the membrane, specifically through nanosheet junctions, is currently one of the main challenges in GO membranes. Shortening this distance is one of the most effective ways to improve membrane's permeability. To this end, some treatments (e.g., thermal treatments) can be used to open pores in GO membranes in a controlled manner. These holey-GO (hGO) membranes have been shown to undergo up to 3.8 times higher water permeability relative to regular GO membranes, despite being up to 4 times thicker [48]. The water path can also be facilitated by an interlayer spacing control, such as the one presented by Dong et al. [49]. They prepared a new kind of stimuli-responsive GO-based membrane with reversible, gas-tunable water permeability: water flux remains higher than that of pure GO under CO_2 stimulation and, upon removal of CO_2 by Ar bubbling, a high rejection rate of dye molecules and $MgCl_2$ is achieved—without noticeable degradation of the water permeability.

Stability is another critical factor for GO/rGO lamellar membranes. Most of the studies focus on creating covalent or non-covalent bonds between adjacent GO/rGO sheets. Different cross-linking molecules and ions, such as polyions, multivalent cations, dopamine, glutaraldehyde, and diamine can borrow structural stability to the membrane. Functionalization by Janus particles has also been used to reinforce these structures. The incorporation of asymmetrically functionalized Janus GO (JGO) sheets in GO films exhibited extraordinary stability in the water at broad pH values even under agitation [50]. These structures present dual interfaces (i.e., hydrophilic and hydrophobic) that interact with the extended π-systems on adjacent nanosheets. The approach also allows for high molecular retention of charged and uncharged dye molecules, rhodamine B, and brilliant blue G compounds, while maintaining water permeability comparable with previously reported GO-based membranes under osmotic pressure.

In addition to the approaches described above, a new trend to better control the properties of GO-based membranes is the combination with other materials forming hybrid structures, aiming to increase both permeability and selectivity. Other 2D materials such as graphitic carbon nitrides (g-C_3N_4) and MXenes can ideally engage with GO membranes to produce enhanced nanofiltration devices.

The combination of GO membranes with g-C_3N_4 aims to improve their anti-pressure ability and, with the addition of other materials (e.g., metal nanoparticles), add catalytic properties. This approach solves a limitation of the pure GO membrane, enabling the degradation of molecules of environmental interest. It also brings forward additional benefits, such as increasing the number of permeable nanochannels, which allows for a twofold water permeability compared with traditional GO membranes [51]. GO/g-C_3N_4 hybrid membranes also feature enhanced stability and rigidity.

Wu et al. [52] developed different GO hybrid membranes with g-C_3N_4 for nanofiltration using glycine as a molecular binder. Contrary to expectations, the carbon nitride group reduced nanochannel dimensions while the glycine led to expanded channels. The hybrid membranes resulted in faster water transport, without compromising the retention of dye molecules.

MXenes are a family of 2D materials based on transition metal carbides and nitrides. Their structure and hydrophilicity allow for fast and selective transport of water [53]. However, the retention of various molecules, such as dyes, is quite poor, for the addition of GO/rGO layers can borrow a significant improvement.

A recent study described the preparation of a hybrid GO/MXene ($Ti_3C_2T_x$) membrane through vacuum filtration [54]. The resultant membrane presents a typical lamellar structure with increased interlayer spacing compared with pristine GO. Both hydrophilicity and increased space provide these membranes with excellent surface wettability to water and organic solvents. For instance, high flow for pure solvents, with emphasis on water (~21 $L \cdot m^{-2} \cdot h^{-1}$), and excellent performance in the retention of dyes (above 90%) were found for aqueous and organic solutions.

Another work on composite membranes based on GO and MXene found an even greater water flux of ~72 $L \cdot m^{-2} \cdot h^{-1} \cdot bar^{-1}$, while the reference GO membrane achieved 6.5 $L \cdot m^{-2} \cdot h^{-1} \cdot bar^{-1}$ under the same experimental conditions [55]. To this end, the authors used a GO/MXene mass ratio of 1/4. Different proportions of GO and MXene led to varied permeabilities and rejection rates. Nevertheless, the rejection of conventional organic dyes of small molecules was found to exceed 99.5%, highlighting their excellent removal efficiency. The overall good performance is a consequence of the increased interlayer spacing and an electrostatic effect due to a decrease in the oxygen-containing functional groups.

As described throughout this section, GO species are materials with many advantages that enable lamellar GO/rGO membranes with excellent properties for water purification. Fully understanding their structure–property relationship is fundamental to improve performance, stability, selectivity, and anti-fouling effectively.

3.2 *GO/rGO 3D Membranes*

3D assembly of GO/rGO monolayers is one of the most promising strategies for nanofiltration technologies [56]. Similar to the lamellar case, the morphology of 3D membranes can be controlled during their synthesis. The hierarchical structures can be divided into two families: the hollow and the 3D porous structures. The former includes nanoshells, nanospheres, nanococoons, and nanocapsules, while the latter is composed of aerogels, hydrogels, sponges, and foams.

3D-GO networks were first obtained by Xu and co-authors [57]. The foam structure obtained by a hydrothermal method showed high capacity and conductance, substantial surface area, and high mechanical and thermal stability. Controlling the amount of oxygen in the reduction process and employing distinct treatment methods are both effective to tune the membrane's adsorptive properties. In fact, these structures have been widely employed in adsorption experiments with promising results [58].

There are countless functionalization options for 3D GO/rGO materials. For instance, the irradiation by ion beams generates wrinkles, folds, and defects in the structure that are a perfect fit for adsorption of naphthalene [59], a carcinogenic agent whose eradication from water is a current challenge. Ji et al. [60] explored the use of three distinct carbohydrates as reducing and spacing agents in 3D-rGO aerogels. They found distinct structural characteristics for each carbohydrate, which can be further tailored for specific pollutant adsorption.

Polydopamine (PDA) can be easily employed as both reducing and functionalization agents. This is possible due to a spontaneous process of polymerization by dopamine and the self-assembly of graphene to form a hydrogel during the hydrothermal process. A high density of PDA functional groups in the graphene surface leads to a high adsorption capacity of a large number of pollutants such as heavy metals, synthetic dyes, and aromatic pollutants [61]. Other functional groups have been added to GO hydrogels in order to enhance their adsorptive properties. For instance, glutaraldehyde accelerates the cross-linking process between PDA-coated GO sheets and polyethyleneimine, resulting in a 3D hydrogel that is formed within only 3 min. This membrane can efficiently adsorb anionic dyes and organic solvents from water [62]. Also, the addition of 1H, 1H, 2H, 2H-perfluorodecanethiol to the PDA-rGO structure can create a sponge for water–oil separation [63].

GO/rGO hydro and aerogels containing bacterial cellulose (BC-GO/rGO) have also been investigated. They can effectively adsorb distinct heavy metals from water, exhibiting high structural stability [64]. Desorption/readsorption experiments with BC/GO demonstrated their excellent recyclability [65]. Additionally, these structures have shown underwater superoleophobicity and underoil superhydrophobicity [66, 67], which is a desirable characteristic for the removal of oils and organic solvents from water. Cellulose acetate nanofibers can be employed to prevent GO nanosheets from self-assembly after reduction. This increases the connectivity between nanopores, creating aerogels able to retain oils and organic dyes [68].

The combination of different reducing agents can lead to unique properties. For instance, a specific adsorbent agent is able to remove either cationic or anionic dyes. However, combining GO with polyvinyl alcohol and using L-cysteine as the reducing agent makes an ultralight aerogel that is able to absorb both species [69], increasing the membrane's scope.

Chitosan is a biopolymer extensively employed to functionalize 3D-GO/rGO hydro and aerogels, leading to materials with high adsorption of heavy metals, neutral and charged dyes, and oils [70–72]. Additionally, GO, chitosan, and cellulose composites can be combined to create porous nanospheres. These structures have been used as an alternative for removing heavy metals from water solutions and even directly from the soil, which broadens their use in environmental treatments [73, 74]. Also, chitosan and activated carbon-functionalized GO flakes have demonstrated promising removal of pharmaceuticals and personal care products from water [75].

The combination of GO/rGO with other biopolymers also leads to interesting adsorption properties. GO/silk fibroin hybrid aerogel has been designed for dye and heavy metal adsorption [76], while carbon nanofiber/GO composite aerogels were used in highly efficient oil absorption prototypes [77].

The hydrothermal polymerization of lignin and sodium alginate (SA) in the presence of GO in an aqueous system was used to create a hydrogel with high adsorption of heavy metals [78]. On the other hand, SA/rGO composite hydrogel cross-linked by Fe^{3+} exhibited an improved dye adsorption performance [79], especially for cationic dyes. The efficient removal is achieved due to a synergetic interaction between GO/rGO and SA. Another advantage of SA is to create 3D printable and mechanically robust hydrogels, as recently obtained from a GO and amino-GO (aGO) non-covalently functionalized with SA [80].

The functionalization is not restricted to carbon-based molecules. Many other materials can be used to get improved properties. Cadmium sulfide (CdS)-functionalized aerogels showed an enhanced absorption of ionic dyes [81], while adding iron nanoparticles to GO hydrogels make it highly efficient to remove organic pollutants [82]. N-doped rGO aerogels showed excellent catalytic degradation of antibiotic contaminants in water [83]. Silver phosphate/GO aerogel microspheres that can photodegrade pharmaceuticals and pesticides were recently obtained [84]. Nd_2O_3 nanoparticles [85] and rGO/rare-earth-metal-oxide [86] aerogels are also promising functionalizations for dye removal.

Polyvinyl alcohol, a biocompatible polymer, bonded with iron oxide can be used as a precursor to prepare an Fe_3O_4@Fe/GO aerogel. This membrane achieved higher antibiotic activity in comparison with other 3D-GO/rGO materials and total organic carbon removal from water [87]. Also, Fe_3O_4@Fe/GO nanocomposites can be employed in the immobilization of pharmaceutical drugs, such as phenazopyridine [88], and heavy metals [89]. It is also possible to create a modified rGO aerogel by exchanging the biopolymer for a polyethyleneimine (PEI). The resulting Fe_3O_4@PEI/rGO aerogel has been shown to effectively remove a polar non-steroidal anti-inflammatory drug from water [90]. An alternative membrane to adsorb similar drugs was obtained through cobalt-based ferrite ($CoFe_2O_4$) functionalization of GO-based nanocomposites [91].

The possibilities of creating GO/rGO-based 3D structures for water purification are numerous, as nanofiller membranes, foams, and sponges for water-pollutant separation. However, it is important to address some concerns about almost all current synthesis routines. A significant issue is the reducing agents. Most of the chemicals employed to reduce the disperse GO/rGO sheets into a 3D structure are toxic. For instance, one of the most used reducing agents is hydrazine (N_2H_4), which exhibits excellent reducing capacity, but it is also a carcinogenic agent related to lung and colon cancer [92] as well as basal cell carcinomas [93]. Another reducing agent widely applied in 3D-GO synthesis is the sodium borohydride ($NaBH_4$), a highly toxic compound that, in contact with water, starts a reaction that releases gases with potential spontaneous ignition. The time required for the synthesis is also a challenge. There are too many steps in current methods, taking several days—or weeks—to get a sample. Besides that, in many cases, templates have to be employed. It is still necessary to create new chemical routines to obtain these membranes, using non-pollutant reducing agents and speeding the process up to achieve large industrial production.

4 Computational Studies

Whether predicting new materials or exploring physical phenomena unlikely to be explained through experiments, computer simulations have decisively advanced materials science. The contribution becomes more evident when we scale down to nanomaterials. They are expected to perform differently from bulk. For instance, graphene exhibits extraordinary electronic and mechanical properties and extremely high thermal conductivity. The latter was observed to even increase with the sample's length. Non-equilibrium molecular dynamics (NEMD) simulations showed that this behavior is a consequence of the 2D nature of phonons in graphene, and to the change of the phonon population at stationary non-equilibrium conditions [94], providing a fundamental understanding of thermal transport in 2D materials. Theory and simulations have also been coupled with experiments to develop more robust and stiffer nanomaterials. This is the case of a recent finite element analysis combined with a two-photon lithography used to produce carbon nanostructures stronger than diamonds [95].

Possible computational approaches for nanofluidics and nanomaterials include quantum mechanics (QM) calculations, molecular dynamics (MD) simulations, and hybrid methods such as QM/MM—a link between quantum and classical world providing relevant chemical and physical insights. At the QM level, electronic interactions are described by solving the Schrödinger wave equation either directly or by employing functionals with approximations at various levels, such as in the Density Functional Theory (DFT). For instance, these methods are used in calculations of the energy barrier of individual molecules permeating nanopores. Unfortunately, a high computational cost associated with the many-body problem makes it unsuitable for systems with more than a few hundred atoms.

MD simulations are computationally cheaper, allowing for direct observation of kinetic processes of much larger systems—including a 64 million atoms HIV capsid [96]—during widely representative timescales up to μs. A typical MD framework is based on a predetermined distribution of atomic positions and partial charges. While the latter does not change during the simulation, the former evolves in time following Newton's equations of motion. MD simulations on 2D membrane separation usually start from two gas or water reservoirs separated by a nanoporous structure. Then, a pressure gradient is applied, so that the solute/solvent flux can be monitored and further analyzed. Intermolecular interactions are described through empirical force fields based on simple potential functions, which means that chemical interactions involving electronic degrees of freedom are neglected.

The best of both worlds would be a combination where atomic motion is determined by Newton's law and interatomic forces calculated using quantum mechanics. Although further theoretical advancements are needed to make QM/MM suitable for membrane separation simulations, QM and MD approaches are regularly used together to predict mechanical properties. While QM calculations based on first principles have long been considered for elastic evaluation of nanomaterials, reactive force fields implemented in MD simulations can also be used to predict failure mechanisms and deformation. Currently, the adaptive intermolecular reactive empirical bond order (AIREBO) [97], a Tersoff-type potential, is one of the most popular reactive force fields used to compute mechanical properties of graphene and various carbon allotropes. The salient feature of this potential is that the bond order term only depends on the local coordination without the need to consider explicit charges and long-range Coulombic interactions, allowing for excellent computational performance as fast Fourier transforms (FFT) calculations are not needed [98]. However, the cutoffs of the switching functions in AIREBO terms must be carefully selected. ReaxFF [99] is another reactive force field used for hydrocarbon systems, where atomic charges are dynamically optimized during the simulations. It makes ReaxFF more transferable and suitable for complex chemical reactions, such as interactions between graphyne and other molecules in various environments.

Other computational approaches, such as those based on finite element methods, have also been employed to study mechanical properties of nanostructures. For instance, numerical simulations based on finite element methods were used together with Raman spectroscopy to determine Young's modulus of single- and bilayer graphene [100]. Interestingly, electronic structure calculations, Monte Carlo, and MD simulations have shown that the oxidation process significantly decreases the in-plane Young's modulus and fracture strength of GO [101]. Especially nanoporous GO has shown stress/strain slope about one-half that of ideal graphene, with a fracture strength of about one-third of ideal graphene. This is particularly interesting because high-resolution TEM observations have shown that reduced GO platelets present quite large holes [102].

Computer simulations have been at the forefront of nanofluidics, contributing to the establishment of measurement protocols and the manufacture of new materials—currently two of the most challenging experimental endeavors. Notably, MD simulations were used over the past decades to predict and explain high water flux rates and ionic selectivities of few-atom-layered materials. For instance, theoretical works anticipated the continuous flow of water through graphene with pore diameters smaller than 1 nm [103]. The same computational framework was used to show that a single-layer graphene with defect pores was able to reject salt and still perform a water flux two to three orders of magnitude higher than that of traditional RO membranes [104]. Enhanced water flow in GO and rGO has also been reported in computational and experimental works [105].

Additionally, the structure of GO membranes has been modified to enhance both water flux and rejection of salt ions and dye molecules. Further theoretical analysis revealed that the fast water transport found in GO membranes can be attributed to the presence of oxidized and pristine parts in GO sheets [106]. The oxidized region causes adjacent microcrystals to separate and inhibits their reunion, allowing water to flow between the GO layers, while the non-oxidized region provides a capillary network that creates high capillary pressure. The smooth surface of the pristine region enables water to flow under this pressure without friction—previous simulations already showed the high slip length of graphene nanochannels to largely increase water molecules' speed [107]. The underlying mechanism can then be understood as a capillary-like pressure due to a combination of pristine regions (where water molecules can move faster) and oxidized regions where hydrogen bond formation is a driving force for water to move through the membrane. Other structural factors, such as different kinds of defects, valleys, folds, cavitations, and voids, also subsidize the reported high water flux through GO films [108].

4.1 Graphyne-Based Membranes

An emblematic contribution of computer simulations to membrane science is the prediction of graphyne by Baughman et al. [23]. They described a lattice of benzene rings (sp^2 hybridized) linked together with acetylenic bonds ($-C{\equiv}C-$,

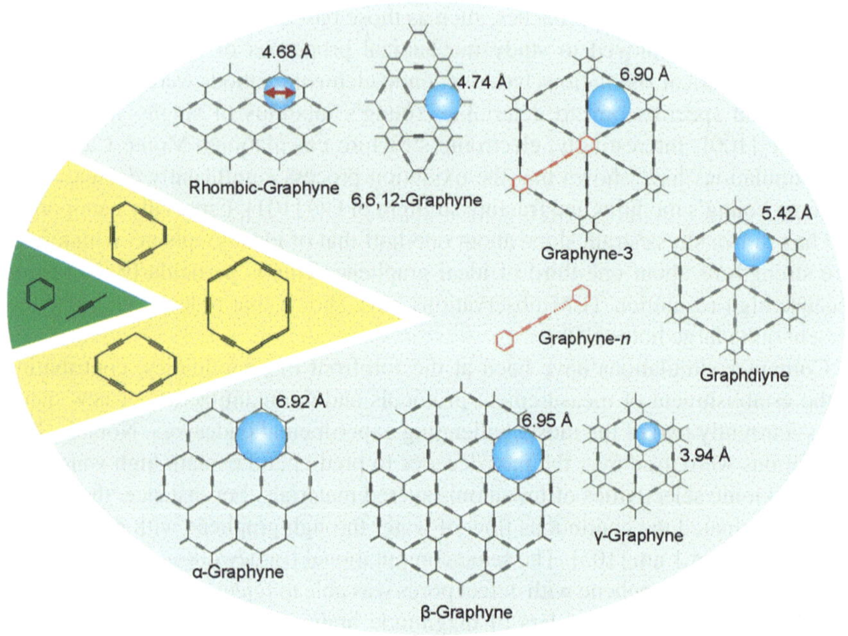

Fig. 3 The graphyne puzzle: building blocks inside the green region are assembled to form graphynes (gray region) with different nanopore structures (yellow region). Adapted with permission from Qiu et al. [109]

sp hybridized)—smallest green sector region of Fig. 3. This architecture is now referred to as γ-graphyne. It has a hexagonal symmetry similar to that of graphene. The length of the acetylenic linkages can be different, leading to the graphyne-*n* structures, in which *n* indicates the number of acetylenic bonds (highlighted in red in Fig. 3). Special attention is devoted to $n = 2$, also known as graphdiyne, the first successfully synthesized graphyne [25]. Further addition of alkyne units leads to graphyne-3, an up-and-coming candidate for water desalination. In summary, it is possible to adjust the nanopore size (yellow region in Fig. 3) just by adding more acetylenic bonds to the structure.

Different arrangements of sp- and sp^2-hybridized carbon atoms lead to the remaining symmetries shown in Fig. 3, namely, α- and β-graphynes. For desalination purposes, the interest is on the intrinsic nanopores resulting from these symmetries—the blue hollow spheres in Fig. 3. MD simulations have exhaustively shown that water permeability increases as the number of acetylenic linkages in graphyne increases. This is expected since geometric factors, most of the time, dictate permeation rates in 2D nanoporous membranes [110]. The γ-graphyne-3 membrane has the smallest pores allowing for water permeation [111, 112]. This structure showed

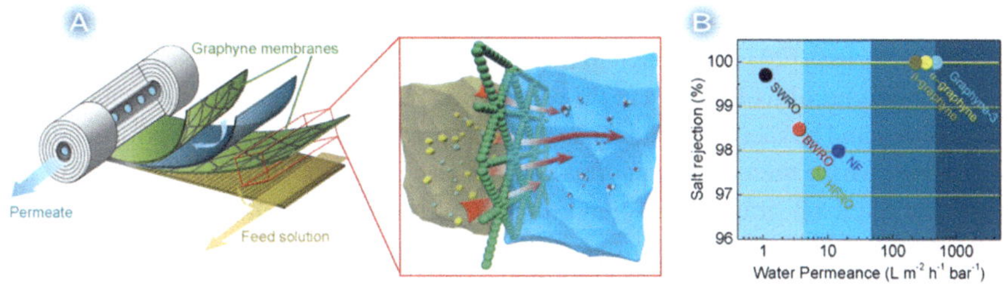

Fig. 4 **a** Prototypical graphyne-based RO membrane and **b** performance comparison between graphynes and conventional commercial RO membranes, such as polymeric seawater RO (SWRO), brackish water RO (BWRO), high-flux RO (HFRO), and nanofiltration (NF). Adapted with permission from Qiu et al. [109] and Xue et al. [113]

two folds increase in the net water flux compared with CNTs with similar diameters, 27.5 and 13.5 ns^{-1} [111], respectively. Similar flow rates were found in β-graphyne membranes, whereas some α-graphynes exhibited reduced water flux [113]. In the case of γ-graphyne-3, water can form a small single file hydrogen-bonded chain reducing the free energy barrier at the pore entrance—only 2.3 k_BT compared with 3.5 k_BT in CNT membranes. Simulations also revealed size-dependent quantized transportation of water across graphyne membranes, where discrete water flow transitioned into continuous flow in γ-graphyne-7 [114].

Current commercial RO plants are typically based on spiral-wound membranes, as depicted in Fig. 4a, where graphyne can be sandwiched between functional layers. The combination of an excellent mechanical strength with well-defined open pores makes graphyne very selective. Precisely, γ-graphyne-3 is predicted to reject 100% of contaminants, such as $CuSO_4$, CCl_4, C_6H_6, and a wide variety of ions, such as Na^+, K^+, Mg^{2+}, Ca^{2+} and Cl^-. The same applies to α- and β-graphyne membranes with similar pore diameters. Figure 4b (water permeance versus salt rejection) shows why these structures are so promising as water purification membranes.

The unique architecture grants graphyne robust mechanical stability, with high strength and stiffness. MD-based biaxial mechanical tensile tests [112] on a series of γ-graphyne-*n* membranes revealed ultimate stress and strain falling between 16.7 and 32.3 GPa and 1.2 and 2.7%, respectively, much higher than conventional polymer-based membranes and comparable to those of CNTs. In particular, γ-graphyne-3 has exhibited very promising mechanical properties, which further stresses its role as an ideal candidate for desalination membranes. Even when mechanical deformations are imposed on γ-graphyne-3 membranes, water permeability is enhanced and ionic rejection is kept unchanged [109].

All desalination results regarding graphyne come from computational simulations because only graphdiyne (γ-graphyne-2) films have been produced so far. The success of graphyne-based desalination membrane depends on the fabrication of large-area graphyne materials—first in the laboratory and eventually in a scalable way. Top-down approaches (e.g., chemical or mechanical exfoliation) are off the table since no bulk phase is experimentally available for graphynes. The alternative is to build up from the molecular scale, using suitable precursor molecules. In 2010, Li and co-workers [25] implemented this bottom-up method to synthesize for the first time graphdiyne films with 1 μm thick. They used a cross-coupling reaction on Cu foil surfaces and hexaethynylbenzene (HEB) as precursors to produce uniform multi-layer films. Since then, a tremendous effort has been made to improve graphdiyne synthesis. Even a β-graphdiyne-containing thin film with a thickness of about 25 nm was fabricated [115]. However, experimental efforts have failed to produce structures other than graphdiynes. This is of particular concern because the best theoretical desalination results have been obtained for γ-graphyne-3, while graphdiyne is impermeable. Improved surface engineering of substrates, rational design, and synthesis of new precursors as well as the development of fundamentally new preparation technologies are needed to realize γ-graphyne-3 and others experimentally. Theoretical works can also contribute to guiding microscopic processes, clarifying mechanisms involved in the growth of large-area graphynes.

5 Conclusions, Challenges, and Perspectives

The reduction of GO is a delicate process, which can produce environmentally undesirable residues. Conventional strategies include thermal, chemical, and solvothermal reduction. Despite thermal annealing being highly effective, it produces a large amount of CO and CO_2 as a result of the elimination of oxygen (from the oxygen-containing functional groups) and carbon (from the pristine graphene regions in GO). Thermal methods are usually very energy-consuming because of the elevated temperatures involved. On the other hand, chemical reducing agents can be more efficient, but many chemical reagents are toxic and non-environmentally friendly. The search for green chemical reagents (e.g., ascorbic acid (vitamin C), sugars, green tea) is a current challenge that is bringing together large theoretical and experimental groups. Hydrothermal reduction is also an emerging, green route that can produce stable rGO suspensions in graphene hydrogels.

It is the transition from bulk to a nanoconfined fluid that makes possible every membrane-based separation plant. Thus, separation efficiency in GO/rGO membranes is closely related to their nano or sub-nanometer channels. If the membrane is not stable enough, the channels do not hold and the entire process is compromised. For instance, the interlayer channels can become wider when converting from dry to hydrated states, in a process called swelling—which is more prominent in GO rather than rGO. On the other hand, external pressure often compacts the membrane, especially lamellar membranes. In this scenario, the pathways available for water molecules to cross the membrane are diminished and there goes the membrane's permeability. One alternative is the use of guest materials such as nanotubes, molecules, or ions. These cross-linkers tend to amplify attractive forces between host nanosheets, which significantly contribute to their structural stability. The host material can even be another 2D nanosheet. For example, pristine graphene and rGO with GO nanosheets can be blended to reduce intersheet repulsion and strengthen the π–π attraction, yielding exceptional stability [116]. The search for new host materials is, therefore, a current area of research that demands further investigation.

From design to the experimental realization of large-scale GO/rGO membranes, computational studies are helping to understand and improve protocols, methods, and applications. If the production of such membranes is still a challenge, it would be impracticable without the guidance of atomic-scale simulations. Only at this level, we can understand some of the factors driving water permeance and pollutant rejection rates. Simulations have also been used as predictive tools for new and improved materials, determining their structural resistance and selectivity properties even before experiments—saving both scientists' money and time. However, these simulations can be quite expensive, using substantial computational resources. Data mining algorithms, such as machine learning (ML) and deep learning (DL), can also extract useful knowledge on new materials, geometries, and functionalizations, creating structure–property maps that play a key role in accelerated materials discovery [117, 118]. Some of its advantages are (i) the huge database with several properties of countless materials and (ii) the speed with which this huge amount of data can be processed to find the best candidates. Inverse design, where data-driven materials with pre-defined target properties are proposed, has emerged as an important numerical tool in recent years by shading some light in hidden information of materials [119]. GO/rGO-based membranes can certainly take advantage of these powerful tools to find a more efficient and safe path.

Despite the extraordinary advancements in the last few years, there is one final barrier to overcome: the industrial scale production. As we have shown, this horizon is not so far for GO/rGO membranes. However, there are still some obstacles. For lamellar GO/rGO, the construction is often by self-stacking nanosheets. Ideally, this method allows for a simple assembly and facilitates the scaling up both by the thickness and surface area. But the cost associated with building up uniform, well-defined layered structures can be disadvantageous in large-scale scenarios. In this way, 3D GO/rGO can stand as a cheaper option. The problem is the scaling, which is still far from industrial needs. One possible solution is to find a material that can combine the low cost of 3D GO/rGO and the self-assembly control from lamellar structures, the last great challenge.

References and Future Readings

1. WWAP and UNESCO (2019) The United Nations world water development report 2019: leaving no one behind, United Nations Educational, Scientific and Cultural Organization
2. Eliasson J (2015) The rising pressure of global water shortages. Nature 517:6
3. WWAP and UNESCO (2017) The United Nations world water development report, 2017: Wastewater: an untapped resource; executive summary, United Nations Educational, Scientific and Cultural Organization

4. Ali I, Alharbi OML, Tkachev A, Galunin E, Burakov A, Grachev VA (2018) Water treatment by new-generation graphene materials: hope for bright future. Environ Sci Pollut Res 25:7315–7329
5. Köhler MH, Bordin JR, de Matos CF, Barbosa MC (2019) Water in nanotubes: the surface effect. Chem Eng Sci 203:54–67
6. Radushkevich LE (1952) The carbon structure formed by the thermic decomposition of carbon monoxide in iron VK. Sov J Phys Chemis 26:88
7. Oberlinv A, Endo M, Koyama T (1976) Filamentous growth of carbon through benzene decomposition. J Cryst Grow 32:335
8. Iijima S (1991) Helical microtubules of graphitic carbon. Nature 354:56
9. Umemoto K, Saito S, Berber S, Tománek D (2001) Carbon foam: spanning the phase space between graphite and diamond. Phys Rev B 64:
10. Novoselov KS, Geim AK, Morozov SV, Jiang D, Zhang Y, Dubonos SV, Grigorieva IV, Firsov AA (2004) Electric field effect in atomically thin carbon films. Science 306(5696):666–669
11. Gaim AK (2009) Graphene: status and prospects. Science 324:1530–1534
12. United Nations (2015) Mine tailings storage: Safety is no accident, Report summary
13. Bradl H (2005) Heavy metals in the environment: origin, interaction and remediation. Elsevier
14. Smith AT, LaChance AM, Zeng S, Liu B, Sun L (2019) Synthesis, properties, and applications of graphene oxide/reduced graphene oxide and their nanocomposites. Nano Mater Scie 1:31–47
15. Whitesides GM, Grzybowski B (2002) Self-assembly at all scales. Science 295:2418–2421
16. Xi Y-H, Hu J-Q, Liu Z, Xie R, Ju X-J, Wang W, Chu L-Y (2016) Graphene oxide membranes with strong stability in aqueous solutions and controllable lamellar spacing. ACS Appl Mater Interfaces 8:15557–15566
17. Zhao S, Zhu H, Wang H, Rassu P, Wang Z, Song P, Rao D (2019) Free-standing graphene oxide membrane with tunable channels for efficient water pollution control. J Hazard Mater 366:659–668
18. Hegab HM, Zou L (2015) Graphene oxide-assisted membranes: Fabrication and potential applications in desalination and water purification. J Membr Sci 484:95–106
19. Wei Y, Zhang Y, Gao X, Ma Z, Wang X, Gao C (2018) Multilayered graphene oxide membranes for water treatment: a review. Carbon 139:964–981
20. Hiew BYZ, Lee LY, Lee XJ, Thangalazhy-Gopakumar S, Gan S, Lim SS, Pan G-T, Yang TC-K, Chiu WS, Khiew PS (2018) Review on synthesis of 3D graphene-based configurations and their adsorption performance for hazardous water pollutants. Process Saf Environ Prot 116:262–286
21. Liu H, Qiu H (2020) Recent advances of 3D graphene-based adsorbents for sample preparation of water pollutants: a review. Chem Eng J 393:
22. Li Y, Xu L, Liu H, Li Y (2014) Graphdiyne and graphyne: from theoretical predictions to practical construction. Chem Soc Rev 43:2572–2586
23. Baughman RH, Eckhardt H, Kertesz M (1987) Structure-property predictions for new planar forms of carbon: layered phases containing sp^2 and sp atoms. J Chem Phys 87:6687–6699
24. Kang J, Wei Z, Li J (2019) Graphyne and its family: recent theoretical advances. ACS Appl Mater Interfaces 11:2692–2706
25. Li G, Li Y, Liu H, Guo Y, Li Y, Zhu D (2010) Architecture of graphdiyne nanoscale films. Chem Commun 46:3256–3258
26. Li J, Chen Y, Gao J, Zuo Z, Li Y, Liu H, Li Y (2019) Graphdiyne sponge for direct collection of oils from water. ACS Appl Mater Interfaces 11:2591–2598
27. Wang F, Zuo Z, Shang H, Zhao Y, Li Y (2019) Ultrafastly interweaving graphdiyne nanochain on arbitrary substrates and its performance as a supercapacitor electrode. ACS Appl Mater Interfaces 11:2599–2607
28. Singh RK, Kumar R, Singh DP (2016) Graphene oxide: strategies for synthesis, reduction and frontier applications. RSC Adv. 6:64993–65011
29. Shamaila S, Sajjad AKL, Iqbal A (2016) Modifications in development of graphene oxide synthetic routes. Chem Eng J 294:458–477

30. Pei S, Wei Q, Huang K, Cheng H-M, Ren W (2018) Green synthesis of graphene oxide by seconds timescale water electrolytic oxidation. Nat Commun 9:145
31. Lyu J, Wen X, Kumar U, You Y, Chen V, Joshi RK (2018) Separation and Purification Using GO and r-GO Membranes. RSC Adv 8:23130–23151
32. Thakur S, Karak N (2015) Alternative methods and nature-based reagents for the reduction of graphene oxide: a review. Carbon 94:224–242
33. Erickson K, Erni R, Lee Z, Alem N, Gannett W, Zettl A (2010) Determination of the local chemical structure of graphene oxide and reduced graphene oxide. Adv Mater 22:4467–4472
34. Banhart F, Kotakoski J, Krasheninnikov AV (2011) Structural defects in graphene. ACS Nano 5:26–41
35. Yousefi N, Lu X, Elimelech M, Tufenkji N (2019) Environmental performance of graphene-based 3D macrostructures. Nat Nanotech 14:107–119
36. Zhu Y, Ji H, Cheng H-M, Ruoff RS (2017) Mass production and industrial applications of graphene materials. Natl Sci Rev 5:90–101
37. Sun M, Li J (2018) Graphene oxide membranes: functional structures, preparation and environmental applications. Nano Today 20:121–137
38. Li H, Song Z, Zhang X, Huang Y, Li S, Mao Y, Ploehn HJ, Bao Y, Yu M (2013) Ultrathin, molecular-sieving graphene oxide membranes for selective hydrogen separation. Science 342:95–98
39. Tsou C-H, An Q-F, Lo S-C, Guzman MD, Hung W-S, Hu C-C, Lee K-R, Lai J-Y (2015) Effect of microstructure of graphene oxide fabricated through different self-assembly techniques on 1-butanol dehydration. J Membr Sci 477:93–100
40. Liu G, Jin W (2019) Chapter 2: Graphene-based membranes. In: Graphene-based membranes for mass transport applications. The Royal Society of Chemistry, pp 14–42
41. Guan K, Zhao D, Zhang M, Shen J, Zhou G, Liu G, Jin W (2017) 3D nanoporous crystals enabled 2D channels in graphene membrane with enhanced water purification performance. J Membr Sci 542:41–51
42. Hu M, Mi B (2014) Layer-by-layer assembly of graphene oxide membranes via electrostatic interaction. J Membr Sci 469:80–87
43. Zhu H, Sun P (eds) (2019) Graphene-based membranes for mass transport applications. The Royal Society of Chemistry
44. Thebo KH, Qian X, Zhang Q, Chen L, Cheng H-M, Ren W (2018) Highly stable graphene-oxide-based membranes with superior permeability. Nat Commun 9:1486
45. Yu H, He Y, Xiao G, Fan Y, Ma J, Gao Y, Hou R, Yin X, Wang Y, Mei X (2020) The roles of oxygen-containing functional groups in modulating water purification performance of graphene oxide-based membrane. Chem Eng J 389:
46. Yang E, Ham M-H, Park HB, Kim C-M, Song J, Kim IS (2018) Tunable semi-permeability of graphene-based membranes by adjusting reduction degree of laminar graphene oxide layer. J Membr Sci 547:73–79
47. Kumar S, Garg A, Chowdhuri A (2019) Sonication effect on graphene oxide (GO) membranes for water purification applications. Mater Res Express 6:085620
48. Buelke C, Alshami A, Casler J, Lin Y, Hickner M, Aljundi IH (2019) Evaluating graphene oxide and holey graphene oxide membrane performance for water purification. J Membr Sci 588:
49. Dong L, Fan W, Tong X, Zhang H, Chen M, Zhao Y (2018) A CO_2-responsive graphene oxide/polymer composite nanofiltration membrane for water purification. J Mater Chem A 6:6785–6791
50. Kim C-M, Hong S, Li R, Kim IS, Wang P (2019) Janus graphene oxide-doped, lamellar composite membranes with strong aqueous stability. ACS Sustain Chem Eng 7:7252–7259
51. Ran J, Pan T, Wu Y, Chu C, Cui P, Zhang P, Ai X, Fu C-F, Yang Z, Xu T (2019) Endowing g-C_3N_4 membranes with superior permeability and stability by using acid spacers. Angew Chem Int Ed 58:16463–16468
52. Wu Z, Gao L, Wang J, Zhao F, Fan L, Hua D, Japip S, Xiao J, Zhang X, Zhou S-F, Zhan G (2020) Preparation of glycine mediated graphene oxide/g-C_3N_4 lamellar membranes for nanofiltration. J Membr Sci 601:

53. Ding L, Wei Y, Wang Y, Chen H, Caro J, Wang H (2017) A two-dimensional lamellar membrane: MXene nanosheet stacks. Angew Chem Int Ed 56:1825–1829
54. Wei S, Xie Y, Xing Y, Wang L, Ye H, Xiong X, Wang S, Han K (2019) Two-dimensional graphene Oxide/MXene composite lamellar membranes for efficient solvent permeation and molecular separation. J Membr Sci 582:414–422
55. Liu T, Liu X, Graham N, Yu W, Sun K (2020) Two-dimensional Mxene incorporated graphene oxide composite membrane with enhanced water purification performance. J Membr Sci 593:
56. Nardecchia S, Carriazo D, Ferrer ML, Gutiérrez MC, del Monte F (2013) Three dimensional macroporous architectures and aerogels built of carbon nanotubes and/or graphene: synthesis and applications. Chem Soc Rev 42:794–830
57. Xu Y, Sheng K, Li C, Shi G (2010) Self-assembled graphene hydrogel via a one-step hydrothermal process. ACS Nano 4:4324–4330
58. Chen X, Lai D, Yuan B, Fu M-L (2019) Tuning oxygen clusters on graphene oxide to synthesize graphene aerogels with crumpled nanosheets for effective removal of organic pollutants. Carbon 143:897–907
59. Liu Y, Bai J, Yao H, Li G, Zhang T, Li S, Zhang L, Si J, Zhou R, Zhang H (2020) Embryotoxicity assessment and efficient removal of naphthalene from water by irradiated graphene aerogels. Ecotoxicol Environ Saf 189:
60. Ji C, Xu M, Bao S, Cai C, Lu Z, Chai H, Yang F, Wei H (2013) Self-assembly of three-dimensional interconnected graphene-based aerogels and its application in supercapacitors. J Colloid Interface Sci 407:416–424
61. Gao H, Sun Y, Zhou J, Xu R, Duan H (2013) Mussel-inspired synthesis of polydopamine-functionalized graphene hydrogel as reusable adsorbents for water purification. ACS Appl Mater Interf 5:425–432
62. Xu J, Du P, Bi W, Yao G, Li S, Liu H (2020) Graphene oxide aerogels co-functionalized with polydopamine and polyethylenimine for the adsorption of anionic dyes and organic solvents. Chem Eng Res Des 154:192–202
63. Cheng Y, Barras A, Lu S, Xu W, Szunerits S, Boukherroub R (2020) Fabrication of super-hydrophobic/superoleophilic functionalized reduced graphene oxide/polydopamine/PFDT membrane for efficient oil/water separation. Sep Purif Technol 236:116240
64. Yu H, Hong H-J, Kim SM, Ko HC, Jeong HS (2020) Mechanically enhanced graphene oxide/carboxymethyl cellulose nanofibril composite fiber as a scalable adsorbent for heavy metal removal. Carbohydrate Polym 240:116348
65. Mensah A, Lv P, Narh C, Huang J, Wang D, Wei Q (2019) Sequestration of Pb(II) ions from aqueous systems with novel green bacterial cellulose graphene oxide composite. Materials 12:218
66. Zhang S, Liu G, Gao Y, Yue Q, Gao B, Xu X, Kong W, Li N, Jiang W (2019) A facile approach to ultralight and recyclable 3D self-assembled copolymer/graphene aerogels for efficient oil/water separation. Sci Total Environ 694:
67. Luo H, Xie J, Wang J, Yao F, Yang Z, Wan Y (2018) Step-by-step self-assembly of 2D few-layer reduced graphene oxide into 3D architecture of bacterial cellulose for a robust, ultralight, and recyclable all-carbon absorbent. Carbon 139:824–832
68. Xiao J, Lv W, Song Y, Zheng Q (2018) Graphene/nanofiber aerogels: performance regulation towards multiple applications in dye adsorption and oil/water separation. Chem Eng J 338:202–210
69. Xiao J, Lv W, Xie Z, Song Y, Zheng Q (2017) l-cysteine-reduced graphene oxide/poly (vinyl alcohol) ultralight aerogel as a broad-spectrum adsorbent for anionic and cationic dyes. J Mater Sci 52:5807–5821
70. Hu J, Zhu J, Ge S, Jiang C, Guo T, Peng T, Huang T, Xie L (2020) Bio-compatible, hydrophobic and resilience graphene/chitosan composite aerogel for efficient oil–water separation. Surf Coat Technol 385:
71. Luo J, Fan C, Xiao Z, Sun T, Zhou X (2019) Novel graphene oxide/carboxymethyl chitosan aerogels via vacuum-assisted self-assembly for heavy metal adsorption capacity. Colloids Surf A Physicochem Eng Asp 578:123584

72. Huang T, Shao Y-W, Zhang Q, Deng Y-F, Liang Z-X, Guo F-Z, Li P-C, Wang Y (2019) Chitosan-cross-linked graphene oxide/carboxymethyl cellulose aerogel globules with high structure stability in liquid and extremely high adsorption ability. ACS Sustain Chem Eng 7:8775–8788
73. Zhao L, Yang S, Yilihamu A, Ma Q, Shi M, Ouyang B, Zhang Q, Guan X, Yang S-T (2019) Adsorptive decontamination of Cu^{2+}-contaminated water and soil by carboxylated graphene oxide/chitosan/cellulose composite beads. Environ Res 179:108779
74. Zhao L, Guan X, Yu B, Ding N, Liu X, Ma Q, Yang S, Yilihamu A, Yang S-T (2019) Carboxylated graphene oxide-chitosan spheres immobilize Cu^{2+} in soil and reduce its bioaccumulation in wheat plants. Environ Int 133:105208
75. Delhiraja K, Vellingiri K, Boukhvalov DW, Philip L (2019) Development of highly water stable graphene oxide-based composites for the removal of pharmaceuticals and personal care products. Ind Eng Chem Res 58:2899–2913
76. Wang S, Ning H, Hu N, Huang K, Weng S, Wu X, Wu L, Liu J, Alamusi (2019) Preparation and characterization of graphene oxide/silk fibroin hybrid aerogel for dye and heavy metal adsorption. Compos Part B: Eng 163:716–722
77. Lin Y-Z, Zhong L-B, Dou S, Shao Z-D, Liu Q, Zheng Y-M (2019) Facile synthesis of electrospun carbon nanofiber/graphene oxide composite aerogels for high efficiency oils absorption. Environ Int 128:37–45
78. Zhou F, Feng X, Yu J, Jiang X (2018) High performance of 3D porous graphene/lignin/sodium alginate composite for adsorption of Cd (II) and Pb (II). Environ Sci Pollut Res 25:15651–15661
79. Xiao D, He M, Liu Y, Xiong L, Zhang Q, Wei L, Li L, Yu X (2020) Strong alginate/reduced grapheme oxide composite hydrogels with enhanced dye adsorption performance. Polym Bull 77:6609–6623
80. Liu S, Bastola AK, Li L (2017) A 3D printable and mechanically robust hydrogel based on alginate and graphene oxide. ACS Appl Mater Interf 9(47):41473–41481
81. Wei X-N, Ou C-L, Fang S-S, Zheng X-C, Zheng G-P, Guan X-X (2019) One-pot self-assembly of 3D CdS-graphene aerogels with superior adsorption capacity and photocatalytic activity for water purification. Powder Technol 345:213–222
82. Neskoromnaya EA, Burakov AE, Babkin AV, Burakova IV, Melezhik AV, Mkrtchyan ES (2020) Development of sorption materials based on iron(III)-chloride-modified graphene oxide for selective removal of organic pollutants from aquatic media. Fullerenes Nanotubes Carbon Nanostruct 28:521–525
83. Wang J, Duan X, Dong Q, Meng F, Tan X, Liu S, Wang S (2019) Facile synthesis of N-doped 3D graphene aerogel and its excellent performance in catalytic degradation of antibiotic contaminants in water. Carbon 144:781–790
84. Liu Y, Yang D, Shi Y, Song L, Yu R, Qu J, Yu Z-Z (2019) Silver phosphate/graphene oxide aerogel microspheres with radially oriented microchannels for highly efficient and continuous removal of pollutants from wastewaters. ACS Sustainable Chem Eng 7:11228–11240
85. Pan L, Liu S, Oderinde O, Li K, Yao F, Fu G (2018) Facile fabrication of graphene-based aerogel with rare earth metal oxide for water purification. Appl Surf Sci 427:779–786
86. Zhang Y, Li K, Liao J (2020) Facile synthesis of reduced-graphene-oxide/rare-earth-metal-oxide aerogels as a highly efficient adsorbent for Rhodamine-B. Appl Surf Sci 504:144377
87. Zhuang Y, Wang X, Zhang L, Dionysiou DD, Shi B (2019) Fe-chelated polymer templated graphene aerogel with enhanced Fenton-like efficiency for water treatment. Environ Sci Nano 6:3232–3241
88. Karimi-Maleh H, Shafieizadeh M, Taher MA, Opoku F, Kiarii EM, Govender PP, Ranjbari S, Rezapour M, Orooji Y (2020) The role of magnetite/graphene oxide nano-composite as a high-efficiency adsorbent for removal of phenazopyridine residues from water samples, an experimental/theoretical investigation. J Mol Liq 298:112040
89. Neolaka YA, Lawa Y, Naat JN, Riwu AA, Iqbal M, Darmokoesoemo H, Kusuma HS (2020) The adsorption of Cr(VI) from water samples using graphene oxide-magnetic (GO-Fe_3O_4) synthesized from natural cellulose-based graphite (kusambi wood or Schleichera oleosa): Study of kinetics, isotherms and thermodynamics. J Mater Res Technol 9:6544–6556

90. Li N, Chen J, Shi Y-P (2019) Magnetic polyethyleneimine functionalized reduced graphene oxide as a novel magnetic sorbent for the separation of polar non-steroidal anti-inflammatory drugs in waters. Talanta 191:526–534
91. Tran TV, Nguyen DTC, Le HT, Vo D-VN, Nanda S, Nguyen TD (2020) Optimization, equilibrium, adsorption behavior and role of surface functional groups on graphene oxide-based nanocomposite towards diclofenac drug. J Environ Sci 93:137–150
92. Ritz B, Zhao Y, Krishnadasan A, Kennedy N, Morgenstern H (2006) Estimated effects of hydrazine exposure on cancer incidence and mortality in aerospace workers. Epidemiology 2006:154–161
93. Aigner BA, Darsow U, Grosber M, Ring J, Plötz SG (2010) Multiple basal cell carcinomas after long-term exposure to hydrazine: case report and review of the literature. Dermatology 221:300–302
94. Xu X, Pereira LFC, Wang Y, Wu J, Zhang K, Zhao X, Bae S, Bui CT, Xie R, Thong BH, Loh KP, Donadio D, Li B, Özyilmaz B (2014) Length-dependent thermal conductivity in suspended single-layer graphene. Nat Commun 5:3689
95. Crook C, Bauer J, Izard AG, de Oliveira CS, de Souza e Silva JM, Berger JB, Valdevit L (2020) Plate-nanolattices at the theoretical limit of stiffness and strength. Nat Commun 11:1579
96. Perilla JR, Schulten K (2017) Physical properties of the HIV-1 capsid from all-atom molecular dynamics simulations. Nat Commun 8:15959
97. Brenner DW, Shenderova OA, Harrison JA, Stuart SJ, Ni B, Sinnott SB (2002) A second-generation Reactive Empirical Bond Order (REBO) potential energy expression for hydrocarbons. J Phys: Condens Matter 14:783–802
98. Yeo J, Jung GS, Martín-Martínez FJ, Beem J, Qin Z, Buehler MJ (2019) Multiscale design of graphyne-based materials for high-performance separation membranes. Adv Mater 31:1805665
99. Srinivasan SG, van Duin ACT, Ganesh P (2015) Development of a ReaxFF potential for carbon condensed phases and its application to the thermal fragmentation of a large fullerene. J Phys Chem A 119:571–580
100. Lee J-U, Yoon D, Cheong H (2012) Estimation of young's modulus of graphene by Raman spectroscopy. Nano Lett 12:4444–4448
101. Paci JT, Belytschko T, Schatz GC (2007) Computational studies of the structure, behavior upon heating, and mechanical properties of graphite oxide. J Phys Chem C 111:18099–18111
102. Gómez-Navarro C, Meyer JC, Sundaram RS, Chuvilin A, Kurasch S, Burghard M, Kern K, Kaiser U (2010) Atomic structure of reduced graphene oxide. Nano Lett 10:1144–1148
103. Suk ME, Aluru NR (2010) Water transport through ultrathin graphene. J Phys Chem Lett 1:1590–1594
104. Cohen-Tanugi D, Grossman JC (2012) Water desalination across nanoporous graphene. Nano Lett 12:3602–3608
105. Cho KM, Lee H-J, Nam YT, Kim Y-J, Kim C, Kang KM, Ruiz Torres CA, Kim DW, Jung H-T (2019) Ultrafast-selective nanofiltration of an hybrid membrane comprising laminated reduced graphene oxide/Graphene oxide nanoribbons. ACS Appl Mater Interf 11:27004–27010
106. Wei N, Peng X, Xu Z (2014) Understanding water permeation in graphene oxide membranes. ACS Appl Mater Interf 6:5877–5883
107. Falk K, Sedlmeier F, Joly L, Netz RR, Bocquet L (2010) Molecular origin of fast water transport in carbon nanotube membranes: superlubricity versus curvature dependent friction. Nano Lett 10:4067–4073
108. Chen B, Jiang H, Liu X, Hu X (2017) Molecular insight into water desalination across multilayer graphene oxide membranes. ACS Appl Mater Interfaces 9:22826–22836
109. Qiu H, Xue M, Shen C, Zhang Z, Guo W (2019) Graphynes for water desalination and gas separation. Adv Mater 31:1803772
110. Köhler MH, Bordin JR, Barbosa MC (2019) Ion flocculation in water: from bulk to nanoporous membrane desalination. J Mol Liq 277:516–521
111. Kou J, Zhou X, Chen Y, Lu H, Wu F, Fan J (2013) Water permeation through single-layer graphyne membrane. J Chem Phys 139:064705

112. Lin S, Buehler MJ (2013) Mechanics and molecular filtration performance of graphyne nanoweb membranes for selective water purification. Nanoscale 5:11801–11807
113. Xue M, Qiu H, Guo W (2013) Exceptionally fast water desalination at complete salt rejection by pristine graphyne monolayers. Nanotechnology 24:505720
114. Zhu C, Li H, Zeng XC, Wang EG, Meng S (2013) Quantized water transport: ideal desalination through graphyne-4 membrane. Sci Rep 3:3163
115. Li J, Xie Z, Xiong Y, Li Z, Huang Q, Zhang S, Zhou J, Liu R, Gao X, Chen C, Tong L, Zhang J, Liu Z (2017) Architecture of β-graphdiyne-containing thin film using modified glaser-hay coupling reaction for enhanced photocatalytic property of TiO_2. Adv Mater 29:1700421
116. Morelos-Gomez A, Cruz-Silva R, Muramatsu H, Ortiz-Medina J, Araki T, Fukuyo T, Tejima S, Takeuchi K, Hayashi T, Terrones M, Endo M (2017) Effective NaCl and dye rejection of hybrid graphene oxide/graphene layered membranes. Nat Nanotechnol 12:1083–1088
117. Motevalli B, Sun B, Barnard AS (2020) Understanding and predicting the cause of defects in graphene oxide nanostructures using machine learning. J Phys Chem C 124:7404–7413
118. Wan J, Jiang J-W, Park HS (2020) Machine learning-based design of porous graphene with low thermal conductivity. Carbon 157:262–269
119. Noh J, Gu GH, Kim S, Jung Y (2020) Machine-enabled inverse design of inorganic solid materials: promises and challenges. Chem Sci 11:4871–4881

Graphene Oxide and Reduced Graphene Oxide as Nanofillers in Membrane Separation

Siamak Pakdel, Sima Majidi, Jafar Azamat, and Hamid Erfan-Niya

Abstract Due to the excellent physiochemical properties of graphene oxide (GO) and its family members, these materials have been extensively used for water purification. This chapter presents and discusses the recent development of GO and reduced graphene oxide (rGO) composites for membrane filtration. At first, we highlight the current synthesis methods of polymer–GO/rGO nanocomposites such as solvent processing, *in situ* polymerization, and melt processing. Also, some novel preparation methods of nanocomposites are discussed. Then, we compare the recent experimental works on the fabrication and testing of these nanocomposite membranes with classical membranes. Finally, Characterization techniques of nanocomposite membranes including spectral characterization, analysis of the membrane surface roughness, morphological study, and measurement of contact angle, as well as the thermal and mechanical properties of the nanocomposite membranes are investigated.

Keywords Nanofiller · Graphene oxide · Reduced graphene oxide · Nanocomposite · Membrane

1 Introduction

An ideal membrane material should be capable of being modified to get the desired performance. Recently, some ideal nanomaterials, such as titanium dioxide [1], nanosilver [2], aquaporins [3], nanosilica [4, 5], graphene oxide (GO), and carbon nanotubes (CNTs) [6–8] have been used to develop membranes with ideal performance and structure [9–11]. Among them, GO is the new class of membrane material that was discovered in 2012 [12], offers superior properties in comparison with ceramic and polymeric membranes in terms of high hydrophilicity, outstanding

S. Pakdel · S. Majidi · H. Erfan-Niya (✉)
Faculty of Chemical and Petroleum Engineering, University of Tabriz, 51666-16471 Tabriz, Iran
e-mail: herfan@tabrizu.ac.ir

J. Azamat
Department of Basic Sciences, Farhangian University, Tehran, Iran

R. Das (ed.), *Two-Dimensional (2D) Nanomaterials in Separation Science*,
Springer Series on Polymer and Composite Materials,
https://doi.org/10.1007/978-3-030-72457-3_5

colloidal stability in water, nanopores' generation, and excellent mechanical behavior [13, 14]. GO is an oxidized form of graphene that is considered to be the thinnest and strongest material with an extended one-atom sheet of sp^2-bonded carbon atoms [15–17]. Moreover, reduced GO (rGO) is another important derivative of graphene, with similar properties as GO that has been used in membrane separation technology [18]. The structure of the GO is supposed to be a network of sp^2-hybridized –C–C– atoms, with approximately 55% of sp^3 C–O bonds in the form of carboxyl, hydroxyl, or epoxy groups [19]. These polar oxygen units facilitated aqueous dispersion and also brought positive consequences in terms of declining sheet conductivity. Hence, an effective removal of these oxygen species from GO for production of rGO is important in many applications, such as linear and non-linear conductive polymer nanocomposite fabrication [20–23].

Alternatively, GO is a better candidate for constructing free-standing and polymer–GO hybrid membranes for water separation processes because of the wide availability of reactive surface sites and the layered structure. GO nanosheets display great antifouling capacity that is a significant desired property in the field of water desalination processes [24, 25]. GO-based thin membranes display promising qualities in the field of water-permeable membrane to be employed in the desalination procedure [26, 27]. GO has also revealed to be effective in allowing the permeation of water, while subsequently blocking penetration of other liquids, vapors, or gases [28].

Ideally, the membranes should be mechanically and chemically stable against high pressures over a long period of operation, while maintaining their desired water flux and ion rejection characteristics. The manufacturing processes of thin-film composite (TFC) membranes, which consist of a substrate and an interfacial polymerized skin layer, have been developed for water purification [29–31]. In these membranes, GO can be used as nanofillers to improve the membrane permselectivity [32].

The nanofillers such as graphene, GO, and rGO have strong interaction with the polymer chain as well as monotonic good dispersion in the polymer matrix. This results in increasing the mechanical strength and thermal stability of the membrane. Moreover, the hydrophilic functional groups bound onto GO improve the wetting properties of the hydrophobic polymer membrane [33]. The GO sheets as two-dimensional (2D) nanofillers can effectively modify the physicochemical properties of the polymer matrix due to their large surface area and functional groups as well as their intrinsic mechanical and thermal stability [34, 35]. However, pristine graphene is not compatible with organic polymers, GO sheets containing epoxide, hydroxyl, diol, carboxyl, and ketone functional groups are known to modify the interactions of GO sheets and polymer matrix and consequently make them compatible [36]. In the past 5 years, numerous studies have been carried out to fabricate polymer–GO nanocomposite membranes [37–40]. Alternatively, several approaches to preparing stacked graphene-based nanosheets or graphene-based polymer nanocomposites have been discovered for their potential applications in the water treatment area. One of these methods is to coat ultrathin, a few-layered graphene/GO/rGO sheets on polymeric substrates [24, 41–43]. The manufactured composite membranes all displayed upgraded filtration performance. For instance, Han et al. [41] synthesized ultrathin graphene nanofiltration (NF) membrane using vacuum filtration of rGO suspension. Likewise, Xu et al. fabricated a loaded GO–TiO_2 membrane through vacuum filtration of GO–TiO_2 composite nanosheets [40, 43]. A variety of polymer–GO composites, such as poly (vinyl acetate)–GO, polyurethane–GO, Nafion–GO, poly (vinyl alcohol)–GO, polybenzimidazole–GO, and poly(allylamine)/GO have been thoroughly investigated [44–47].

In summary, according to literature reports, it is expected that the polymer–GO/rGO nanocomposite membranes could be an outstanding technology for water purification. Some of these interesting works are reviewed in this chapter as a reflection for future works. This chapter will also present the fabrication methods and characterization techniques of polymer–GO/rGO composite membranes.

2 Preparation Methods of Polymer–GO/rGO Nanocomposite

GO surface functional groups, such as epoxide, hydroxyl, carbonyl, and carboxyl groups, make it as a potential candidate for preparing nanocomposites [48, 49]. Most of polymer–GO/rGO composites have been prepared by three common procedures of solvent processing, *in situ* polymerization, and melt processing [50] and, also, other new preparation methods, such as Pickering emulsions and spin coating which are investigated in the next sections.

2.1 Solvent Processing

Among the reported methods for synthesizing polymer–GO/rGO nanocomposites, solvent processing is one of the most current procedures, because of the well dispersion of GO in many solvents [51]. This method is based on a solvent system in which the polymer is solubilized and graphene or modified graphene layers are allowed to swell [52]. GO is directly dispersed in water or various organic solvents such as chloroform, acetone, toluene, dimethylformamide (DMF), and tetrahydrofuran (THF). The sp^2 carbon network of GO sheets can be recovered during the dispersion by reducing the GO sheets using specific solvent [53–55]. After mixing GO and polymer with solvent, the polymer–rGO composites are prepared through reducing GO either at the same time as solvent mixing or after the blending of polymer and GO. Up to now, a variety of polymer–rGO composites have been prepared through this approach [51, 56–59].

Uniform dispersion of graphene/GO/rGO in the solvent solution is important before mixing them with the polymer matrix. For instance, Zhang et al. [60] homogeneously dispersed the rGO in organic media without the addition of complex dispersing agents or the requirement of functionalized GO. In this experiment, the aqueous colloids of rGO sheets were prepared in an alkaline solution. The produced rGO aggregates could be easily redispersed in suitable organic media as individual nanosheets by using NaCl salts as flocculants with the help of sonication, which is an economical and facile technique for the large-scale production of graphene/GO/rGO. The homogeneous dispersion of graphene/GO/rGO in organic solvents facilitates solution-processing method for new composites' development. The covalent [61–63] and non-covalent [64–67] modifications are the strategies that have been incorporated toward the homogeneous dispersion of graphene in various solvents. However, the reaction condition for the grafting modification is usually complex, while for the non-covalent methods, the stabilizing agents are difficult to be removed completely [60]. It should be considered that the organic solvents are generally adsorbed between the graphene layers that adversely impact the performance of the prepared nanocomposites [68]. Solution blending of polymer and GO is often carried out in laboratory preparation [69], because large amounts of organic solvents are required to be used in the preparation especially when the nanocomposite is used for industrial production [68]. Also, it is difficult to completly remove the large quantities of the utilized solvent that makes it not eco-friendly [70].

2.2 *In Situ Polymerization*

In situ polymerization method is often used to prepare graphene-filled polymer composites, such as those with epoxy [71–75]. In this method, graphene/GO/rGO is mixed with pre-polymers or monomers [50]. First, graphene or modified graphene is swollen inside the liquid monomer. Then an appropriate initiator is diffused, and polymerization is begun either by radiation or heat [74, 76]. In epoxy–graphene composite, at first, the graphene-based filler is mixed with epoxy resins by applying high-shear forces and then a curing agent is added to initiate the polymerization [53, 77, 78]. *In situ* polymerization of monomer in the presence of GO fillers can efficiently improve the interfacial bonding strength between GO and polymers. Besides, it improves the dispersion status of GO nanosheets in the polymer matrix [79], but this technique could be carried out only on the laboratory scale [69].

Wang and coworkers [80] introduced an efficient approach of *in situ* polymerization in order to fabricate GO/polyimide (PI) composites with greater performances. NH_2-functionalized GO (ODA-GO) was a versatile starting platform for the *in situ* preparation of composite films via the embedding of polyamic acid onto the GO surface. The graphene sheets, bearing grafted polymer layers, exhibit great dispersibility and compatibility with the polymer matrix and also create strong bonds with the PI polymer chains. The functionalized graphene nanosheets strengthen the PI composites greatly by incorporating the ultrahigh contact area and strong interfacial interactions of the PI matrix. The PI–ODA-GO films, prepared by *in situ* polymerization, exhibit around a 6.4-times increase in tensile modulus, and 240% improvement in tensile strength, for a very low GO content (0.3 wt%). Furthermore, the bulk dielectric constant is decreased from 3.3, for the neat PI film, to just 2.0 for the PI–ODA-GO composite film. Consequently, in situ polymerization to produce PI–ODA-GO films provides an effective method to develop freestanding, lightweight, and strong polyimide–GO composite materials with low dielectric constants and shows promise for a wide range of applications.

Researchers have reported an *in situ* polymerization step that was combined with chemical grafting modification for fabricating poly(vinyl alcohol)-grafted GO/poly(vinyl alcohol) nanocomposites [81]. The poly(vinyl alcohol)-g-GO sheets displayed great dispersion and compatibility with the poly(vinyl alcohol) matrix and also created strong interfacial interactions with the poly(vinyl alcohol) chains of the matrix [68].

2.3 Melt Processing

Melt processing is the most eco-friendly and economical method [53] and is appropriate for mass production of GO composites for industrial applications [82], but the high viscosity of polymer melts and the high temperature used in this process always leads to GO flocculation [83]. Besides, the high shear forces used in melt compounding can often cause the breakage of the filler materials such as the graphene-based nanosheets and CNTs [84]. However, these preparation methods are less or more restricted in applications, they demonstrate their sole advantages to fabricate polymer–GO composite materials [69].

In melt processing, no solvent is needed due to utilizing both high-temperature melting and high-shear forces to mix the filler and matrix materials and combining the graphene or modified graphene with the polymer matrix is done in the molten state [36]. In this procedure, the polymer and graphene/GO/rGO are directly mixed in a twin-screw extruder, and then the parameters of the extruder screw, such as the screw speed, blending time, and temperature are regulated [68].

In general, most of the polymer–GO nanocomposites are prepared by using polypropylene (PP), polyurethane, poly (ethylene terephthalate), poly (ether ketone), Polystyrene (PS), styrene–ethylene/butylene–styrene triblock copolymers [45, 85–87]. Melt blending is a promising method for the preparation of commercial nanocomposite products because of its processing without using organic solvents [68].

2.4 Pickering Emulsions

Pickering emulsion is the preparation of polymer nanocomposites via the self-assembly of nanoparticles at liquid–liquid interface [88]. It is a promising approach for the design and preparation of well-defined nanostructured materials [89–91]. Self-assembly of the materials at the liquid–liquid interface is thermodynamically promising because of the decrease in the interfacial energy of the system upon the formation of a new interface between the liquid and particle phases instead of the liquid–liquid interface [92, 93]. Pickering emulsions generally have less toxicity, less susceptibility to foaming, and lower cost in comparison with common emulsions stabilized by surfactants [94]. The properties of Pickering emulsions are determined by the size of the solid particles and their amphiphilicity. The amphiphilicity of GO makes it perform like a colloidal surfactant. Xie et al. [95] produced PS/GO composite particles by Pickering emulsion stabilized by GO. The results demonstrated that the pH and the size of the GO particles affect the stability of Pickering emulsions and the final morphology of the particles, respectively. Small GO particles were appropriate for Pickering emulsion polymerization to synthesize the composite particles. The prepared composite particles could have a wide range of potential applications such as in environmental remediation, catalysts, sensors, and energy storage. Gudarzi et al. [93] reported a novel method based on Pickering emulsion polymerization that promises well dispersion and increases loading. The main idea of this method is to use a high affinity of GO for assembly at the liquid–liquid interface. A guideline for confirming stable hybrid colloids of polymer–GO with a proper polymer particle size has been presented. Then, a system of polymethyl methacrylate (PMMA)–GO has been selected, and the nanocomposites have been produced by Pickering emulsion polymerization to examine the theory. The PMMA monomer was employed to test the feasibility of the process and up to 5 wt% of GO has been loaded successfully. This high loading of GO paves the way for the industrial-scale production of graphene-based masterbatches that can be used in downstream industries via melt mixing. Solvent- and soap-free characteristics increase the quality of the nanocomposite and make the method environmentally friendly.

2.5 Spin Coating

The homogeneous thin composite films can be synthesized by spin-coating technique in the range of micrometer to nanometer thickness. This method includes numerous stages such as fluid dispense, spin up, stable fluid outflow, spin-off, and evaporation. First, the material is deposited on the turntable and then spin up and spin-off are carried out in sequence while the evaporation is occurred during the process. The solution is dispersed on the turntable via centrifugal force, at a high velocity in a range of 1000–8000 rpm [96, 97], which results in leaving a uniform and thin layer. After this stage, drying of the applied layer occurs in which high volatile components are removed from the substrate because of the evaporation or simply drying. The thickness of the synthesized composite film is controlled by the speed of rotation and the viscosity of the coating solution [98]. Lue et al. [99] fabricated Nafion/GO composite membranes for direct liquid fuel cell applications with the purpose of reducing fuel permeability and extra-improving cell performance. Nafion 212 (N212) composites consist of various GO contents that were fabricated using spin coating methods. The GO filling in the prepared composites had a positive relationship with the water uptake, ionic conductivity, and ion exchange capacity. The spin-coated composite membranes had a higher width to thickness ratio, which is the effective aspect, leading to the higher permeability. The prepared composite with 0.067% loaded GO showed double peak power densities compared to the pristine N212 in direct formic acid, ethanol, and methanol fuel cells. The exact mass fraction of the GO on the composite membranes was obtained to be 2.53 wt%. The well-aligned thin top layer of GO contributed to the ideal fuel cell performance.

3 Characterization of Nanocomposite Membranes

The prepared nanocomposite membranes are characterized by various instrumental methods to study the change in physiochemical properties and structure of the membranes, which is caused by embedding GO or rGO into them. The Scanning Electron Microscope (SEM), Transmission Electron Microscope (TEM), X-ray Diffractometer (XRD), and Atomic Force Microscope (AFM) are typically used for the morphology characterization, whereas X-ray Photoelectron Spectroscope (XPS) and Fourier transform infrared (FTIR) spectroscope are commonly applied for functional group detection. Consequently, in order to study the effect of the surface charge of the nanocomposite membrane, Zeta potential analyses are used. The contact angle studies are employed to evaluate the surface hydrophilicity of the nanocomposite membranes. Thermogravimetric analysis (TGA) can also be used to measure the weight loss and to detect the possible functionalized state of nanocomposite membrane. At last, the mechanical stability of the membranes is measured by tensile tests.

3.1 Membrane Morphology Analysis

(a) **SEM Analysis**

The cross-section morphologies and structure of the prepared nanocomposite membranes are examined through SEM analyses. To make the membranes electrically conducting, the membrane samples are cut into small pieces and then submerged in liquid nitrogen. Then sputtering of conductive metals (e.g. gold, platinum, cadmium, etc.) is used to coat on the cross-sectional layer. The acquisition of cross-sectional images of the membranes is done in very high vacuum conditions [100–103]. Ganesh et al. [38] recorded the cross-sectional SEM images of the prepared membranes that are shown in Fig. 1. Herein, polysulfone (PSF) membrane demonstrated the characteristic asymmetric structure of dense top layer followed by macro-voids. The loading of GO into the PSF polymer matrix has caused significant changes in macro-void structure. Indeed, the hydrophilic basis of the GO is led to extended porosity as well as changes in the macro-voids. As a result, the GO has played an important role in modification of membrane morphology.

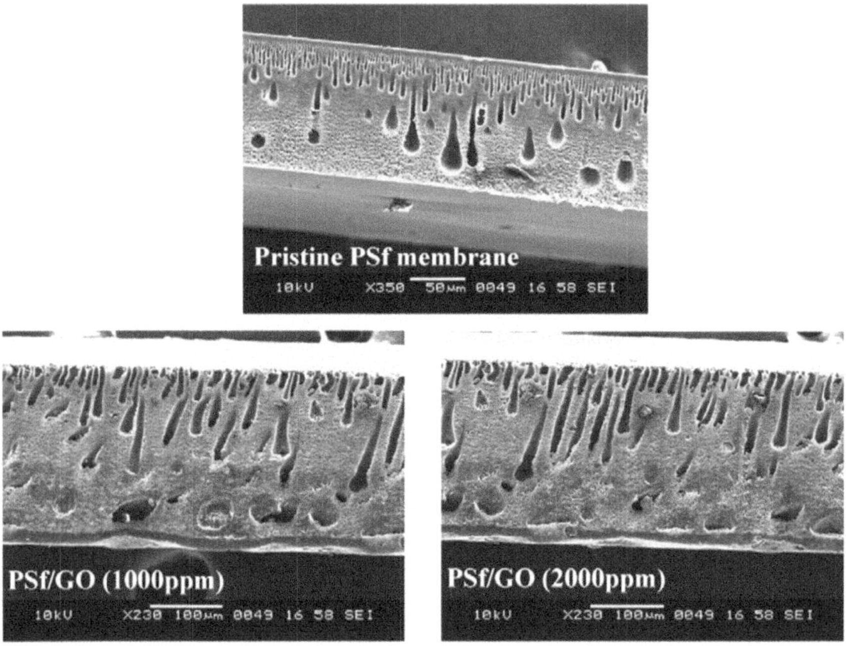

Fig. 1 Cross-sectional SEM images of pristine PSF and PSF–GO mixed matrix membranes. The figure is adapted with permission from Elsevier [38]

(b) **TEM Analysis**

One of the other techniques to examine the morphology of the synthesized nanocomposite and study the distribution of nanofiller in mixed matrix membrane is to use TEM. To prepare the TEM samples, the membranes are cut into very low thickness sections by the microtome method [40, 104]. Then, the images are taken to obtain the TEM cross-section. Wan et al. [104] incorporated GO into the four polymer matrix (poly(L-lactic acid) (PLLA), poly(ε-caprolactone) (PCL), polystyrene (PS), and high-density polyethylene (HDPE)) to study the reinforcement behavior of the prepared nanocomposite membranes. TEM images obtained from GO and the four polymers' dispersion are presented in Fig. 2, which indicates flexible and well dispersed of GO in PLLA, PCL, PS, and HDPE polymers.

(c) **XRD Analysis**

XRD analysis, as a non-destructive test method, is one of the microstructural analysis methods used for the identification of crystallinity of polymers. In reality, textural features of synthesized membranes are characterized by XRD analysis on XRD devices. The emission current and accelerating voltage are set. Consequently, data with Cu-Kα radiation are collected in a certain range of 2θ [103, 105, 106]. Strankowski et al. [106] fabricated a new polyurethane (PU) nanocomposite by

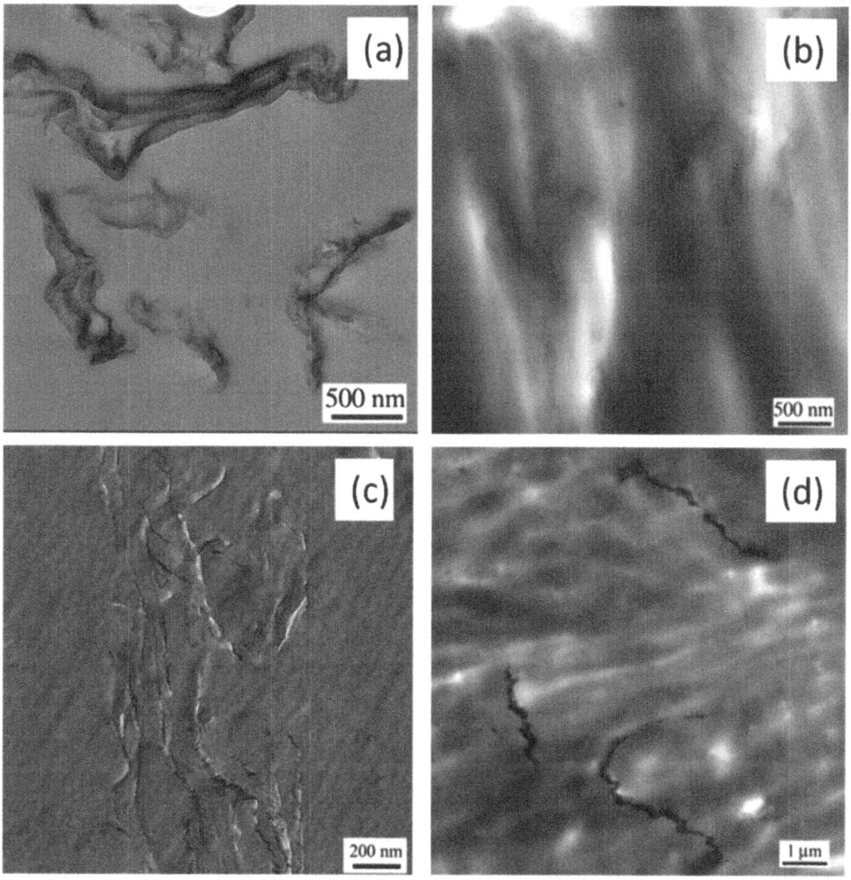

Fig. 2 TEM images of various polymers–GO nanocomposites with 1.0 wt% GO concentration: **a** PLLA–GO, **b** PCL–GO, **c** PS–GO, and, with 0.5 wt% GO concentration: **d** HDPE–GO. The figure is adapted with permission from the Royal Society of Chemistry [104]

embedding a thermally rGO to modify the matrix of PU elastomers. Investigation of nanocomposites by applying XRD method indicated notable improvement in sample properties. XRD spectra for neat PU and PU–rGO are shown in Fig. 3, which indicates a maximum peak around 21°. For PU nanocomposites, XRD pattern demonstrates a similar pattern comparing with PU composite due to the predominant amount of polymer materials versus the quantity of loaded rGO. By increasing the rGO content, the intensity reduces from 850 to 650 a.u., which is concluded that interactions between rigid segments are being more disrupted with increasing rGO loading and leads to more amorphous and disordered nanocomposite structure. It should also be noted that the intensity discrepancy between 2 wt% of rGO loading and higher content is because of the overloading of composite with the rGO as nanofiller.

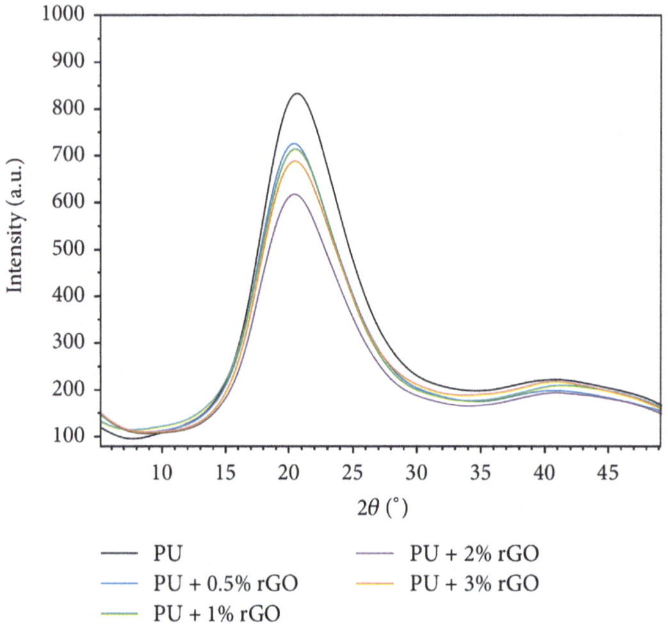

Fig. 3 XRD spectra of PU nanocomposites. The figure is adapted with permission from Hindawi Publishing Corporation[106]

(d) AFM Analysis

Surface topographies of the composite membranes were analyzed using AFM instrument. Quantitative information could be extracted from the AFM three-dimensional topographies as taken in the semi-contact mode, and the root mean square (RMS) roughness was recorded [39, 40, 107]. Yin et al. [40] used AFM measurements to study the surface roughness of the prepared mixed matrix membranes as shown in Fig. 4. The obtained results showed that the incorporation of GO into the TFC during the phase inversion process has an impact on the membrane surface roughness. The RSM values decreased at first with an increase in GO loading, due to the interruption of GO nanosheets on the growth of leaf-like structure during the nanocomposite fabrication process, while the observed increment of roughness at GO concentration of 0.02 wt% may be caused by the surface-located clusters form due to the aggregation of GO.

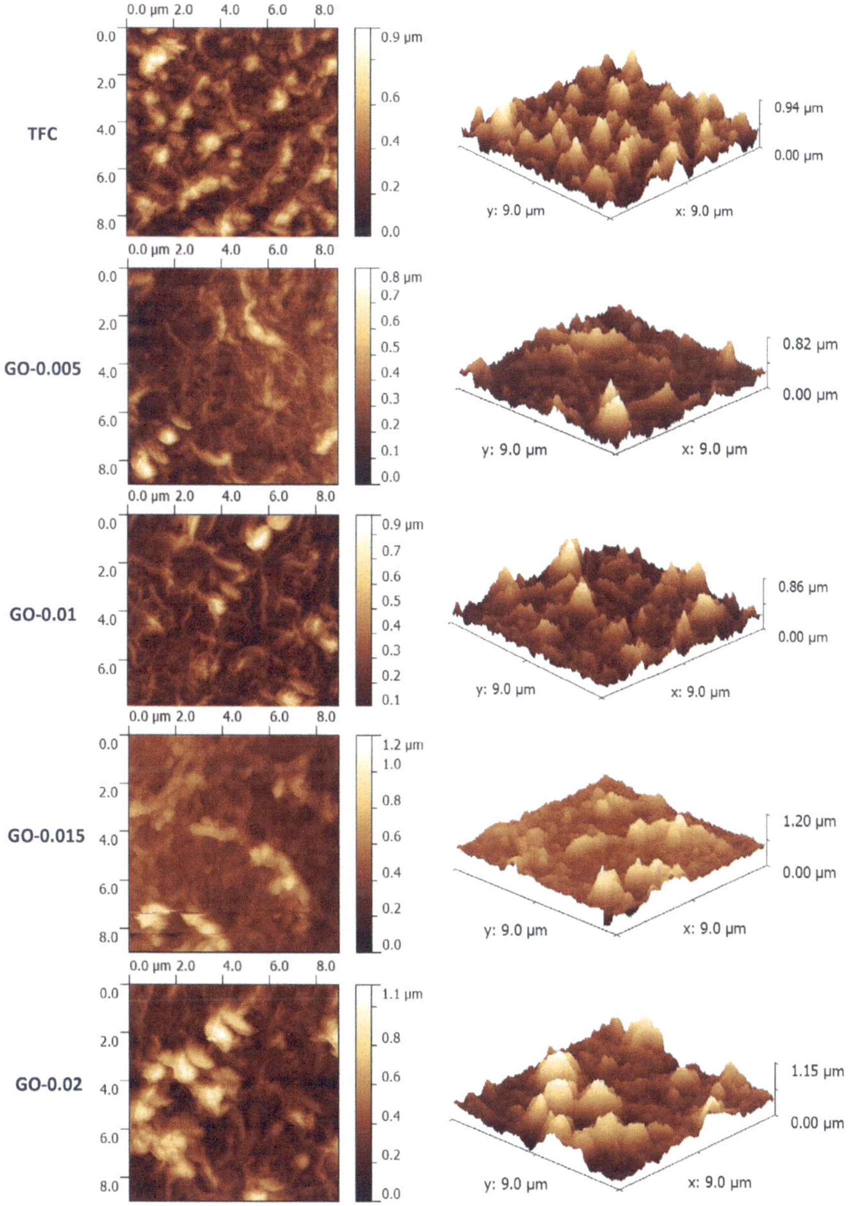

Fig. 4 AFM analysis of membrane surface with a scan area of 9 × 9 μm. The figure is adapted with permission from Elsevier [40]

3.2 Functional Groups of Membrane Analyses

(a) FTIR Analysis

Fourier transform infrared (FTIR) spectroscopy is used to identify the presence of GO or rGO and their interaction with polymer membrane matrix. In other words, the attenuated total reflectance (ATR) FTIR spectra are applied to study the membrane matrix surface morphology changes due to characteristic nanofiller content [100, 102, 106]. Bano et al. [108] used ATR FTIR spectra to confirm the interaction of GO with the polyamide (PA) membrane and the developing of thin-film nanocomposite (TFN).The FTIR spectra of PA and PA–GO TFN 2 are presented in Fig. 5. As seen in Fig. 5, the characteristic absorption bands in PA–GO composite membrane demonstrate the fact that incorporation of GO into the PA membrane does not affect the structural feature of the base membrane. Additionally, compared with the bare PA membrane, the modified PA–GO composite membrane depicted an intense and wider absorption band at 3367 cm^{-1}, which attributes to the improvement of surface hydrophilicity.

(b) XPS Analysis

XPS spectra measurement is conducted to investigate the surface chemistry information of materials. In other words, oxidation of the nanocomposites can be confirmed by XPS spectrum. The spectrum is collected as the intensity (number of counts per second) versus binding energy [107, 109]. Österholm et al. [110] developed one-step electrochemical polymerization method for incorporation of GO into polypyrrole (PPy) and poly(3,4-ethylenedioxythiophene) (PEDOT) without any additional dopants. To study the effect of GO loading into the polymers, reference films without GO were prepared using the same monomer concentration in 0.1 M KCl (PPy–Cl)

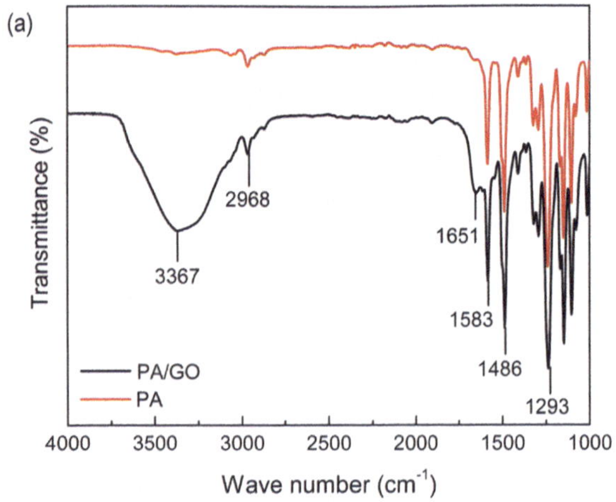

Fig. 5 FTIR spectra of PA TFC and PA–GO TFN 2 membranes. The figure is adapted with permission from Royal Society of Chemistry [108]

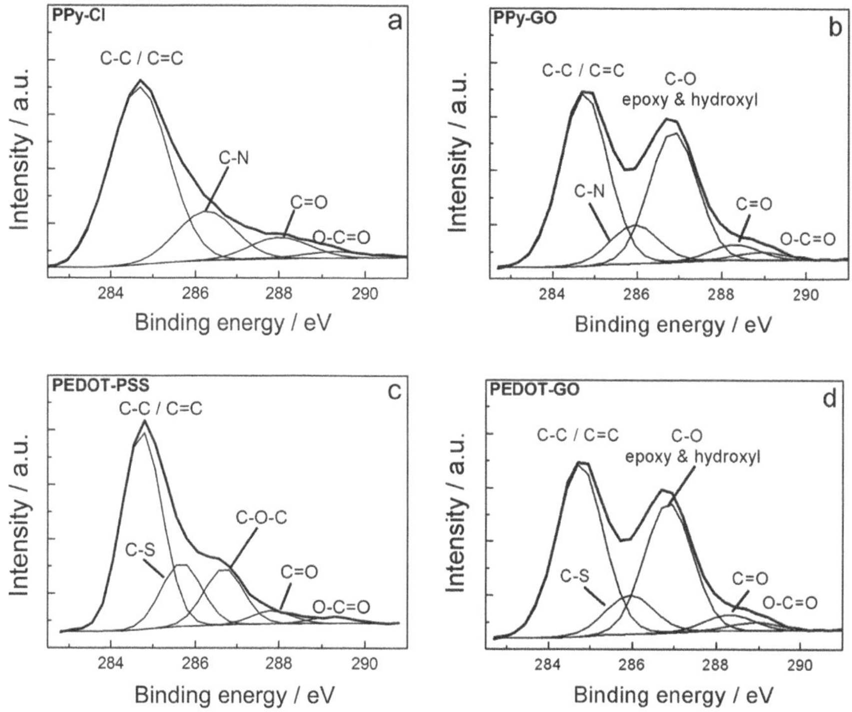

Fig. 6 C 1s XPS spectra of **a** PPy–Cl, **b** PPy–GO, **c** PEDOT–PSS, and **d** PEDOT–GO. The figure is adapted with permission from Elsevier [110]

and in 0.1 M NaPSS (PEDOT–PSS). They applied XPS to verify the embedding of GO into the polymers. The C 1s core-level spectra of the nanocomposite films and their respective reference films are depicted in Fig. 6. An additional strong peak is observed in XPS spectra of PPy–GO and PEDOT–GO compared with that of PPy–Cl and PEDOT–PSS, which is a clear evidence to the successful incorporation of GO into the polymer films.

3.3 Evaluation of Membrane Surface Charge

(a) **Zeta Potential Analysis**

The surface charge of the nanocomposite membranes is studied using zeta potential analysis. The zeta potential of the membrane surface is determined using an electrokinetic analyzer [102, 108]. Bano et al. [108] indicated that the incorporation of GO into the PA polymer induces a negative charge on the surface of the nanocomposite membranes (Fig. 7). This was justified using zeta potential investigation. The

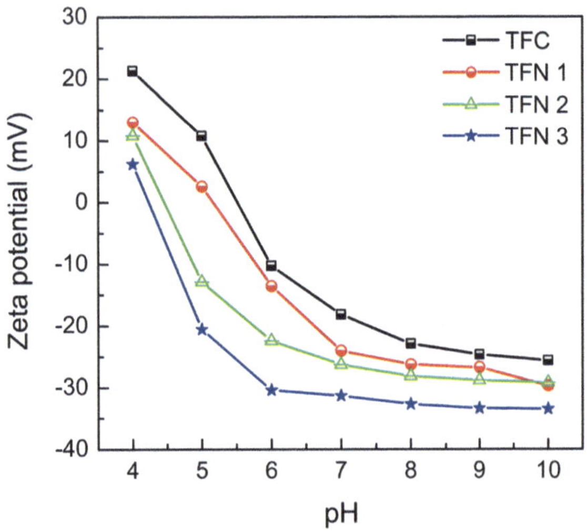

Fig. 7 Surface charge of the membranes at different GO content over pH range. The figure is adapted with permission from Royal Society of Chemistry [108]

GO was incorporated into PA in 0.1, 0.2, and 0.3 wt% concentration to arrange TFN 1, TFN 2, and TFN 3, respectively. Besides, at higher pH, the membrane with high GO content demonstrated a more negative surface charge.

3.4 Measurement of Contact Angle to Evaluate the Surface Hydrophilic Behavior

The contact angle measurement on the skin surfaces of the composite membranes was carried out with contact angle measuring instrument to assess the surface hydrophilicity, which plays an important role in the performance of the membranes. In general, membranes with hydrophilic surface generate high water flux in comparison with hydrophobic membranes. The sessile drop method to determine the water contact angles was applied in some researches [40, 105, 107], while contact angle goniometer could also be used [39, 101].

3.5 Thermal Stability Investigation of the Membranes

(a) **TGA Analysis**

To evaluate the thermal stability of the membranes, TGA was employed by some researchers [93, 103, 107, 111]. In this analysis, the thermal degradation process is monitored through the change of mass as a function of temperature. Strankowski et al. [111] showed that the doping of rGO into microporous PU (MPU) matrix leads to

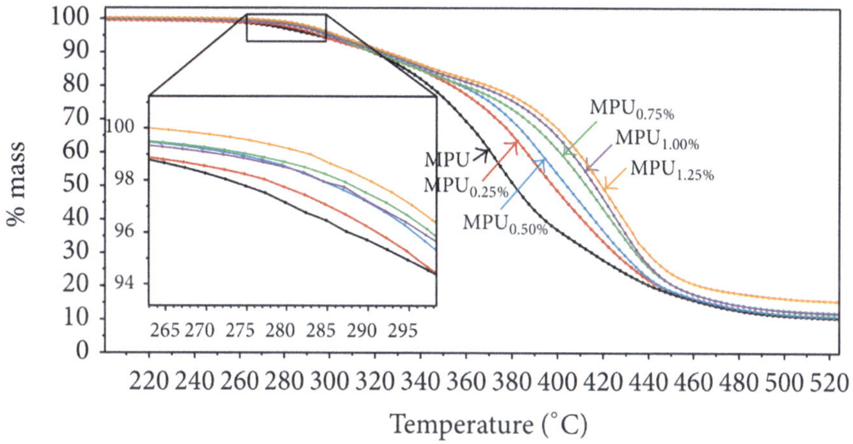

Fig. 8 TGA thermograms and % mass versus temperature for MPU matrix and MPU–rGO nanocomposites. The figure is adapted with permission from Royal Society of Chemistry [111]

Table 1 Temperatures corresponding to 2, 5, and 10% weight loss and maximum degradation rate (T_{max}) (±1°C) obtained from the first derivative of TGA signal [111]

Samples/$_{\%wt}$	$T_{2\%}$/°C	$T_{5\%}$/°C	$T_{10\%}$/°C	T_{max}/°C
MPU	273	294	319	373
$MPU_{0.25\%}$	278	296	320	398
$MPU_{0.50\%}$	284	299	320	399
$MPU_{0.75\%}$	287	301	320	419
$MPU_{1.0\%}$	284	302	323	424
$MPU_{1.25\%}$	290	304	326	424

enhancement of the MPU–rGO nanocomposite thermal stability in comparison with the bare MPU. TGA of the MPU matrix and MPU–rGO nanocomposites are shown in Fig. 8, and the results acquired from the evaluation of TGA curves are presented in Table 1. The rGO is a stable material, which is stiffer than polyurethane matrix. In reality, limited chain mobility can be achieved by employing this nanofiller, which affects on thermal stability of nanocomposite [112].

3.6 *Mechanical Property Evaluation of the Membranes*

Measurement of mechanical stability of the membranes is performed using universal tensile tests [93, 105, 111]. Gudarzi and Sharif [93] carried out tensile tests of the prepared membranes to investigate the influence of GO incorporation on the mechanical properties of the PMMA–GO nanocomposite membranes, which were prepared

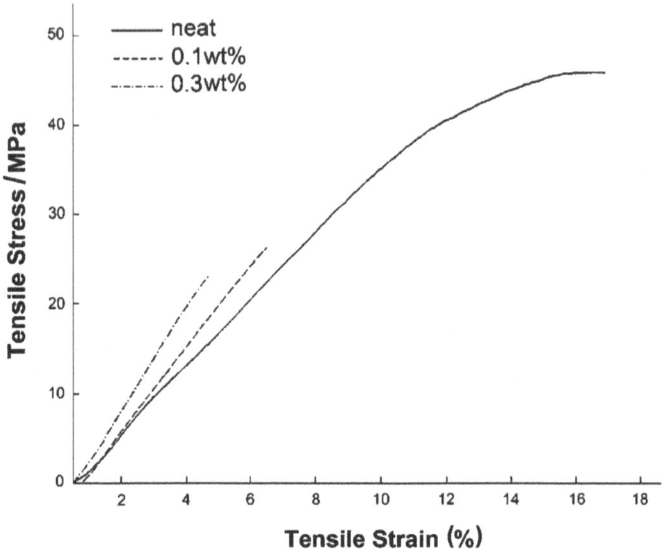

Fig. 9 Comparison between the stress–strain plot of neat PMMA and PMMA–GO mixed matrix. The figure is adapted with permission from Royal Society of Chemistry [93]

by melt mixing. As seen in Fig. 9, a considerable difference in tensile properties is observed by loading a very low concentration of GO.

4 GO- and rGO-Based Nanocomposite Membranes for Water Purification

The amazing characteristics of graphene derivatives like GO and rGO have presented a promising role in developing of nanocomposite membranes [18, 52, 113, 114]. On the other hand, modifying conventional membranes by incorporating GO- and rGO-based materials is a novel approach for water purification. Recently, considerable efforts have been done in the incorporation of GO and rGO into polymer matrix membranes as nanofillers, in order to improve water permeability and salt rejection. GO and rGO have proven to be effective fillers in polymer nanocomposite membrane because of their ideal properties and dispersibility in polymer matrixes [115, 116]. Ganesh et al. [38] prepared GO by oxidation of graphite applying $KMnO_4$ as an oxidizing agent. Then prepared GO was incorporated into PSF-based matrix to prepare PSF–GO mixed matrix membrane. They used wet-phase inversion method to prepare the PSF–GO membrane. To study the effect of the GO nanofiller, 1000 and 2000 ppm of GO were loaded in the fabrication of nanocomposite membrane. The prepared GO and PSF–GO were characterized using some instrumental methods such as SEM, AFM, FTIR, etc. The results revealed that the impregnated GO into PSF

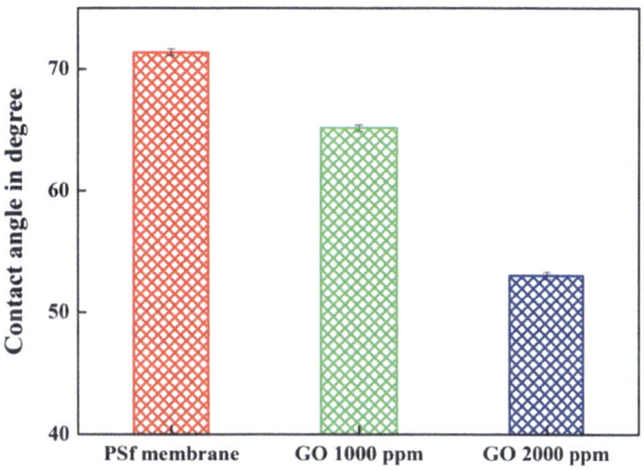

Fig. 10 Water contact angle of different membranes. The figure is adapted with permission from Elsevier [38]

matrix could improve mechanical strength and physical properties of the host PSF polymer. To understand the hydrophilicity of the membrane surface, water contact angle and also water uptake of the membrane were measured. Figure 10 shows the water contact angle of membranes. As shown in Fig. 10, decreasing trend of contact angle is seen by increasing the doped GO in the polymer matrix, which demonstrates enhancement of the hydrophilic property of the membrane. This can be explained by the behavior of GO during the formation of membrane that in the hydrophilic GO moves toward the top surface of the mixed matrix membrane. Moreover, water uptake results indicated that by increasing the loaded GO loading in the membrane, the liquid uptake increases, which verify the enhancement in hydrophilic sites in the mixed matrix.

Pure water flux and salt rejection of the fabricated membrane were also studied. As is seen in Fig. 11, water flux of the membrane increases with increasing the loaded

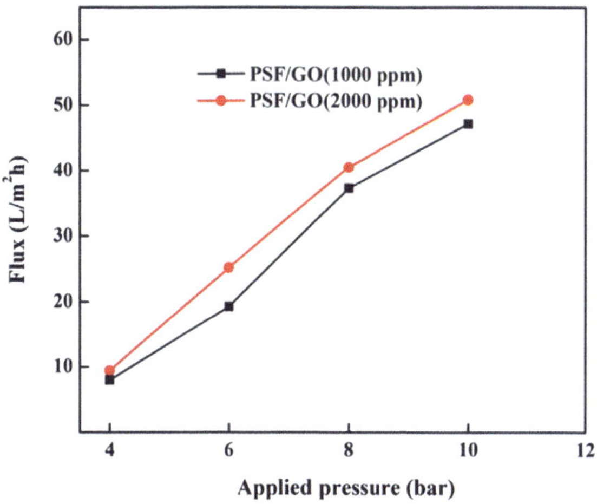

Fig. 11 Pure water flux of the membrane at different applied pressures. The figure is adapted with permission from Elsevier [38]

Table 2 Contact angle of PVDF and PVDF/GO membranes [100]

Membrane	Contact angle (deg)
PVDF	72.6 ± 1.5
PVDF/GO-0.5	70.5 ± 1.1
PVDF/GO-1	64.2 ± 0.8
PVDF/GO-2	60.5 ± 1.8

GO in the membrane at different applied pressures. Indeed, impregnation of GO into the polymer matrix led to more hydrophilic sites, which facilities the sorption of water into the membrane [117]. Salt rejection test was conducted by evaluating the rejection of 1000 ppm of Na_2SO_4 and NaCl solutions. The rejection results showed an increasing trend with increasing GO content in the polymer matrix. Membrane with 2000 ppm of GO doping demonstrated a maximum of 72% Na_2SO_4 rejection at 4 bar applied pressure.

The effect of the GO impregnation into the polyvinylidene fluoride (PVDF) membranes was reported by Zhao et al. [100]. In this study, GO was obtained from natural graphite powder using modified Hummers' method [118]. Then, PVDF applied as the bulk material, N,N-dimethylacetamide (DMAc) as the solvent, GO as the nanofiller and deionized water as the non-solvent coagulation bath to prepare the PVDF/GO membranes via immersion precipitation phase inversion technique. The GO nanosheets in the concentration range from 0 to 2 wt% were incorporated into PVDF membranes, and the nanocomposites were defined as PVDF (as reference), PVDF/GO-0.5, PVDF/GO-1, and PVDF/GO-2. The contact angle of the membranes was measured as presented in Table 2. By increasing the impregnated GO from 0 to 2 wt%, the contact angle reduced from 72.6 ± 1.5° to 60.5 ± 1.8°, which resulted in the improvement of surface hydrophilicity. The ultrafiltration system was applied to study the water permeation flux of the membranes. The pure water flux (J_{w1}) and permeation flux (J_p) through the membranes are shown in Fig. 12, which in

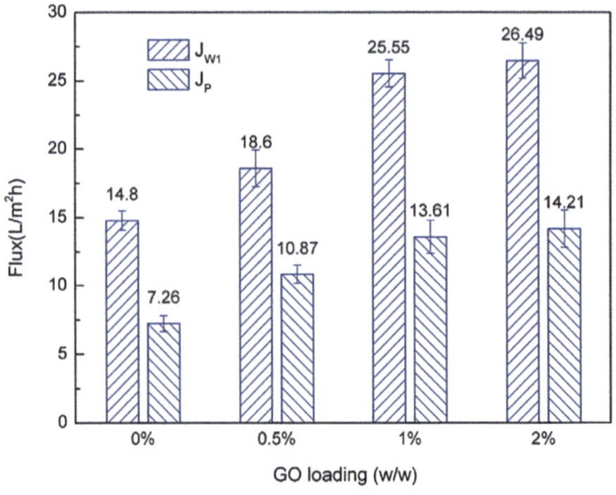

Fig. 12 Effect of GO content on water permeation flux. The figure is adapted with permission from Elsevier [100]

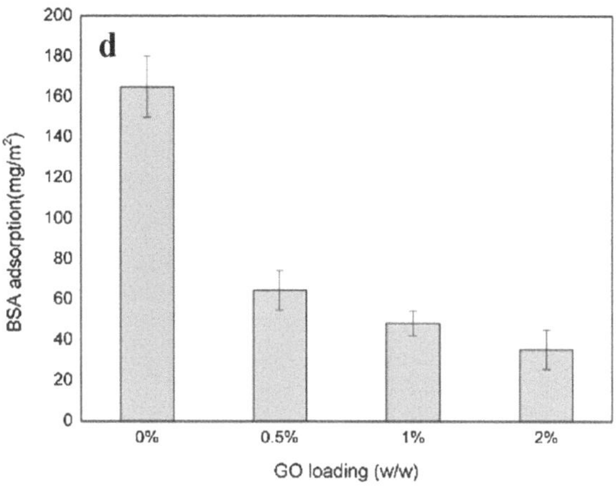

Fig. 13 Effect of GO content on adsorbed BSA. The figure is adapted with permission from Elsevier [100]

both terms increase with an increase in the GO concentration. The results depend on the hydrophilicity enhancement of the membranes with increasing the GO content, which led to more water molecules penetrate through the membranes [119]. Moreover, this hydrophilic property increasing can make the membranes more fouling resistant. To examine the fouling characteristics of the membranes, bovine serum albumin (BSA) was applied as the model protein. As can be seen from Fig. 13, the adsorbed protein by the membranes decreases with incorporation of GO into the PVDF, which indicates the fouling resistance potential of the added GO.

Zinadini et al. [39] reported that the doped GO improves the properties of bare polyethersulfone (PES), on the other hand, the novel mixed matrix possesses excellent fouling resistance and hydrophilic ability. The GO was prepared from natural graphite powder using Hummers' method [120]. The phase inversion method induced by the immersion precipitation technique was applied to prepare the asymmetric PES–GO mixed matrix. The GO nanoplates were blended in PES in 0.1, 0.5, and 1.0 wt% concentration to fabricate M1, M2, and M3, respectively. The measurement of water contact angle was conducted to study the hydrophilic property of the membrane surface. From Fig. 14, the contact angle reduced remarkably with incorporation of GO into the membranes. The decrease in contact angle could be responsible for the hydrophilicity improvement of the blending membrane. As shown in Fig. 14, the trend of pure water flux is in accordance with the improvement of hydrophilic property. Furthermore, the best practice of mixed matrix membranes was found in GO concentration of 0.5 wt%. They examined the membrane dye rejection by evaluating direct red 16 retention. The results revealed that the rejection potential of the GO blended membranes is higher than that of the bare PES membrane (Fig. 15). Indeed, the negative surface charge of the membranes can be induced by acidic functional groups of the GO [121], which caused retention with negative dye. Reduction in dye retention in M2 membrane can be ascribed to the membrane pore radius enhancement. To investigate the fouling resistance of the membranes, the membranes were fouled by 8000 ppm of powder milk solution, and the flux recovery ratio (FRR) of

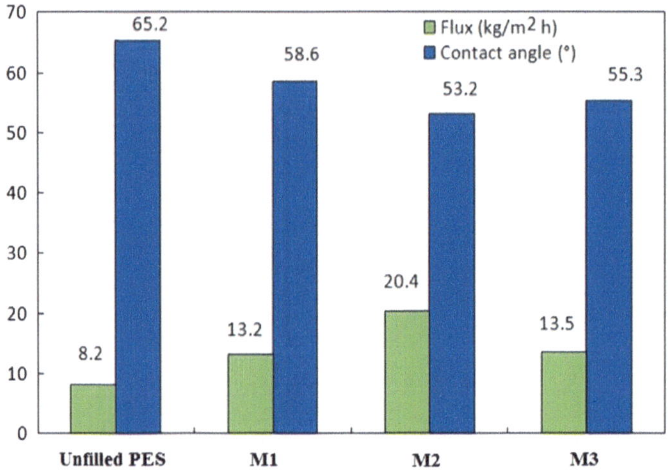

Fig. 14 Contact angle and pure water flux (after 60 min) of the prepared membranes. The figure is adapted with permission from Elsevier [39]

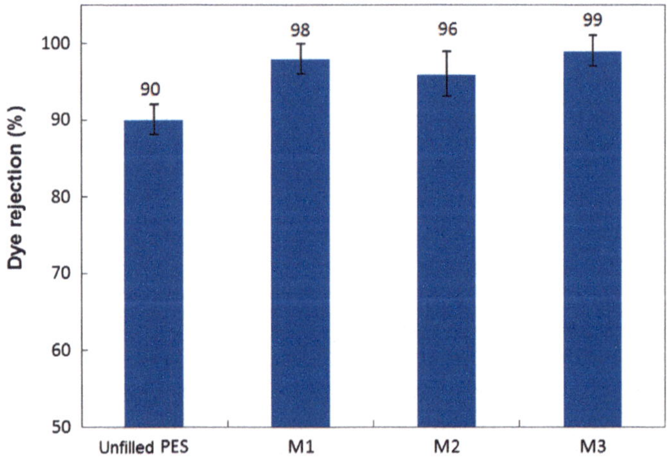

Fig. 15 Dye retention performance of the PES/GO nanofiltration membranes (0.4 MPa, pH = 6.0 ± 0.1, after 60 min filtration).The figure is adapted with permission from Elsevier [39]

the membranes was determined, which is presented in Fig. 16. The impregnation of GO into the PES membrane can considerably enhance the antifouling property of the modified PES membranes compared with the unfilled one.

Bano et al. [108] demonstrated that impregnation of GO into a TFC-PA can effectively increase its hydrophilicity, which resulted in its flux and antifouling property improvement [118, 120]. GO was incorporated into PA polymer using interfacial polymerization to develop TFN-PA/GO matrix membrane. The GO was loaded into

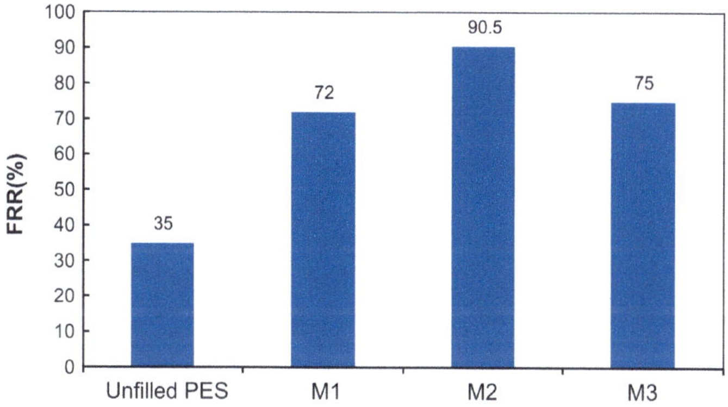

Fig. 16 Water flux recovery ratio of the reported membranes after protein fouling. The figure is adapted with permission from Elsevier [39]

Table 3 Contact angle values of the TFC and TFN membranes [108]

Sample	GO (wt%)	Contact angle (°)
TFC	0	88 ± 3
TFN 1	0.1	66 ± 2
TFN 2	0.2	65 ± 3
TFN 3	0.3	60 ± 1

PA in 0.1, 0.2, and 0.3 wt% concentration to arrange TFN 1, TFN 2, and TFN 3, respectively. The results of the contact angle are presented in Table 3. Polyamide TFC membranes are relatively hydrophobic [122] with a higher contact angle, but reducing the trend of contact angle with increasing GO content indicated enhancement in the hydrophilicity. The effect of GO on the performance of the membranes was also studied. As is seen in Fig. 17, there is an optimum point for membrane water flux. In other words, the water flux enhances with increasing concentration of GO up to 0.2 wt% and then begins reducing. Three different salt solutions (2000 ppm of NaCl, $MgSO_4$, and Na_2SO_4) were used to examine the salt rejection, which only a negligible change was observed in the rejection term of the membrane. The modified matrix membrane also indicated excellent anti-fouling properties against BSA and humic acid (HA), which may be attributed to the hydrophilic properties increasing [123].

Pal et al. [105] developed polysulfone-nanostructured rGO (PS/nRGO)-based nanocomposite UF membranes using the in-house-synthesized nRGO as reinforcing material with the variation of nRGO from 1 to 8 w/w%. Asymmetric flat sheet type of PS/nRGO composite membrane with a very thin and dense skin layer was prepared via non-solvent-induced phase inversion method and characterized by various instrumental techniques. The membranes made under similar conditions named as PS-nRGO UF-1, PS-nRGO UF-2, PS-nRGO UF-3, and PS-nRGO UF-4 for using 1, 2, 4, and 8 w/w% of nRGO, respectively. The hydrophilic property of the membranes,

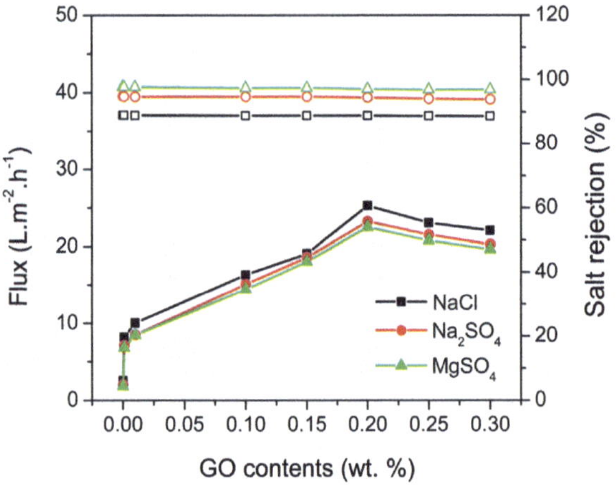

Fig. 17 Flux and salt rejection of PA membranes with different GO contents. The figure is adapted with permission from Royal Society of Chemistry [108]

as represented by the water contact angle values obtained on each of the membrane surface, is shown in Fig. 18. As is seen in Fig. 18, nRGO-incorporated Ps membranes demonstrate more hydrophilic surface compared with PS ultrafiltration membrane. In reality, the presence of oxygenated functionalities in nRGO supports the increase of hydrophilic behavior of the skin surface of PS/nRGO composite membranes [38]. They also investigated the performance of the membranes in terms of pure water

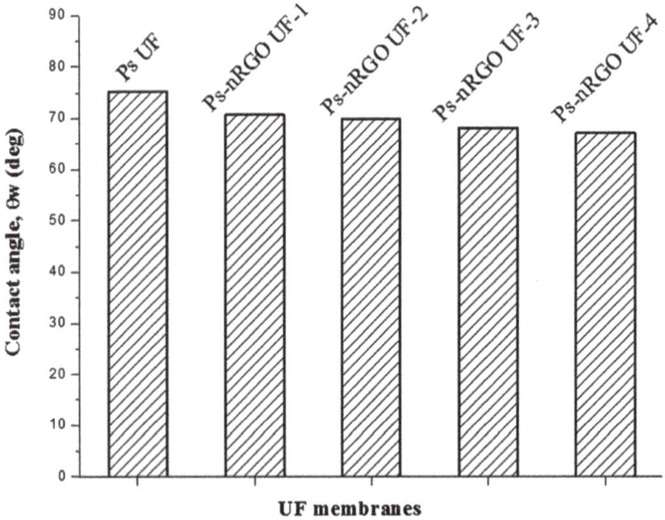

Fig. 18 Variation in contact angles (water) of Ps ultrafiltration and Ps-nRGO composite membranes. The figure is adapted with permission from Royal Society of Chemistry [105]

Table 4 Thermal features and tensile properties of Ps UF and Ps–nRGO composite UF membranes [105]

	Thermal features		Mechanical features	
Membrane code	T_d (°C)	Mass loss (%)	TS (MPa)	EB (%)
Ps UF	507.8	57.61	2.94 ± 0.16	36.83
Ps-nRGO UF-1	513.5	51.56	3.33 ± 0.11	32.41
Ps-nRGO UF-2	512.9	53.61	3.42 ± 0.07	30.04
Ps-nRGO UF-3	509.7	54.01	3.12 ± 0.19	34.01
Ps-nRGO UF-4	510.4	52.55	2.84 ± 0.24	38.18

permeability and solute rejection. Measurements were carried out using polyethylene glycol (PEG) with an average molecular weight of 35 kDa and polyethylene oxide (PEO) with average molecular weight of 100 kDa. The studies showed that an optimum loading of nRGO (2 w/w%) into the PS matrix leads to membranes with high solute rejection without comprising the flux reduction as compared with the PS ultrafiltration membrane. Moreover, the optimum loading of nRGO resulted in a membrane with better thermal and mechanical stability, which is presented in Table 4.

In another study, TFN membrane was developed by Yin et al. [40]. In this work, GO was prepared applying a modified Hummers' method [120] from graphite flakes and used as nanofiller to produce the TFN membrane. Then, GO nanosheets were embedded into the ployamide (PA) layer using the in situ interfacial polymerization process at 0–0.02 wt% of GO. Results indicated that by GO content enhancement, the contact angle of TFN membranes slightly reduced from 60.4 ± 2.5° to 55.4 ± 1.7°, which led to increasing hydrophilicity of the membrane. The water flux and salt rejection of the membrane were also investigated. The membrane water permeance was improved under optimal loaded GO concentration as demonstrated in Fig. 19. To evaluate the salt rejection of the membrane, 2000 mg/L of Na_2SO_4 and NaCl were tested. From Fig. 19, the rejection of Na_2SO_4 and NaCl reduced slightly, which shows the negligible effect of incorporated GO into the membrane on salt rejection. In addition, in order to study the influence of the employed nanofiller, they compared their results with other studies as shown in Table 5, which indicated relatively good effect of the added GO. Furthermore, the stability of the synthesized membrane was investigated by testing selective samples for 72 h filtration time, which showed acceptable stability under examining conditions.

Fryczkowska et al. [101] reported an innovative technique to incorporate rGO into the polyacrylonitrile (PAN) membranes. In this work, GO was prepared from graphite powder by the modified Hummers' method [120]. Then, GO thermal reduction method was performed to prepare rGO. The phase inversion method was applied to fabricate the PAN/rGO membranes from the dispersion of rGO in a solution of PAN dissolved in DMF. Structural studies confirmed the good dispersion of rGO into PAN membranes. To investigate the effect of rGO concentration on the performance of fabricated nanocomposite, a wide range of rGO concentrations (from 0.11% to 29.4% w/w) were examined (Table 6). The comparison of the contact angle (Table 6)

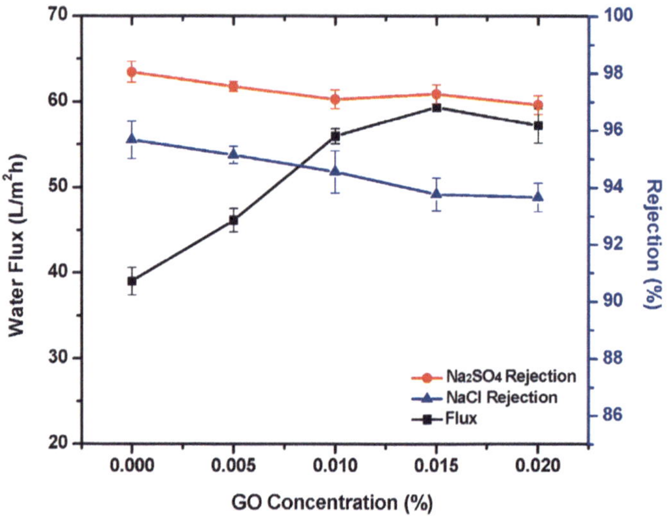

Fig. 19 Water flux and salt rejection of TFN membranes at different CO concentrations. The figure is adapted with permission from Elsevier [40]

Table 5 Performance of TFN membranes [40]

Nanofiller	Loading (wt%)	Water permeability (L/m^2 h psi)	NaCl rejection	Reference
Zeolite	0.4% in hexane	0.0943	93.9 ± 0.3%	Jeong et al. [124]
Silica	0.6% in water	0.06	95.1%	Javad and Singh [125]
MCM-41	0.05% in hexane	0.155	97.9 ± 0.3%	Yin et al. [126]
MWNTs	0.1% in water	0.121	~ 90%	Zhao et al. [127]
GO	0.015% in hexane	0.198	93.8 ± 0.6%	This study [40]

Table 6 The composition of the prepared membranes [101]

Type of membrane	Amount of rGO (g)	Amount of PAN (g)	Concentration of rGO (% w/w)	Concentration of PAN (% w/w)	Contact angle (deg)
0	0	12.0	0	100/0	52.0 ± 2.8
A	0.0135	12.0	0.11	99.89	49.5 ± 2.7
B	0.027	12.0	0.22	99.78	47.3 ± 4.1
C	0.054	12.0	0.45	99.55	48.2 ± 2.7
D	0.1	12.0	0.83	99.27	45.2 ± 2.6
E	0.5	12.0	4.0	96.0	40.7 ± 3.3
F	1.0	12.0	7.7	92.3	42.9 ± 4.3
G	2.0	12.0	14.3	85.7	45.8 ± 4.0
H	5.0	12.0	29.4	70.6	47.3 ± 4.6

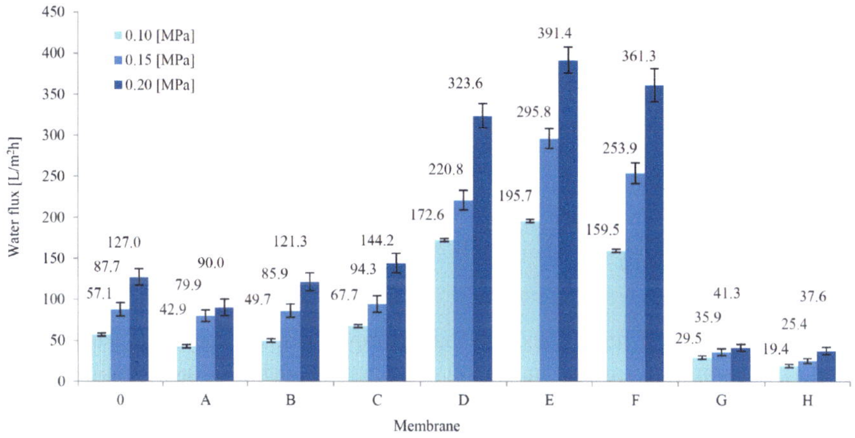

Fig. 20 Values of volumetric permeate flux for the prepared membranes. The figure is adapted with permission from MDPI, Basel, Switzerland [101]

of bare PAN membrane with the impregnated membranes indicated that the introduction of rGO into the PAN membranes decreases the values of contact angle, which resulted in the improvement of membrane hydrophilicity. To clarify, the addition of small portions of rGO (up to 4% w/w) into polymer matrix led to the decrement in the contact angle. The volumetric permeate flux was determined under operating pressures of 0.1, 0.15, and 0.2 MPa to study the transport properties of the membranes as shown in Fig. 20. The highest value of this term was obtained in membrane E, which contained 4.0% w/w of rGO. This result is consistent with their finding about hydrophilic property based on contact angle.

Evaluation of the fouling properties of the membranes by the solution of BSA at a concentration of 0.1 g/L was also carried out. A reduction in the volumetric permeate flux under the effect of BSA was seen for all the membranes. The largest decrease was 84%, which was recorded for bare PAN membrane, while the smallest decrease was 25% for membrane; Table 6(E). Indeed, the results demonstrated that a small incorporation of rGO into the PAN polymer leads to an effective antifouling property of the PAN/RGO membranes. Furthermore, Fryczkowska et al. [101] studied the biostatic properties of the membranes by exposing the samples to bacteria and fungi, i.e., the Gram-positive *S. aureus* and Gram-negative *E. coli* and *C. albicans*, which are capable of causing infections in humans. The obtained results verified the biostatic properties of PAN/RGO membranes by preventing the development of bacteria and fungi on their surface.

The oxidation process can introduce oxygen-containing functional groups into the basic plane of GO [128], which led to GO and rGO, which suggest eminent potential as nanofillers to synthesize nanocomposite materials with superior chemical stability, strong hydrophilicity, and excellent antifouling properties in comparison with the base polymer [33, 113, 129]. As a result, by modifying the polymeric matrix with even a small amount of GO or rGO, we can acquire nanocomposites with distinctive mechanical and physical properties and outstanding membrane performance [102, 130, 131]. Furthermore, the rGO is characterized by a low number of oxygen functional groups, which signifies that rGO possesses a better compatibilization influence on the mixed matrix membrane compared with pure graphene and can be easily dispersed in the composite [132, 133].

5 Research Gaps

According to several researches, GO nanocomposite membranes have displayed good separation performance in several applications such as water treatment, gas separation, and fuel cell. However, the stability of GO composite membrane has not been tested at a wide scale and future study is necessary for ensuring whether GO/rGO would not be leached from the membrane during operation. Adding crosslinker to bind the GO inter-sheets and polymer matrix could increase the stability of the membrane. Another important factor is to consider the optimization of the GO loading into the polymer matrix [122]. Despite the several preparation methods for fabricating the graphene-based nanocomposite membrane, new avant-grade methods are expected to be developed for decreasing the preparation cost and also be friendly to the ecosystem [102]. The design and fabrication of new polymer–GO nanocomposite membrane should be encouraged to further leveraging the nanocomposite membrane performance for effective water purification. It would replace our dependency on traditional hydrophobic membranes, which have low water permeability, selectivity, and lifetime.

6 Conclusions

GO and rGO have proven to be efficient nanofillers in polymer nanocomposite membranes due to their ideal material properties and dispersibility in polymer matrix. Being a low-cost material and their availability for mass production, GO/rGO has been widely used in water purification system. They improved the mechanical, electrical, and thermal properties of nanocomposites, which are better than many types of traditional filler materials. These advantages of GO/rGO fillers originate from their large surface area for efficient heat/electrical conduction and load transfer as well as the 2D functionable surfaces for the strong matrix/filler interactions. Among the various nanocomposite preparation methods, *in situ* polymerization and melt processing are the popular methods of the polymer–GO/rGO nanocomposite preparation. A significant number of polymer–GO/rGO nanocomposites have been successfully fabricated using these methods. The results of experimental works reveal that the dispersion of the nanofiller in the membrane matrix and process conditions can positively influence the final properties of nanocomposites. The appropriate incorporation of GO and rGO into a polymer matrix can remarkably improve the physical, mechanical, and hydrophilic properties of the host polymer at extremely low concentrations. Also, polymer–GO/rGO nanocomposite membranes have been found to improve the water permeability of the studied systems, which reduces energy consumption and operating costs of the system. In addition, these nanofillers improve the antifouling performance of some membranes that leads to decrease the membrane cleaning costs during recycling. In reality, these improvements of the membrane could be attributed to powerful interfacial interactions between graphene/GO/rGO and polymer matrix.

References and Future Readings

1. Rahimpour A et al (2011) TiO2 entrapped nano-composite PVDF/SPES membranes: Preparation, characterization, antifouling and antibacterial properties. Desalination 278(1–3):343–353
2. Yin J et al (2013) Attachment of silver nanoparticles (AgNPs) onto thin-film composite (TFC) membranes through covalent bonding to reduce membrane biofouling. J Membr Sci 441:73–82
3. Zhao Y et al (2012) Synthesis of robust and high-performance aquaporin-based biomimetic membranes by interfacial polymerization-membrane preparation and RO performance characterization. J Membr Sci 423:422–428
4. Jin L et al (2012) Synthesis of a novel composite nanofiltration membrane incorporated SiO2 nanoparticles for oily wastewater desalination. Polymer 53(23):5295–5303
5. Das R, Khayet M (2019) Nanotechnology Based Platforms for Efficient Water Desalination. Desalination 451:1–1
6. Kim E-S et al (2012) Development of nanosilver and multi-walled carbon nanotubes thin-film nanocomposite membrane for enhanced water treatment. J Membr Sci 394:37–48
7. Das R et al (2015) Covalent functionalization schemes for tailoring solubility of multi-walled carbon nanotubes in water and acetone solvents. Sci Adv Mater 7(12):2726–2737
8. Das R (2017) Nanohybrid catalyst based on carbon nanotube. Carbon nanostructures. Springer International Publishing AG. https://doi.org/10.1007/978-3-319-58151-4_2
9. Wei Y et al (2018) Multilayered graphene oxide membranes for water treatment: a review. Carbon 139:964–981
10. Banerjee P et al (2018) Membrane technology, in carbon nanotubes for clean water. Springer, pp 127–150
11. Das R et al (2017) Recent advances in nanomaterials for water protection and monitoring. Chem Soc Rev 46(22):6946–7020
12. Nair R et al (2012) Unimpeded permeation of water through helium-leak–tight graphene-based membranes. Science 335(6067):442–444
13. Xu Q et al (2015) Graphene and graphene oxide: advanced membranes for gas separation and water purification. Inorg Chem Front 2(5):417–424
14. Mohammad A, Asiri AM (2017) Inorganic pollutants in wastewater: methods of analysis, removal and treatment. Mater Res Forum LLC
15. Abbott's IE (2007) Graphene: exploring carbon flatland. Phys Today 60(8): 35
16. Novoselov KS, Geim A (2007) The rise of graphene. Nat. Mater 6(3):183–191
17. Si Y, Samulski ET (2008) Synthesis of water soluble graphene. Nano Lett 8(6):1679–1682
18. Lyu J et al (2018) Separation and purification using GO and r-GO membranes. RSC Adv 8(41):23130–23151
19. Lerf A et al (1998) Structure of graphite oxide revisited. J Phys Chem B 102(23):4477–4482
20. Wang Z et al (2012) Graphene oxide filled nanocomposite with novel electrical and dielectric properties. Adv Mater 24(23):3134–3137

21. Pang H et al (2010) An electrically conducting polymer/graphene composite with a very low percolation threshold. Mater Lett 64(20):2226–2229
22. Gómez H et al (2011) Graphene-conducting polymer nanocomposite as novel electrode for supercapacitors. J Power Sour 196(8):4102–4108
23. Guex LG et al (2017) Experimental review: chemical reduction of graphene oxide (GO) to reduced graphene oxide (rGO) by aqueous chemistry. Nanoscale 9(27):9562–9571
24. Hu M, Mi B (2013) Enabling graphene oxide nanosheets as water separation membranes. Environ Sci Technol 47(8):3715–3723
25. Mahmoud KA et al (2015) Functional graphene nanosheets: The next generation membranes for water desalination. Desalination 356:208–225
26. Cohen-Tanugi D, Grossman JC (2012) Water desalination across nanoporous graphene. Nano Lett 12(7):3602–3608
27. Hosseini M, Azamat J, Erfan-Niya H (2019) Water desalination through fluorine-functionalized nanoporous graphene oxide membranes. Mater Chem Phys 223:277–286
28. Sun P et al (2013) Selective ion penetration of graphene oxide membranes. ACS Nano 7(1):428–437
29. Cadotte J et al (1980) A new thin-film composite seawater reverse osmosis membrane. Desalination 32:25–31
30. Das R (2019) Polymeric materials for clean water. Springer
31. Xu G-R et al (2019) Two-dimensional (2D) nanoporous membranes with sub-nanopores in reverse osmosis desalination: Latest developments and future directions. Desalination 451:18–34
32. Ali ME et al (2016) Thin film composite membranes embedded with graphene oxide for water desalination. Desalination 386:67–76
33. Jin F et al (2013) High-performance ultrafiltration membranes based on polyethersulfone–graphene oxide composites. Rsc Adv 3(44):21394–21397
34. Dreyer DR et al (2010) Graphite oxide. Chem Soc Rev 39:228–240
35. Choi BG et al (2012) Enhanced transport properties in polymer electrolyte composite membranes with graphene oxide sheets. Carbon 50(15):5395–5402
36. Kuilla T et al (2010) Recent advances in graphene based polymer composites. Prog Polym Sci 35(11):1350–1375
37. Wu H, Tang B, Wu P (2014) Development of novel SiO2–GO nanohybrid/polysulfone membrane with enhanced performance. J Membr Sci 451:94–102
38. Ganesh B, Isloor AM, Ismail AF (2013) Enhanced hydrophilicity and salt rejection study of graphene oxide-polysulfone mixed matrix membrane. Desalination 313:199–207
39. Zinadini S et al (2014) Preparation of a novel antifouling mixed matrix PES membrane by embedding graphene oxide nanoplates. J Membr Sci 453:292–301
40. Yin J, Zhu G, Deng B (2016) Graphene oxide (GO) enhanced polyamide (PA) thin-film nanocomposite (TFN) membrane for water purification. Desalination 379:93–101
41. Han Y, Xu Z, Gao C (2013) Ultrathin graphene nanofiltration membrane for water purification. Adv Func Mater 23(29):3693–3700
42. Joshi R, et al (2014) Precise and ultrafast molecular sieving through graphene oxide membranes. Science 343(6172):752–754
43. Xu C et al (2013) Graphene oxide–TiO2 composite filtration membranes and their potential application for water purification. Carbon 62:465–471
44. Bao Q et al (2010) Graphene–polymer nanofiber membrane for ultrafast photonics. Adv Func Mater 20(5):782–791
45. Kim H, Miura Y, Macosko CW (2010) Graphene/polyurethane nanocomposites for improved gas barrier and electrical conductivity. Chem Mater 22(11):3441–3450
46. Choi BG et al (2010) Solution chemistry of self-assembled graphene nanohybrids for high-performance flexible biosensors. ACS Nano 4(5):2910–2918
47. Satti A, Larpent P, Gun'ko Y (2010) Improvement of mechanical properties of graphene oxide/poly (allylamine) composites by chemical crosslinking. Carbon 48(12):3376–3381

48. Yu X et al (2017) Fabrication technologies and sensing applications of graphene-based composite films: advances and challenges. Biosens Bioelectron 89:72–84
49. Fan Y (2017) Synthesis of graphene and graphene-based composite membrane
50. Ng CY et al (2019) Fabrication of graphene-based membrane for separation of hazardous contaminants from wastewater In: Graphene-based nanotechnologies for energy and environment. Elsevier, pp 267–291
51. Liu Y, Feng J (2017) An attempt towards fabricating reduced graphene oxide composites with traditional polymer processing techniques by adding chemical reduction agents. Compos Sci Technol 140:16–22
52. Stankovich S et al (2006) Graphene-based composite materials. Nature 442(7100):282–286
53. Hussain F et al (2006) Polymer-matrix nanocomposites, processing, manufacturing, and application: an overview. J Compos Mater 40(17):1511–1575
54. Liang J et al (2009) Molecular-level dispersion of graphene into poly (vinyl alcohol) and effective reinforcement of their nanocomposites. Adv Func Mater 19(14):2297–2302
55. Wang W-P, Pan C-Y (2004) Preparation and characterization of polystyrene/graphite composite prepared by cationic grafting polymerization. Polymer 45(12):3987–3995
56. Wu SL, Shi TJ, Zhang LY (2016) Latex co-coagulation approach to fabrication of polyurethane/graphene nanocomposites with improved electrical conductivity, thermal conductivity, and barrier property. J Appl Polym Sci 133(11)
57. Zhu Y et al (2010) Exfoliation of graphite oxide in propylene carbonate and thermal reduction of the resulting graphene oxide platelets. ACS Nano 4(2):1227–1233
58. Song P et al (2011) Fabrication of exfoliated graphene-based polypropylene nanocomposites with enhanced mechanical and thermal properties. Polymer 52(18):4001–4010
59. Yan DX et al (2015) Structured reduced graphene oxide/polymer composites for ultra-efficient electromagnetic interference shielding. Adv Func Mater 25(4):559–566
60. Zhang C et al (2012) A novel approach for transferring water-dispersible graphene nanosheets into organic media. J Mater Chem 22(23):11748–11754
61. Xu LQ et al (2010) Dopamine-induced reduction and functionalization of graphene oxide nanosheets. Macromolecules 43(20):8336–8339
62. Yang H et al (2009) Covalent functionalization of chemically converted graphene sheets via silane and its reinforcement. J Mater Chem 19(26):4632–4638
63. Shan C et al (2009) Water-soluble graphene covalently functionalized by biocompatible poly-L-lysine. Langmuir 25(20):12030–12033
64. Geng J et al (2010) Preparation of graphene relying on porphyrin exfoliation of graphite. Chem Commun 46(28):5091–5093
65. Zu S-Z, Han B-H (2009) Aqueous dispersion of graphene sheets stabilized by pluronic copolymers: formation of supramolecular hydrogel. J Phys Chem C 113(31):13651–13657
66. Ren L et al (2010) A smart pH responsive graphene/polyacrylamide complex via noncovalent interaction. Nanotechnology 21(33):335701
67. Choi E-Y et al (2010) Noncovalent functionalization of graphene with end-functional polymers. J Mater Chem 20(10):1907–1912
68. Chen W et al (2018) A critical review on the development and performance of polymer/graphene nanocomposites. Sci Eng Compos Mater 25(6):1059–1073
69. Zhang L et al (2018) Preparation of polymer/graphene oxide nanocomposites by a two-step strategy composed of in situ polymerization and melt processing. Compos Sci Technol 154:1–7
70. Barroso-Bujans F et al (2010) Permanent adsorption of organic solvents in graphite oxide and its effect on the thermal exfoliation. Carbon 48(4):1079–1087
71. Yu A et al (2007) Graphite nanoplatelet−epoxy composite thermal interface materials. J Phys Chem C 111(21):7565–7569
72. Xiao X, Xie T, Cheng Y-T (2010) Self-healable graphene polymer composites. J Mater Chem 20(17):3508–3514
73. Rafiee MA et al (2010) Fracture and fatigue in graphene nanocomposites. Small 6(2):179–183

74. Liang J et al (2009) Electromagnetic interference shielding of graphene/epoxy composites. Carbon 47(3):922–925
75. Huang X et al (2012) Graphene-based composites. Chem Soc Rev 41(2):666–686
76. Zheng W, Lu X, Wong SC (2004) Electrical and mechanical properties of expanded graphite-reinforced high-density polyethylene. J Appl Polym Sci 91(5):2781–2788
77. Lee WD, Im SS (2007) Thermomechanical properties and crystallization behavior of layered double hydroxide/poly (ethylene terephthalate) nanocomposites prepared by in-situ polymerization. J Polym Sci Part B: Polym Phys 45(1):28–40
78. Hsueh H-B, Chen C-Y (2003) Preparation and properties of LDHs/polyimide nanocomposites. Polymer 44(4):1151–1161
79. Verdejo R et al (2011) Graphene filled polymer nanocomposites. J Mater Chem 21(10):3301–3310
80. Wang J-Y et al (2011) Preparation and properties of graphene oxide/polyimide composite films with low dielectric constant and ultrahigh strength via in situ polymerization. J Mater Chem 21(35):13569–13575
81. Zhang S et al (2018) Enhanced tensile strength and initial modulus of poly (vinyl alcohol)/graphene oxide composite fibers via blending poly (vinyl alcohol) with poly (vinyl alcohol)-grafted graphene oxide. J Polym Res 25(3):65
82. Chen B et al (2008) A critical appraisal of polymer–clay nanocomposites. Chem Soc Rev 37(3):568–594
83. Song SH et al (2010) Physical and thermal properties of acid-graphite/styrene-butadiene-rubber nanocomposites. Korean J Chem Eng 27(4):1296–1300
84. Dasari A, Yu Z-Z, Mai Y-W (2009) Electrically conductive and super-tough polyamide-based nanocomposites. Polymer 50(16):4112–4121
85. Wakabayashi K et al (2008) Polymer-graphite nanocomposites: effective dispersion and major property enhancement via solid-state shear pulverization. Macromolecules 41(6):1905–1908
86. Tewatia A et al (2017) Characterization of melt-blended graphene–poly (ether ether ketone) nanocomposite. Mater Sci Eng B 216:41–49
87. You F et al (2014) In situ thermal reduction of graphene oxide in a styrene–ethylene/butylene–styrene triblock copolymer via melt blending. Polym Int 63(1):93–99
88. Pickering SU (2001) Cxcvi.—emulsions. J Chem Soc Trans 91(1907):2001–2021
89. Böker A et al (2007) Self-assembly of nanoparticles at interfaces. Soft Matter 3(10):1231–1248
90. Wang D, Duan H, Möhwald H (2005) The water/oil interface: the emerging horizon for self-assembly of nanoparticles. Soft Matter 1(6):412–416
91. San Miguel A et al (2010) Smart colloidosomes with a dissolution trigger. Soft Matter 6(14):3163–3166
92. Dinsmore A et al (2002) Colloidosomes: selectively permeable capsules composed of colloidal particles. Science 298(5595):1006–1009
93. Gudarzi MM, Sharif F (2011) Self assembly of graphene oxide at the liquid–liquid interface: a new route to the fabrication of graphene based composites. Soft Matter 7(7):3432–3440
94. Read E et al (2004) Effect of varying the oil phase on the behavior of pH-responsive latex-based emulsifiers: demulsification versus transitional phase inversion. Langmuir 20(18):7422–7429
95. Xie P et al (2013) Pickering emulsion polymerization of graphene oxide-stabilized styrene. Colloid Polym Sci 291(7):1631–1639
96. Kumar A, Nanda D (2019) Methods and fabrication techniques of superhydrophobic surfaces. In: Superhydrophobic polymer coatings. Elsevier. pp 43–75
97. Senez V, Thomy V, Dufour R (2014) Nanotechnologies for synthetic super non-wetting surfaces. nanotechnologies for synthetic super non-wetting surfaces, pp 1–12
98. Yilbas BS, Al-Sharafi A, Ali H (2019) Self-cleaning of surfaces and water droplet mobility. Elsevier
99. Lue SJ et al (2015) Novel bilayer well-aligned Nafion/graphene oxide composite membranes prepared using spin coating method for direct liquid fuel cells. J Membr Sci 493:212–223

100. Zhao C et al (2013) Effect of graphene oxide concentration on the morphologies and antifouling properties of PVDF ultrafiltration membranes. J Environ Chem Eng 1(3):349–354
101. Fryczkowska B et al (2020) The influence of graphene addition on the properties of composite rGO/PAN membranes and their potential application for water disinfection. Membranes 10(4):58
102. Wang N et al (2012) Self-assembly of graphene oxide and polyelectrolyte complex nanohybrid membranes for nanofiltration and pervaporation. Chem Eng J 213:318–329
103. Jamil N et al (2019) Mixed matrix membranes incorporated with reduced graphene oxide (rGO) and zeolitic imidazole framework-8 (ZIF-8) nanofillers for gas separation. J Solid State Chem 270:419–427
104. Wan C, Chen B (2012) Reinforcement and interphase of polymer/graphene oxide nanocomposites. J Mater Chem 22(8):3637–3646
105. Pal A et al (2015) Reinforcement of nanostructured reduced graphene oxide: a facile approach to develop high-performance nanocomposite ultrafiltration membranes minimizing the trade-off between flux and selectivity. RSC Adv 5(58):46801–46816
106. Strankowski M et al (2016) Polyurethane nanocomposites containing reduced graphene oxide, FTIR, Raman, and XRD studies. J Spectrosc
107. Jiang Y et al (2019) Graphene oxides as nanofillers in polysulfone ultrafiltration membranes: shape matters. J Membr Sci 581:453–461
108. Bano S et al (2015) Graphene oxide modified polyamide nanofiltration membrane with improved flux and antifouling properties. J Mater Chem A 3(5):2065–2071
109. Wan C, Frydrych M, Chen B (2011) Strong and bioactive gelatin–graphene oxide nanocomposites. Soft Matter 7(13):6159–6166
110. Österholm A et al (2012) Electrochemical incorporation of graphene oxide into conducting polymer films. Electrochim Acta 83:463–470
111. Strankowski M et al (2016) Thermal and mechanical properties of microporous polyurethanes modified with reduced graphene oxide. Int J Polym Sci
112. Chen SQ, Wang Y (2010) Microwave-assisted synthesis of a Co 3 O 4–graphene sheet-on-sheet nanocomposite as a superior anode material for Li-ion batteries. J Mater Chem 20(43):9735–9739
113. Dreyer DR et al (2010) The chemistry of graphene oxide. Chem Soc Rev 39(1):228–240
114. Soldano C, Mahmood A, Dujardin E (2010) Production, properties and potential of graphene. Carbon 48(8):2127–2150
115. Yoo BM et al (2014) Graphene and graphene oxide and their uses in barrier polymers. J Appl Polym Sci 131(1)
116. Smith AT et al (2019) Synthesis, properties, and applications of graphene oxide/reduced graphene oxide and their nanocomposites. Nano Mater Sci 1(1):31–47
117. Roh IJ, Greenberg AR, Khare VP (2006) Synthesis and characterization of interfacially polymerized polyamide thin films. Desalination 191(1–3):279–290
118. Kovtyukhova NI et al (1999) Layer-by-layer assembly of ultrathin composite films from micron-sized graphite oxide sheets and polycations. Chem Mater 11(3):771–778
119. Kausar A (2019) Applications of polymer/graphene nanocomposite membranes: a review. Mater Res Innov 23(5):276–287
120. Hummers Jr WS, Offeman RE (1958) Preparation of graphitic oxide. J Am Chem Soc 80(6):1339–1339
121. Dimiev AM, Alemany LB, Tour JM (2013) Graphene oxide. Origin of acidity, its instability in water, and a new dynamic structural model. ACS Nano **7**(1):576–588
122. Perreault F, Tousley ME, Elimelech M (2014) Thin-film composite polyamide membranes functionalized with biocidal graphene oxide nanosheets. Environ Sci Technol Lett 1(1):71–76
123. Wang Z et al (2012) Novel GO-blended PVDF ultrafiltration membranes. Desalination 299:50–54
124. Jeong B-H et al (2007) Interfacial polymerization of thin film nanocomposites: a new concept for reverse osmosis membranes. J Membr Sci 294(1–2):1–7

125. Jadav GL, Singh PS (2009) Synthesis of novel silica-polyamide nanocomposite membrane with enhanced properties. J Membr Sci 328(1–2):257–267
126. Yin J et al (2012) Fabrication of a novel thin-film nanocomposite (TFN) membrane containing MCM-41 silica nanoparticles (NPs) for water purification. J Membr Sci 423:238–246
127. Zhao H et al (2014) Improving the performance of polyamide reverse osmosis membrane by incorporation of modified multi-walled carbon nanotubes. J Membr Sci 450:249–256
128. Allen MJ, Tung VC, Kaner RB (2010) Honeycomb carbon: a review of graphene. Chem Rev 110(1):132–145
129. Koinuma M et al (2012) Photochemical engineering of graphene oxide nanosheets. J Phys Chem C 116(37):19822–19827
130. Moghaddasi A (2019) Polymer-Matrix nanocomposite membranes for water treatment
131. de Lannoy C-F, Soyer E, Wiesner MR (2013) Optimizing carbon nanotube-reinforced polysulfone ultrafiltration membranes through carboxylic acid functionalization. J Membr Sci 447:395–402
132. Jing Q et al (2015) Chemical functionalization of graphene oxide for improving mechanical and thermal properties of polyurethane composites. Mater Des 85:808–814
133. Geim, A.K., Graphene: status and prospects. Science 324(5934):1530–1534

Porous Graphene Membranes for Solute Separation via Reverse Osmosis and Electrodialysis

Chengzhen Sun, Mei Liu, Hassan, and Bofeng Bai

Abstract Graphene, the thinnest material known to science, is a very promising candidate for separation membranes with ultra-high molecular permeance. It has been widely demonstrated that nanoporous graphene membranes have a great potential for solute separation. In this chapter, the recent advances on the nanoporous graphene membranes for the application in water purification via reverse osmosis and electrodialysis are both reviewed. The separation mechanisms and fabrication methods of this atomically thick membrane are especially discussed by highlighting the representative theoretical and experimental works. Currently, the studies on the porous graphene membranes via electrodialysis are relatively limited comparing to those of reverse osmosis. It is expected that more and more researchers are attracted to this frontier research field to make the two-dimensional membranes for water purification a reality by overcoming the challenges faced currently.

Keywords Porous graphene · Membrane · Water purification · Reverse osmosis · Electrodialysis

1 Introduction

Porous graphene has been proved by researchers among the most suitable and efficient water purification membranes in different desalination and water treatment technologies due to its encouraging results [1–3]. It has a two-dimensional (2D) hexagonal structure [4], stable chemical structure [5], high mechanical strength [6], high water flux and high selectivity with suitable pore size and specific pore chemical properties [4, 7, 8]. Pristine graphene sheets are impermeable to the smallest atoms like helium, but they allow protons to pass through [9]. By removing certain carbon atoms, the formation of selective nanopore makes graphene a promising molecular

C. Sun (✉) · M. Liu · Hassan · B. Bai
State Key Laboratory of Multiphase Flow in Power Engineering, Xi'an Jiaotong University, Shaanxi 710049, China
e-mail: sun-cz@xjtu.edu.cn

R. Das (ed.), *Two-Dimensional (2D) Nanomaterials in Separation Science*, Springer Series on Polymer and Composite Materials,
https://doi.org/10.1007/978-3-030-72457-3_6

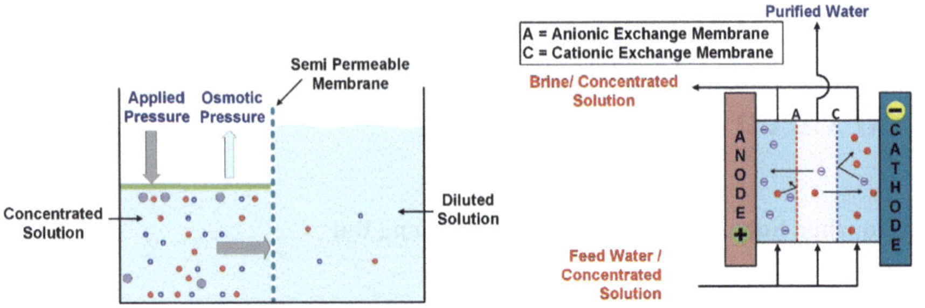

Fig. 1 Schematic diagram for RO (left) and electrodialysis (right) water purification technologies

sieve in separation science [10, 11]. With its wide application in various fields [12–17], it has entered a new era of research with excellent separation performances including reverse osmosis (RO) and electrodialysis [18].

RO has been the most popular technology for both water desalination and treatment so far [19, 20]. It is a pressure-driven technology, in which a semi-permeable membrane separates a dilute solution from a concentrated solution under the applied pressure on the concentrated side (see Fig. 1). As the solvent of the concentrated solution flows through the membrane, the concentrated solution becomes more concentrated, which requires a higher driving pressure to resist the higher osmotic pressure π (N/m^2), which is defined by the Van's Hoff equation:

$$\pi = C_s RT \tag{1}$$

where C_s is the sum of molalities of total solute ions (mol/m^3), R is the ideal gas constant and T is the absolute temperature of the system. The osmotic pressure magnitude is very large even in a very dilute solution [21]. The applied pressure must be larger than the osmotic pressure of the concentrated solution to prevent the freshwater from reverting to the feed water. A semi-permeable membrane should have a significant selectivity to allow the passage of solvents (e.g. water molecules) and to reject solutes (e.g. salt ions). A good RO membrane should possess high permeability, high selectivity and higher mechanical strength with minimum thickness. Thus, these requirements make nanoporous graphene (NPG) membrane an ideal candidate for desalination and a strong alternative to the conventional RO membranes [22].

Electrodialysis with ion-exchange membranes is considered to be one of the most cost-effective desalination techniques. It is the procedure in which ions are transported selectively through ion exchange membranes using an external electric field or electrostatic effect of anions and cations (Fig. 1). Electrodialysis and its corresponding processes are all based on electrochemical potential and ion exchange membranes. In electrodialysis, feed concentrated solution is entered in the multi membrane cells with applied pressure and an external electric field is provided through the anode and cathode. Electrodes attract ions with opposite charge and repel ions with the same charge to the other electrode, which likewise attracts them, and ion exchange membranes selectively allow anions or cations to pass through. Electrodialysis, on the other hand, has some advantages and some limitations compared with conventional water desalination techniques (e.g. RO). For example, it requires minimum water pretreatment. However, electrodialysis with conventional membranes can only be operated for brackish/lower salt concentrated water due to their higher thickness and lower permeability. Owing to the high water permeability of the NPG membrane and its remarkable ion selectivity, it has a great potential to overcome these limitations of conventional membranes [3, 23].

The selective ion transport through functionalized graphene nanopores was detected in 2008 by Sint et al. [24], and they proposed that functionalized graphene nanopores with certain size, charge, and chemical property possessed high selectivity for different ions in aqueous solution. In 2010, Suk and Aluru [25] performed molecular dynamics (MD) simulations and showed that the NPG membrane possessed a high water flux rate compared to the conventional RO membranes. The NPG membrane was proposed as a promising candidate for water desalination by Cohen-Tanugi and Grossman [1] in 2012 based on the MD simulation results, and later on, they continued the work on graphene-based water desalination membranes in which they reported that graphene exhibited a high salt rejection rate along with higher water permeability compared to the conventional RO membranes [26]. Afterward in 2014, O'Hern et al. [2] conducted experiments in this novel study and measured the transport rates of various ions across a single-layer NPG. In 2015, Surwade et al. [27] presented the promising salt rejection rate of nearly 100% using single-layer NPG with high water flux. In 2016, Rollings et al. [28] experimentally measured the ion selectivity of graphene nanopore using electrodialysis technique and especially the cation–anion selectivity at different voltage levels with different pore sizes. They indicated that a comparatively large porous graphene membrane can exhibit phenomenal ion selectivity. Zhang et al. [29] also demonstrated that graphene can reject salt ions effectively with energy conservation under low applied electric potential and lower driving pressure.

In this chapter, we review the theoretical and experimental works regarding NPG membranes for water purification using RO and electrodialysis techniques. We summary the research achievements and discussions in these representative works for the NPG membranes in the area of water purification. This chapter shows that the NPG membrane has tremendous potential for water purification with a good selective ionic transportability and high water flux. We anticipate that this chapter will not only provide an exposure to the researchers in the research field of solute separation but also navigate and lead toward the new applications of NPG membranes in related areas like heavy metal removing [30], reverse electrodialysis [31], gas separation [32].

2 Porous Graphene as Reverse Osmosis Membranes

(a) **Experimental works**

Following the first theoretical demonstration of NPG water purification membranes by Cohen-Tanugi et al. [1] in 2012, several experimental works were successively conducted by different groups. Surwade et al. [27] measured the water permeation and ion rejection rate through a NPG membrane with tunable nanopores created via the oxygen plasma etching process on the ambient-pressure CVD graphene and achieved a nearly 100% salt rejection rate for the dissolved ions (e.g. K^+, Na^+, Li^+ and Cl^-) and a high water permeation rate. O'Hern et al. [33] investigated the salt rejection rates of a defect-free NPG membrane and observed a high rejection rate of multivalent ions and small molecules; and this membrane can achieve a high water permeation flux. Subsequently, they [2] measured the transport rates of different ions through NPG membranes fabricated by ion bombardment and oxidative etching; the pore diameter was on the order of 0.40 nm and the pore density exceeded 10^{12} cm^{-2} (Fig. 2). Prior to these two studies, the effect of intrinsic defects on the permeation of ions through the CVD graphene was evaluated by O'Hern et al. [34] and they found that different ions presented a size-selective transport through the graphene with intrinsic defects of sizes 1–15 nm. Kim et al. [35] evaluated the effects of oxidation degree of graphene on the water purification efficiency, and the results showed that the water permeation rate through the NPG membranes increased proportionally with the increase of the driving pressure. They also found that the permeated concentration of Na^+ through the NPG membranes decreased by 67.7% and 64.5% for the oxidation degree of 28.1% and 53.9%, respectively, compared to a commercial polyamide membrane. Rollings et al. [28] proved that the NPG membranes exhibited a high selectivity (about 100) of K^+ cations over Cl^- anions and that the transport rates of monovalent cations were 5 times higher than those of divalent cations. Surprisingly, they observed that the large pores with diameters of about 20 nm still presented a high K^+/Cl^- selectivity, meaning that the requirement on the precise control of the pore sizes to selectively transport ions was not high. Kafiah et al. [36] fabricated the microscale NPG membranes based

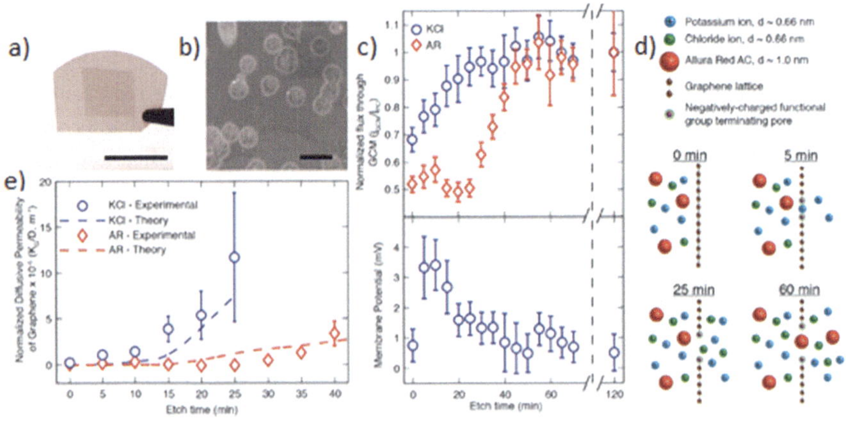

Fig. 2 Experimental transport measurements through NPG membranes. This figure is directly reproduced with permission from the literature [2]

on monolayer CVD graphene and the measurement results showed that the rejection rates of KCl ions were 57 and 40% for two different membrane samples. However, the rejection rates can be increased up to 84% by sealing the intrinsic defects on graphene through interfacial polymerization. Li et al. [37] prepared the NPG membranes with the carboxyl groups functionalized pores by using the irradiation with swift heavy ions and investigated their ionic sieving performance; they found that the permeation rate of ions had a great dependence on the temperature and H^+ concentration in the solutions and that the external electric fields can enhance the permeation flux of ions through the NPG membranes. A four-layer NPG membrane with nanopores generated by using metal oxide nanoparticles at high temperature was fabricated by Wei et al. [38] and they indicated that the water flux can be up to 4600 $L \cdot m^{-2} \cdot h^{-1}$ for the pores of sizes 50 nm and of density 1.0×10^7 cm^{-2} at the driving pressure of 0.2 bar. In short, the macroscopic NPG sheets have been synthesized and demonstrated to be competent to the ionic separations [2, 27, 39], which is a big step toward the industrial applications of graphene-based membranes. Moreover, the pore sizes of the NPG sheets can be precisely controlled and the high ion rejection and water permeation flux can be achieved [40].

The group of Prof. Karnik at the Massachusetts Institute of Technology made a significant contribution to the development of NPG water purification membranes [41–46]. They fabricated large-scale atomically thick NPG membranes and demonstrated their potentials for purifying kinds of ions and solutes. Meanwhile, they concluded that the single-layer graphene membranes can withstand ultrahigh applied pressure by monitoring the failure of centimeter-scale NPG membranes on porous supports. Although the graphene-based membranes work very well in laboratory, many challenges remain to make the cutting-edge membranes practicable in industry. The main challenges for the industrial-scale graphene-based membranes include the fabrication of large-scale graphene-based membranes with high mechanical stability and free of intricate defects and artificial damages, the generation of high-density pores in the graphene sheets with precisely controlled sizes and compositions, and other general challenges, such as supporting, blocking, fouling and concentration polarization. Most recently, the group of Prof. Quan Yuan at Wuhan University and the group of Prof. Xiangfeng Duan at the University of California Los Angeles cooperatively demonstrated a large-area hybrid membrane with graphene-nanomesh and single-walled carbon nanotubes [47]. Their research indicated that such membranes exhibited a high transport rate of water and a high rejection of ions owing to the high-density selective nanopores in the graphene nanomesh. Meanwhile, these membranes could well maintain the structural integrity and had a high mechanical strength under the support of microscopic single-walled carbon nanotubes. Excitingly, the hybrid membranes demonstrated in this work exhibited the advantages of both NPG and graphene oxide membranes, namely high molecular and ionic permeance owing to the permeable pores in the graphene nanomesh and they are easy to scale-up with the help of macroscopic single-walled carbon nanotube networks. Additionally, these novel membranes overcame the key challenges faced currently for the graphene-based membranes—fabrication of large-area graphene sheets with high-density selectively permeable pores and high mechanical strength. This work further demonstrates that the graphene-based membranes are really practicable by gradually overcoming the challenges toward the industrial applications. All of these state-of-the-art works offered a proof that the NPG membranes can serve as a novel and high-efficient membrane for water purification, along with a high water permeance and high ionic rejection rate thanks to its atomic thickness and excellent ionic sieving effects. We hope that the ongoing efforts on the graphene-based membranes can not only make the 2D separation membranes feasible but also facilitate the applications of graphene sheets in energy harvesting and storage, bioprocessing and so on.

(b) **Simulation works**

The simulation works were conducted relatively considerably comparing to the experimental works, especially by using the MD method. The first simulation work was conducted by Cohen-Tanugi and Grossman [1] in 2012, who first demonstrated the concept of NPG membranes for water purification. They demonstrated that the saltwater can be effectively purified by the single-layer NPG sheet and the desalination performance was dependent on the pore size, pore functionalization and driving pressure. Meanwhile, the NPG membranes presented water permeability of several orders of magnitude higher than those of conventional RO water desalination membranes. For the salt rejection rate, the small pores usually exhibited a higher rate, but it presented a decreasing trend at high pressures; the rejection rate of hydrogenated pores can be higher comparing to the hydroxylated pores. Subsequently, Konatham et al. [48] found that the NPGs with pores of diameter $\approx$7.5 Å can effectively reject ions and that the ion rejection rates can be enhanced by the carboxyl groups on the pore rims (Fig. 3). Guerrero-Avilésa and Orellana [49] identified a water permeation flux through the hydroxylated pores agreeing with the experimental measurements and showed that the O-passivation on the hydroxylated pores can enhance the water transport rate owing to the formed hydrogen bonds with the water molecules. Kommu et al. [50] showed a higher salt rejection rate and intermediate water permeance of the NPG membranes with N-functionalized pores while a lower salt rejection rate and a higher water permeance of the NPG membranes with OH-functionalized pores.

After the demonstration of NPG water purification membranes by these pioneering MD simulation works, many simulation studies were performed for more detailed investigations. The effects of chemical functionalization, pore size and other factors on the desalination performance were revealed. Ebrahimi [51] revealed that the curvature of graphene sheets and the shape of iso-surfaces of density affected the rejection rates of ions. Wang et al. [52] displayed that the chemical functionalization on the pore rim can greatly affect the ion rejection rate and that the potential of mean forces can well explain the hierarchy of water permeation rates through the NPG membranes. Chen and Yang [53] indicated that the NPG membranes with pores partially functionalized by hydroxyl groups showed a high salt rejection rate. The desalination performance of NPG membranes was studied by Chogani et al. [54] by considering the membrane flexibility and the result showed that the water permeation flux increased while the salt rejection rate decreased with the increase of the applied

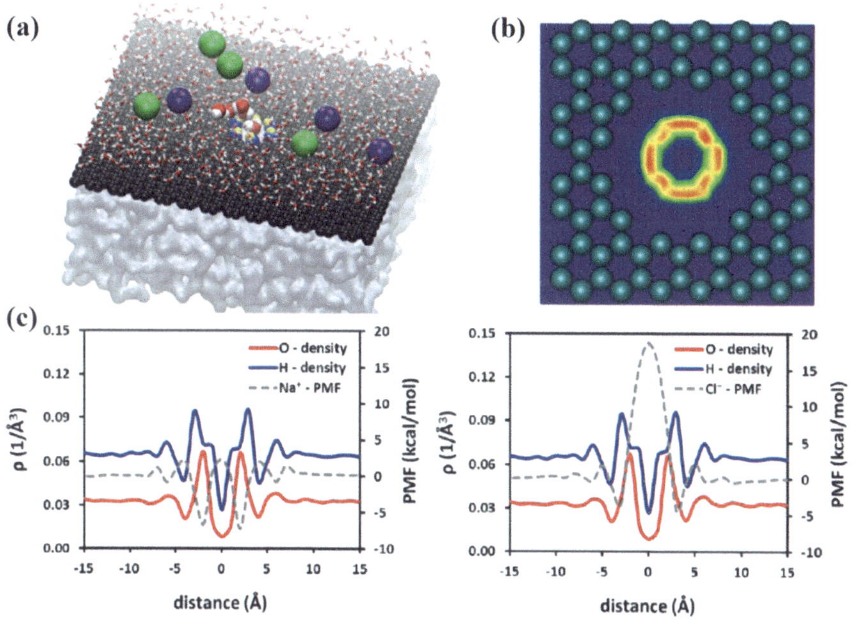

Fig. 3 MD simulations on the NPG membranes. **a** Schematic illustration of the water desalination process, **b** density distribution of water molecules in the graphene nanopores, **c** molecular density and potential of mean force distributions along the direction of vertical to graphene surfaces. Figures are reproduced with permission from [48]

pressure and the pore diameter. In addition, the effects of other liquids on the permeation of water through NPG membranes were studied. Hou et al. [55] concluded that the selective permeation of water and ethanol molecules through NPGs was mainly dependent on the molecular adsorption on hydrophobic surfaces and the molecular trapping on pores. Gravelle et al. [56] showed that the NPG membranes exhibited a counter-intuitive "self-semi-permeability" to water in the presence of ethanol because the adsorption of ethanol in nanopores prevented the water molecules from entering the pores. Shi et al. [57] found that the pore size was very critical for the ethanol/water separation and that the water permeation rates can be enhanced by the hydrophilic functionalization on the pore rims, because of reducing the energy barriers. Darvishi and Foroutan [58] found that the NPGs can be employed to separate water from a gaseous mixture via the preferential adsorption of water molecules on the graphene surfaces. Owing to the ultrahigh permeability of NPG membranes, the pressure requirement and energy consumption for water purification can be reduced. Cohen-Tanugi et al. [59] demonstrated that a tripling in the water permeability can result in a 44% lower pressure requirement or 15% less energy consumption for a seawater RO plant. Furthermore, Cohen-Tanugi and Grossman [26] showed that the NPG membranes still presented a high water permeability at low pressures in the real

RO systems, being comparable with the predicted water permeability at very high pressures.

The desalination performance of multilayer NPG membranes was also investigated by MD simulations. The two-layer NPG membranes with a lower surface tension were shown by Garnier et al. [60] to exhibit a lower water permeation flux compared to the monolayer NPG membranes. Shahbabaei et al. [61] observed that the multilayer NPG membranes with hourglass-shaped pores presented a high water permeation flux with a dependence on the length of water flow path inside the pores and especially that the hydrophilic graphene surfaces could double the water permeation flux owing to the strong hydrogen bond interactions. They [62] also indicated that the water occupancy across the asymmetric pore increased and thus a higher water permeation flux was yielded compared to the symmetric pore owing to the enhanced hydrogen bond interactions. The effects of channel morphology on the water permeation through the bilayer graphene membranes were studied by Liu et al. [63] and the results showed that the water permeation rate greatly depended on the curvature of graphene sheets. The same authors [64] found that the layered water structure inside the bilayer graphene nanochannels was significantly important for the channel size-dependent water transport rate. Based on the understanding of the mechanisms of water transport through multilayer NPG membranes, Jiang et al. [65] established an equivalent one-dimensional (1D) nanochannel model to describe the multilayer graphene membranes. By using MD simulations and continuum fracture mechanics, Cohen-Tanugi and Grossman [66] demonstrated that the NPG membranes can maintain the mechanical integrities under the high pressures in the RO systems with the water flowing through the NPG membranes. The NPG membranes can endure a pressure higher than 57 MPa on a porous substrate with pores of a diameter smaller than 1 μm. They [67] further showed that the desalination performance of multilayer NPGs was as good as the monolayer NPG membranes and that the desalination performance can be regulated by the configurational parameters.

The permeation of ions and water through graphene nanopores are very essential for the understanding of the mechanisms of NPG water purification membranes. Here, the studies on the permeation of water and ions are summarized, although some works have been conducted before the demonstration of NPG water purification membranes. Sint et al. [24] showed from MD simulations that the diverse ions can selectively transport through the graphene nanopores with a high selectivity and a very high ion transport rate (see Fig. 4). Hu et al. [68] investigated the transport of Na^+ and Cl^- ions through graphene nanopores and studied the effects of the hydrodynamic transport, thermal fluctuations as well as electric pressure. Zhao et al. [69] demonstrated that the selectivity of K^+ and Cl^- ions was greatly dependent on the pore size and the partial charges on pore rim. Kang et al. [70] showed that the selectivity of Na^+ and K^+ ions through graphene nanopores was affected by the ionic size and the distance between the functional groups. Suk and Aluru [71] concluded that the concentration and mobility of ions in the graphene nanopores decreased as the pore radius was smaller than 0.9 nm because the layered liquids appeared in the pores. They [25] also found that the graphene nanopores with a bigger diameter can provide a higher water permeation flux comparing to

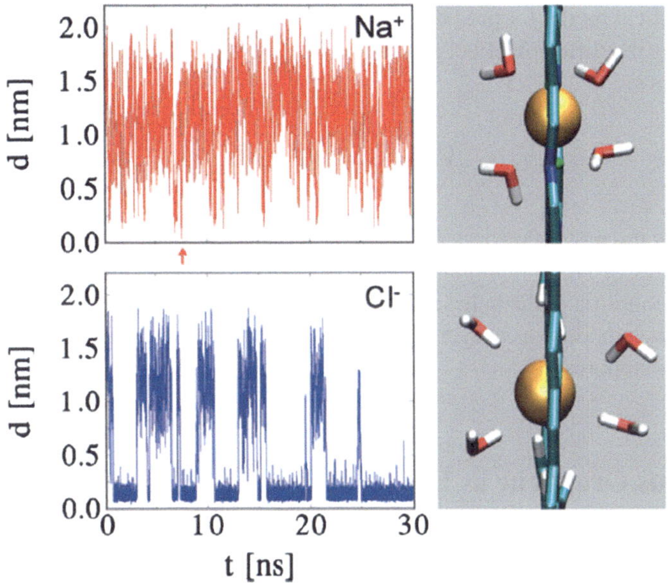

Fig. 4 Ion transport through the graphene nanopores. Time-dependent distance between ions and pore centers. This figure is reproduced with permission from [24]

the carbon nanotubes, and vice versa. Zhu et al. [72] indicated that the graphene nanopores with a diameter smaller than 15 Å showed a water permeation flux with a nonlinear dependence on the pore radius. Muscatello et al. [73] found that the entrance effects and the pathway of water molecules dominated the water permeation through graphene nanopores. These investigations on the transport of water and ions through graphene nanopores can definitely deepen the mechanisms of water purification through NPG-based membranes.

The above-mentioned investigations on the transport of water and ions through graphene nanopores can help building the theoretical models to predict the permeation rates of water and ions through the NPGs. Generally, the water permeation rate through the graphene nanopores can be predicted based on the modified Hagen–Poiseuille equation by considering the entrance/exit effects and the velocity slip etc. The original Hagen–Poiseuille equation is:

$$Q = \frac{\pi R^4 \Delta P}{8\mu L} \tag{2}$$

where ΔP is the pressure drop, Q is the volumetric flow rate, μ is the water viscosity, R is the pore radius, L is the membrane thickness. Suk and Aluru [74] developed a semi-empirical model by modifying the Hagen–Poiseuille equation with the slip length and considering the entrance/exit effects. Walther et al. [75] also developed a model by considering both the friction pressure loss and entrance/exit pressure loss.

The theoretical model for predicting the transport rates of ions can be developed by considering the diffusion effect and the convective effect. The basic formula should be as follows:

$$Q_i = D_i \frac{\Delta c_i}{L} A_p + c_i Q \quad (3)$$

where Q_i is the molar flow rate of ions, Δc_i is the concentration difference between the two sides of NPG membranes, D_i is the diffusion coefficient of ions, c_i is the ionic concentration of the bulk solutions, A_p is the pore area. To our best knowledge, currently such model was not yet established to predict the ion transport rates through graphene nanopores.

3 Porous Graphene as Electrodialysis Membranes

(a) Experimental works

Few comprehensive experiments have been conducted so far in this promising field. Rollings et al. [28] studied the ion selectivity of graphene nanopores in the electrodialysis process, focusing on the relationship between ion selectivity and pore size. They created graphene nanopores by using the electrical pulse method. Their study gave some surprising results that graphene nanopores showed K^+/Cl^- selectivity ratios over 100 and monovalent/divalent cation selectivity up to 5, even when the pores were as large as about 20 nm in diameter. They also showed that the potential of graphene for ion selectivity was underestimated so far due to inappropriate mechanisms, because high throughput, highly selective graphene electrodialysis membranes had no necessity to precisely control the size of the pores (Fig. 5). The promising ion selectivity of graphene for both monovalent and divalent cations and anions even for large pores under high electric field intensity shows that the porous graphene can be effectively used not only as electrodialysis membrane but also for other selective ions transportation fields, especially in reverse electrodialysis. The low requirement on the precise control of pore size makes the membrane synthesis would be more reliable, simple and cost-effective. Unfortunately, the experimental works on the water and ion transport through porous graphene membranes under the electrical fields are very limited currently, and we hope that more and more outstanding experimental measurements can be performed in the near future.

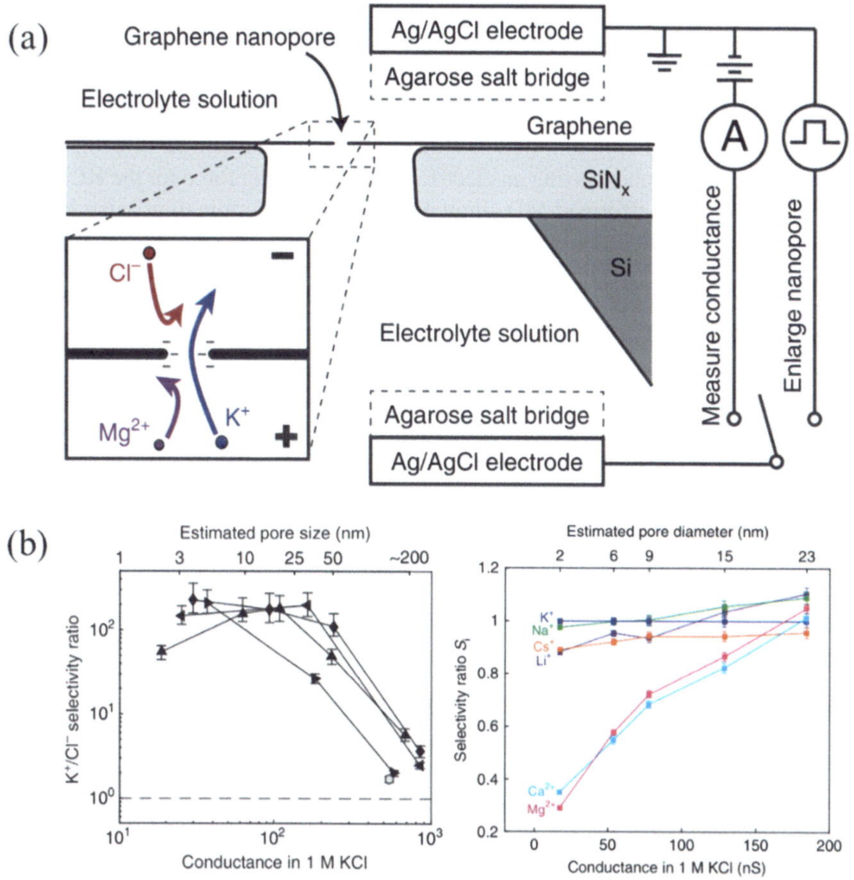

Fig. 5 Ion selectivity of graphene nanopores in electrodialysis process. **a** Schematic of the experimental setup, **b** K^+/Cl^- and inter-cation selectivity ratio as a function of pore size. Figures are reproduced with permission from [28]

(b) Simulation works

Due to the novelty of the NPG membranes, there also have been a limited number of theoretical and simulation works on NPG as electrodialysis membranes so far. Functionalized and charge-modified graphene membranes were used as selective ion transport membranes for the first time by Sint et al. [24]. They used the MD method to simulate the transport of various ions through the graphene membranes with different nitrogen and fluorine modified pores. When the driving electric field was as large as 0.1 V/nm, only Li^+, K^+, Na^+ were allowed to pass through the F–N pore, with a passing ratio of 9:14:33. While H pore only allowed Cl^-, and Br^- ions to pass through. The results showed that ion transport largely depended on ion sizes. Small-sized ions had a high rejection rate due to their small radii and high bonding to the water molecules in their hydration shells while the large radii ions were free from their hydration shells due to their large size without much bonding to the water molecules. Zhao et al. [69] carried out a promising and comprehensive theoretical study on charge-modified NPG as the ion exchange membranes for desalination based on energy conservation. MD simulations were performed to investigate the ion selectivity of graphene using an electric field as a driving force for the KCl solutions. Azamat [76] performed MD simulations for water desalination using bilayer NPG membranes under an electric field. They used a 30 × 30 × 90 Å^3 amorphous cell containing 1600 water molecules including 0.5 M salt and 2 modified NPG membranes with a pore diameter of 6 Å and an area of 30 × 30 Å^2. An external electric field from 0 to 35 V was applied (Fig. 6a). It was found that under the action of electric field, the F-pore and the H-pore of the membrane were preferential selective to Na^+ and Cl^-, respectively. Meanwhile, the higher the electric field, the faster the ions moved from the brine. The sodium ions and the chloride ions met an energy barrier, thus the cations and anions could not pass through the H-pore and F-pore of the membrane, respectively. Zhang et al. [29] also performed MD simulations for desalination using a bilayer NPG membrane nanochannel. The graphene nanochannels of different heights of 7–20 Å have been used under the driving pressure of 0–300 MPa with and without vertical electric fields of different intensities. MD simulations were performed to measure the water flux and salt rejection rate with

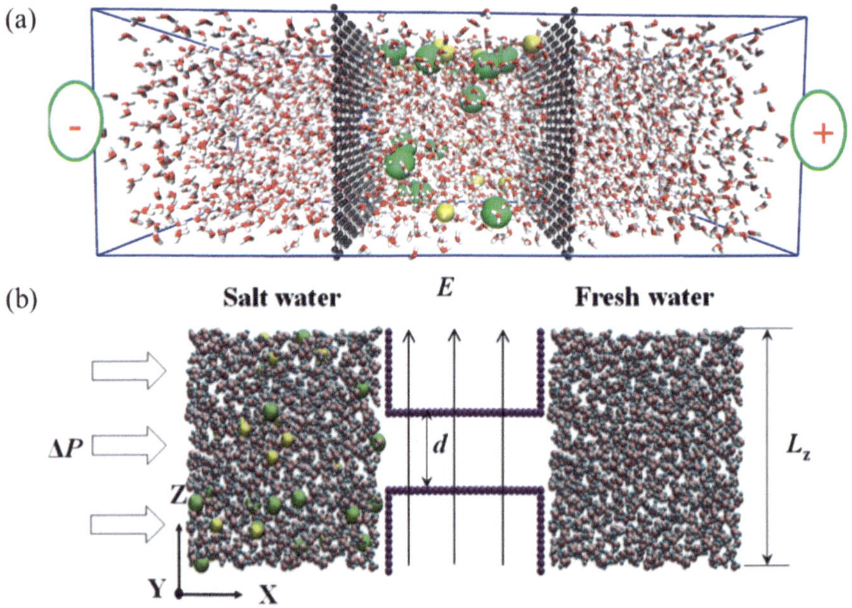

Fig. 6 MD simulation models of electrodialysis membranes. **a** Schematic diagram of the simulation models using bilayer NPG membranes under an electric field. Figure is reproduced with permission from [76], **b** schematic diagram of the simulation models using a bilayer NPG membrane nanochannel. Figure is reproduced with permission from [29]

different pressure, channel height and electric field intensity (Fig. 6b). The results showed that the behaviors of salt ions and water molecules were diverse, namely, the number of water molecules in the nanoscale channel was not affected by the driving pressure, whereas the number of ions was sensitive to the pressure when the channel thickness was small. The salt rejection decreased with the increase of the driving pressure and channel height and it increased with the increase of the vertical electric field. Lohrasebi et al. [3] also used multilayer graphene membranes as selective ion transport membranes under the action of the applied external electric field. They performed MD simulations and measured the ion selectivity, salt rejection rate and water permeability. They used 3 box, 2 bilayer membrane with a rigid piston (for applying pressure), plus 2 external electric fields in the opposite direction applied to the porous graphene membranes. Four levels of external electric fields from 0.1 to 100 mV/Å were applied along with a pore radius of 4 Å and different width levels between bilayer membranes. The pore size was designed according to the hydrated ions so that no ion could pass through the membrane without applying the electric field. The results indicated that the increase of the applied electric field improved ion selectivity and the salt rejection rate was found to be more than 94% under 10 mV/Å.

4 Research Gaps

Graphene and its derivatives are naturally supposed to be a very promising candidate for the separation membranes owing to their thinnest thickness so far, because the membrane permeance, quantifying the permeation abilities of molecules and ions, is generally accepted to be inversely proportional to the thickness of membranes. Such high-efficiency membranes are very crucial for the energy-saving membrane separation technology, which is widely applied in the separation industries such as water purification. Currently, two kinds of graphene-based membranes are proposed for the separation technology; the first is NPG membrane while the second is graphene oxide membrane. The mechanism of NPG membranes is basically the size-sieving effects of the selective nanopores, while that of graphene oxide membranes is the selective molecular and ionic transport through the interlayer channels among the laminated graphene oxide sheets and the defects in the graphene oxide sheets. The molecular permeance of NPG membranes is very high comparing to the graphene oxide membranes with a relatively thick thickness of layered structures. The fabrication process of NPG membranes is very complex although the top-down and bottom-up methods both seem to be effective; while that of graphene oxide membranes is relatively simple. In short, the advantage of NPG-based membranes is their ultra-high molecular permeance and that of graphene oxide membranes is easy of fabrication. People are trying to make the graphene-based membranes have the advantages of both NPG and graphene oxide membranes, namely high molecular permeance and ease of fabrication.

Although the NPG membranes work very well in the laboratory, many challenges are still faced to make these membranes practicable in the industry. The main challenges for the realization of industrial-scale NPG membranes include fabrication of large-scale NPG sheets with a high mechanical stability and free of intricate defects and artificial damages, transfer of graphene sheets to a supporting substrate with no damage, and generation of high-density pores in the graphene sheets with precisely controlled sizes. All of these challenges should be the main concerns of future studies, which are expected to fill the gaps between laboratory study and industrial application of graphene-based membranes. We hope that the continuous efforts on the graphene-based membranes can not only make such separation membranes practicable but also stimulate the applications of graphene sheets in energy harvesting and storage, bioengineering and so on [77], because graphene sheets have a great application prospect in the above fields apart from the separation science and technology. For example, graphene is a good hydrogen production material for its excellent proton conductance and electron transport efficiency; graphene-related materials can be employed in hydrogen storage for their low gravimetric density and strong hydrogen affinity; graphene can be used in supercapacitors for its large specific area and high electrical conductivity; NPG is also very promising as DNA sequencing tools for its intrinsic solid-state nanopores. In summary, graphene-based membranes have shed a bright light, but many efforts must be devoted to solve these challenges faced currently.

5 Conclusions

Graphene has shed light on the two-dimensional high-permeability separation membranes in the fields of water purification due to its atomic thickness. NPG as one of the graphene-based membranes has demonstrated to be very promising for water purification based on the mechanisms of RO and electrodialysis. Recently, many theoretical and experimental works were conducted on the NPG-based membranes for water purification. The separation performances, underlying mechanisms, membrane fabrication methods of these membranes were discussed in detail. In this chapter, we summary the representative experimental and simulation works on the NPG membranes both on the RO and electrodialysis. Apart from the solution separation, the NPG-based membranes are expected to have a great application prospect in other ion-related applications, such as energy harvesting from the selective ion transport through membranes, energy storage based on the supercapacitors with nanostructured materials, etc. For example, the NPG-related materials have shown promises in power generation from the salinity gradient based on the reverse electrodialysis with selective ion transport. It is hoped that the researches on the NPG-based membranes can not only make the two-dimensional separation membranes more and more promising but also promote the developments of other ion-related applications.

Acknowledgments The financial supports from the National Natural Science Foundation of China for the Basic Science Center Program for Ordered Energy Conversion (Number 51888103) and general program (Number 51876169) are highly acknowledged.

References and Future Reading

1. Cohen-Tanugi D, Grossman JC (2012) Water desalination across nanoporous graphene. Nano Lett 12(7):3602–3608
2. O'Hern SC, Boutilier MSH, Idrobo J-C, Song Y, Kong J, Laoui T, Atieh M, Karnik R (2014) Selective ionic transport through tunable subnanometer pores in single-layer graphene membranes. Nano Lett 14(3):1234–1241
3. Lohrasebi A, Rikhtehgaran S (2018) Ion separation and water purification by applying external electric field on porous graphene membrane. Nano Research 11(4):2229–2236
4. Allen MJ Tung VC, Kaner RB (2010) Honeycomb Carbon: a review of graphene. Chem Rev 110(1):132–145
5. Girit CO, Meyer JC, Erni R, Rossell MD, Kisielowski C, Yang L, Park CH, Crommie MF, Cohen ML, Louie SG, Zettl A (2009) Graphene at the edge: stability and dynamics. Science 323(5922):1705–1708
6. Meyer JC, Geim AK, Katsnelson MI, Novoselov KS, Booth TJ, Roth S (2007) The structure of suspended graphene sheets. Nature 446(7131):60–63
7. Sun C, Wen B, Bai B (2015) Recent advances in nanoporous graphene membrane for gas separation and water purification. Sci Bull 60(21):1807–1823
8. Das R, Vecitis CD, Schulze A, Cao B, Ismail AF, Lu X, Chen J, Ramakrishna S (2017) Recent advances in nanomaterials for water protection and monitoring. Chem Soc Rev 46(22):6946–7020
9. Sun C, Zhu S, Liu M, Shen S, Bai B (2019) Selective molecular sieving through a large graphene nanopore with surface charges. J Phys Chem Lett 10(22):7188–7194
10. Xu P, Yang J, Wang K, Zhou Z, Shen P (2012) Porous graphene: properties, preparation, and potential applications. Chin Sci Bull 57(23):2948–2955
11. Kazemi AS, Abdi Y, Eslami J, Das R (2019) Support based novel single layer nanoporous graphene membrane for efficacious water desalination. Desalination 451:148–159
12. Berry V (2013) Impermeability of graphene and its applications. Carbon 62:1–10
13. Das R, Leo BF, Murphy F (2018) The toxic truth about carbon nanotubes in water purification: a perspective view. Nanoscale Res Lett 13(1):183
14. Das R, Ali ME, Hamid SBA, Ramakrishna S, Chowdhury ZZ (2014) Carbon nanotube membranes for water purification: a bright future in water desalination. Desalination 336:97–109
15. Ali ME, Das R, Maamor A, Hamid SBA (2014) Multifunctional Carbon Nanotubes (CNTs): a new dimension in environmental remediation. Adv Mater Res 832:328–332
16. Das R (2017) Carbon nanotube in water treatment. In: Nanohybrid catalyst based on carbon nanotube. Carbon nanostructures. Springer, Cham. https://doi.org/10.1007/978-3-319-58151-4_2
17. Das R (2018) Carbon nanotubes for clean water. Carbon nanostructures. Springer, Cham. https://doi.org/10.1007/978-3-319-95603-9
18. Das R (2017) Advanced membrane materials for desalination: carbon nanotube and graphene. In: Inorganic pollutants in wastewater. Materials Research Forum. https://doi.org/10.21741/9781945291357
19. Liu Q, Xu G-R, Das R (2019) Inorganic scaling in reverse osmosis (RO) desalination: mechanisms, monitoring, and inhibition strategies. Desalination 468:114065

20. Das R (2019) Polymeric materials for clean water. Polymer and composite materials. Springer, Cham. https://doi.org/10.1007/978-3-030-00743-0
21. Chian ESK, Chen JP, Sheng P-X, Ting Y-P, Wang LK (2007) Reverse osmosis technology for desalination. In: Advanced physicochemical treatment technologies. handbook of environmental engineering. Humana Press. https://doi.org/10.1007/978-1-59745-173-4_6
22. Xu G-R, Xu J-M, Su H-C, Liu X-Y, Lu L, Zhao H-L, Feng H-J, Das R (2019) Two-dimensional (2D) nanoporous membranes with sub-nanopores in reverse osmosis desalination: latest developments and future directions. Desalination 451:18–34
23. Strathmann H (2010) Electrodialysis, a mature technology with a multitude of new applications. Desalination 264(3):268–288
24. Sint K, Wang B, Kral P (2008) Selective ion passage through functionalized graphene nanopores. J Am Chem Soc 130(49):16448
25. Suk ME, Aluru NR (2010) Water transport through ultrathin graphene. J Phys Chemis Lett 1(10):1590–1594
26. Cohen-Tanugi D, Grossman JC (2014) Water permeability of nanoporous graphene at realistic pressures for reverse osmosis desalination. J Chem Phys 141(7):074704
27. Surwade SP, Smirnov SN, Vlassiouk IV, Unocic RR, Veith GM, Dai S, Mahurin SM (2015) Water desalination using nanoporous single-layer graphene. Nat Nanotechnol 10(5):459–464
28. Rollings RC, Kuan AT, Golovchenko JA (2016) Ion selectivity of graphene nanopores. Nat Commun 7:11408
29. Hui Zhang BLM-SW, Zhou K, Law AW-K (2017) Transport of salty water through graphene bilayer in an electric field: a molecular dynamics study. Comput Mater Sci 131:100–107
30. Azamat J, Khataee A, Joo SW (2014) Functionalized graphene as a nanostructured membrane for removal of copper and mercury from aqueous solution: a molecular dynamics simulation study. J Mol Graph Model 53:112–117
31. Fu Y, Guo X, Wang Y, Wang X, Xue J (2019) An atomically-thin graphene reverse electrodialysis system for efficient energy harvesting from salinity gradient. Nano Energy 57:783–790
32. Sun C, Bai B (2017) Molecular sieving through a graphene nanopore: non-equilibrium molecular dynamics simulation. Sci Bulln 62(8):554–562
33. O'Hern SC, Jang D, Bose S, Idrobo J-C, Song Y, Laoui T, Kong J, Karnik R (2015) Nanofiltration across defect-sealed nanoporous monolayer graphene. Nano Lett 15(5):3254–3260
34. O'Hern SC, Stewart CA, Boutilier MSH, Idrobo J-C, Bhaviripudi S, Das SK, Kong J, Laoui T, Atieh M, Karnik R (2012) Selective molecular transport through intrinsic defects in a single layer of CVD graphene. ACS Nano 6(11):10130–10138
35. Kim S, Song Y, Ibsen S, Ko S-Y, Heller MJ (2016) Controlled degrees of oxidation of nanoporous graphene filters for water purification using an aqueous arc discharge. Carbon 109:624–631
36. Kafiah FM, Khan Z, Ibrahim A, Karnik R, Atieh M, Laoui T (2016) Monolayer graphene transfer onto polypropylene and polyvinylidenedifluoride microfiltration membranes for water desalination. Desalination 388:29–37
37. Li Z, Liu Y, Zhao Y, Zhang X, Qian L, Tian L, Bai J, Qi W, Yao H, Gao B, Liu J, Wu W, Qiu H (2016) Selective separation of metal ions via monolayer nanoporous graphene with carboxyl groups. Anal Chem 88(20):10002–10010
38. Wei G, Quan X, Chen S, Yu H (2017) Superpermeable atomic-thin graphene membranes with high selectivity. ACS Nano 11(2):1920–1926
39. Celebi K, Buchheim J, Wyss RM, Droudian A, Gasser P, Shorubalko I, Kye J-I, Lee C, Park HG (2014) Ultimate permeation across atomically thin porous graphene. Science 344(6181):289–292
40. Zhao J, He G, Huang S, Villalobos LF, Dakhchoune M, Bassas H, Agrawal KV (2019) Etching gas-sieving nanopores in single-layer graphene with an angstrom precision for high-performance gas mixture separation. Sci Adv 5(1):eaav1851
41. Kidambi PR, Mariappan DD, Dee NT, Vyatskikh A, Zhang S, Karnik R, Hart AJ (2018) A scalable route to nanoporous large-area atomically thin graphene membranes by roll-to-roll

chemical vapor deposition and polymer support casting. ACS Appl Mater Interf 10(12):10369–10378
42. Kidambi PR, Nguyen GD, Zhang S, Chen Q, Kong J, Warner J, Li AP, Karnik R (2018) Facile fabrication of large-area atomically thin membranes by direct synthesis of graphene with nanoscale porosity. Adv Mater 30(49):1804977
43. Kidambi PR, Boutilier MS, Wang L, Jang D, Kim J, Karnik R (2017) Selective nanoscale mass transport across atomically thin single crystalline graphene membranes. Adv Mater 29(19):1605896
44. Wang L, Williams CM, Boutilier MS, Kidambi PR, Karnik R (2017) Single-layer graphene membranes withstand ultrahigh applied pressure. Nano Lett 17(5):3081–3088
45. Jang D, Idrobo J-C, Laoui T, Karnik R (2017) Water and solute transport governed by tunable pore size distributions in nanoporous graphene membranes. ACS Nano 11(10):10042–10052
46. Kidambi PR, Jang D, Idrobo JC, Boutilier MS, Wang L, Kong J, Karnik R (2017) Nanoporous atomically thin graphene membranes for desalting and dialysis applications. Adv Mater 29(33):1700277
47. Yang Y, Yang X, Liang L, Gao Y, Cheng H, Li X, Zou M, Ma R, Yuan Q, Duan X (2019) Large-area graphene-nanomesh/carbon-nanotube hybrid membranes for ionic and molecular nanofiltration. Science 364(6445):1057–1062
48. Konatham D, Yu J, Ho TA, Striolo A (2013) Simulation insights for graphene-based water desalination membranes. Langmuir 29(38):11884–11897
49. Guerrero-Aviles R, Orellana W (2017) Energetics and diffusion of liquid water and hydrated ions through nanopores in graphene: ab initio molecular dynamics simulation. Phys Chem Chem Phys 19(31):20551–20558
50. Kommu A, Namsani S, Singh JK (2016) Removal of heavy metal ions using functionalized graphene membranes: a molecular dynamics study. Rsc Advances 6(68):63190–63199
51. Ebrahimi S (2016) Influence of curvature on water desalination through the graphene membrane with Si-passivated nanopore. Comput Mater Sci 124:160–165
52. Wang Y, He Z, Gupta KM, Shi Q, Lu R (2017) Molecular dynamics study on water desalination through functionalized nanoporous graphene. Carbon 116:120–127
53. Chen Q, Yang X (2015) Pyridinic nitrogen doped nanoporous graphene as desalination membrane: molecular simulation study. J Membr Sci 496:108–117
54. Chogani A, Moosavi A, Rahiminejad M (2016) Numerical simulation of salt water passing mechanism through nanoporous single-layer graphene membrane. Chem Product Process Model 11(1):73–76
55. Hou Y, Xu Z, Yang X (2016) Interface-induced affinity sieving in nanoporous graphenes for liquid-phase mixtures. J Phys Chem C 120(7):4053–4060
56. Gravelle S, Yoshida H, Joly L, Ybert C, Bocquet L (2016) Carbon membranes for efficient water-ethanol separation. J Chem Phys 145(12):124708
57. Shi Q, He Z, Gupta KM, Wang Y, Lu R (2017) Efficient ethanol/water separation via functionalized nanoporous graphene membranes: insights from molecular dynamics study. J Mater Sci 52(1):173–184
58. Darvishi M, Foroutan M (2015) Mechanism of water separation from a gaseous mixture via nanoporous graphene using molecular dynamics simulation. RSC Adv 5(99):81282–81294
59. Cohen-Tanugi D, McGovern RK, Dave SH, Lienhard JH, Grossman JC (2014) Quantifying the potential of ultra-permeable membranes for water desalination. Energy Environ Sci 7(3):1134–1141
60. Garnier L, Szymczyk A, Malfreyt P, Ghoufi A (2016) Physics behind water transport through nanoporous boron nitride and graphene. J Phys Chem Lett 7(17):3371–3376
61. Shahbabaei M, Tang D, Kim D (2017) Simulation insight into water transport mechanisms through multilayer graphene-based membrane. Comput Mater Sci 128:87–97
62. Shahbabaei M, Kim D (2017) Transport of water molecules through noncylindrical pores in multilayer nanoporous graphene. Phys Chem Chem Phys 19(31):20749–20759
63. Liu B, Wu R, Law AW-K, Feng X-Q, Bai L, Zhou K (2016) Channel morphology effect on water transport through graphene bilayers. Sci Rep 6:38583

64. Liu B, Wu R, Baimova JA, Wu H, Law AW-K, Dmitriev SV, Zhou K (2016) Molecular dynamics study of pressure-driven water transport through graphene bilayers. Phys Chem Chem Phys 18(3):1886–1896
65. Jiang G, Wang P, Cheng C, Li D, Liu JZ (2018) An equivalent 1D nanochannel model to describe ion transport in multilayered graphene membranes. Progr Nat Sci Mater Int 28(2):246–250
66. Cohen-Tanugi D, Grossman JC (2014) Mechanical strength of nanoporous graphene as a desalination membrane. Nano Lett 14:6171–6178
67. Cohen-Tanugi D, Lin L-C, Grossman JC (2016) Multilayer nanoporous graphene membranes for water desalination. Nano Lett 16(2):1027–1033
68. Hu G, Mao M, Ghosal S (2012) Ion transport through a graphene nanopore. Nanotechnology 23(39):395501
69. Zhao S, Xue J, Kang W (2013) Ion selection of charge-modified large nanopores in a graphene sheet. J Chem Phys 139(11):114702
70. Kang Y, Zhang Z, Shi H, Zhang J, Liang L, Wang Q, Agren H, Tu Y (2014) Na^+ and K^+ ion selectivity by size-controlled biomimetic graphene nanopores. Nanoscale 6(18):10666–10672
71. Suk ME, Aluru NR (2014) Ion transport in sub-5-nm graphene nanopores. J Chem Phys 140(8):084707
72. Zhu C, Li H, Meng S (2014) Transport behavior of water molecules through two-dimensional nanopores. J Chem Phys 141(18):18C528
73. Muscatello J, Jaeger F, Matar OK, Mueller EA (2016) Optimizing water transport through graphene-based membranes: insights from nonequilibrium molecular dynamics. ACS Appl Mater Interf 8(19):12330–12336
74. Suk M, Aluru NR (2013) Molecular and continuum hydrodynamics in graphene nanopores. RSC Adv 3(24):9365–9372
75. Walther JH, Ritos K, Cruz-Chu ER, Megaridis CM, Koumoutsakos P (2013) Barriers to superfast water transport in carbon nanotube membranes. Nano Lett 13(5):1910–1914
76. Azamat J (2016) Functionalized graphene nanosheet as a membrane for water desalination using applied electric fields insights from molecular dynamics simulations. J Phys Chem C 120(41):23883–23891
77. Sun CZ, Zhou RF, Zhao ZX, Bai BF (2020) Nanoconfined fluids: what can we expect from them? J Phys Chem Lett 11(12):4678–4692

Hexagonal Boron Nitride (h-BN) in Solutes Separation

Sima Majidi, Siamak Pakdel, Jafar Azamat, and Hamid Erfan-Niya

Abstract Thanks to rapid progress in the investigation of two-dimensional (2D) substances, hexagonal boron nitride (h-BN) has shown outstanding properties as filtration membrane and adsorbent for water purification applications. In this chapter, we explain the synthesis methods of h-BN including exfoliation and chemical vapor deposition. And then, the computer simulations in the field of h-BN nanosheets are described, and several recent studies of the membrane filtration of h-BN and pollutant adsorption of h-BN are reviewed from molecular dynamics insights. All findings suggest the promising potential of h-BN for water purification. We further reveal a brief experimental works which investigated the performance of h-BN for separation of solutes. The h-BN not only shows considerable adsorption potential for a wide range of oils, model dyes, and various metallic ions but also reveals high recoverability. To this end, several other new fabrication techniques are described. Despite several researches in this field, there are still research gaps which have been mentioned in the last section of this chapter.

Keywords Hexagonal boron nitride · Membrane · Adsorption · Molecular dynamics · Water purification

1 Introduction

Bulk hexagonal boron nitride (h-BN) is a graphite type structure, in which flat networks of BN hexagons are regularly set [1]. Two-dimensional (2D) h-BN nanosheets are sp^2-hybridized 2D insulators [2, 3] with sub-lattices being inhabited with the same numbers of nitrogen (N) and boron (B) atoms alternately in a honeycomb arrangement [4]. The loading design of interlayer in the h-BN presents

S. Majidi · S. Pakdel · H. Erfan-Niya (✉)
Faculty of Chemical and Petroleum Engineering, University of Tabriz, Tabriz 51666-16471, Iran
e-mail: herfan@tabrizu.ac.ir

J. Azamat
Department of Basic Sciences, Farhangian University, Tehran, Iran

R. Das (ed.), *Two-Dimensional (2D) Nanomaterials in Separation Science*,
Springer Series on Polymer and Composite Materials,
https://doi.org/10.1007/978-3-030-72457-3_7

its B atoms in each successive BN layer sitting under or above the N atoms in the contiguous layers. These obtained properties demonstrate the B–N bonds polarity, especially the partial ionic character of the covalent B–N bonds. Actually, the N and B atoms inside a 2D layer, are connected together by strong B–N covalent bonds, while the 2D layers are linked with each other through the weak van der Waals (vdW) forces [5, 6]. The hexagonal crystal structure of h-BN with the interlayer spacing of 0.33 nm and crystallographic variables of $a = 0.25$ nm and $c = 0.666$ nm empowers admirable interaction with graphene, which indicates its excellent capacity for numerous industrial applications.

Regarding the ultra-flat atomic surface, which is free of dangling bonds and with negligible defects, a small lattice constant mismatch of about 1.7% was obtained [7]. The 2D h-BN has been considered as a supreme outlook for future graphene electronics with greater stability and efficiency [8]. Unlike graphene, the h-BN sheet has a wide gap insulator of 5–6 eV which can be altered by edge passivation with different types of atoms, higher chemical inertness and thermal stability, resistance to oxidation, and good optical properties [9–13]. These make it an interesting material for optoelectronic technologies [14, 15], tunnel devices, and field-effect transistors [16, 17].

2 h-BN Synthesis Methods

Synthesis and processing of h-BN can affect the structure, crystallinity, and characteristics of 2D h-BN nanosheets. However, developing the attractive 2D h-BN with desired structure and functional features for particular applications is still a challenge [18, 19]. Several producing methods have been developed to synthesis mono- and multi-layer h-BN, especially with the aim of producing a high efficiency of considerable lateral size and great quality [4]. The process of 2D h-BN production can be classified into bottom-up and top-down methods. In the bottom-up methods, a film is grown on the surface; and in top-down methods, bulk h-BN crystal is exfoliated to obtain the purpose structure. Generally, taking bulk h-BN, then break the vdW forces between the layers of h-BN and separate the resulting 2D nanosheets is the main idea of top-down methods. These methods mainly involve mechanical and chemical exfoliation methods [20]. Mechanical and liquid-phase exfoliations are two common techniques used to separate single sheets from stacked 2D layered crystals by breaking the weak vdW bonds between the layers. Mechanical exfoliation can lead to sheets with perfectly crystalline structures [15], which the sheets synthesizes via this technique are employed to investigate the inherent properties of substances [21].

2.1 Mechanical Exfoliation

Mechanical exfoliation was the first effort in synthesizing atomic h-BN sheets, in which the BN layer is mechanically cleaved or peeled [22]. The mechanical exfoliation method is low cost, which presents h-BN samples with great quality and even can be performed in most physics or material science laboratories. Indeed, this method has prepared a simple but potent technique to manufacture 2D materials [23–27] which was initially applied to separate graphene in 2004 [25]. Then it has been used for other substrate materials, like h-BN and molybdenum disulfide (MoS_2) exfoliation [23]. The usage of arranged sticky tape to peel off 2D nanosheets is one of the popular mechanical exfoliation procedures [23, 28]. This procedure results in about ten layers or 3.5-nm-thick h-BN nanosheets that could be owing to the strong lip-to-lip interactions of h-BN basal planes [29]. The major disadvantage of mechanical exfoliation procedures is less yield of h-BN nanosheets [21], and the size of the fabricated structures is typically restricted [20]. Other limitations include the lack of control on atomically thin flakes production followed by transferred them at random locations on the substrate. Additionally, thick flakes are transmitted along with thin flakes which are necessary for further characterizations to identify mono- and few-layer flakes [30–32]. Moreover, in the case of transparent 2D substances, like single layer h-BN, the recognition of such thin flakes on the substrate is a challenge and long-term due to its low optical absorption [33].

2.2 Chemical Exfoliation

The liquid solvents, like dichloromethane [34] and dimethylformamide [35] are used to perform the chemical exfoliation techniques. The procedure of these techniques is totally simple, which offers a great yield in comparison to the previous method, mechanical exfoliation [21]. The process of exfoliation follows the below steps:

(I) Self-curling of the sheets is obtained by adsorption of cations at the edges of h-BN surface.
(II) Consecutive curling of BN layer is achieved by entering of anions and cations into the interlayer region.
(III) The surface reaction with hydroxides produces layers with thickness ranging from 2 to 4 nm, which is resulted in direct peeling off from the bulk material.

This technique offers considerable potential advantages including being facile and low cost as well as simply transmittable to usual solvents, like ethanol and water [36]. Du et al. [37] investigated a simple method for chemical exfoliation of h-BN, which was derived from Hummers' method of oxidation and exfoliation of graphite to graphene [38]. The combination of h-BN powder with $KMnO_4$ and H_2SO_4 under heating and then H_2O_2 addition resulted in the exfoliation of h-BN to a small number of single-layer boron nitride nanosheets (BNNSs). Atomic force microscopy was used to measure the thickness, which presented about 1.44 nm for the thinnest obtained BNNS, relating to two layers of h-BN. In addition to the mentioned advantages of this method, the procedure indicated that not only high temperatures, vacuum, or high pressures are not essential but also simple scale up of the method is possible via wet chemical processing [22].

2.3 *Liquid Exfoliation*

The exfoliation of layered compounds in liquids is another possible solution to obtain a great amount of dispersed nanosheets, which involves oxidation, ion intercalation/exchange, or surface passivation by solvents. This allows the methods to give a substantial amount of 2D materials which can be processed by available industrial procedures.

Liquid exfoliation results in the synthesis of thin composites with scalable potential and facile processes using standard technologies. Fabrication of mono- and few-atomic-layer nanosheets from polycrystalline h-BN through liquid exfoliation was firstly reported by Han et al. [39], where 0.2 mg of h-BN was put into 5 mL of 1,2-dicloroethane solution of poly (m-phenyl-enevinylene-co-2,5-dictoxy-p phenylene vinylene) in a 1.2 mg/10 mL ratio and then to breaking up and dispersing the h-BN powder into few layered h-BN upon sonication for an hour. Subsequently, many other solvents have been verified to extend the nanosheet formation. One of the highly polar solvents is N, N-dimethylformamide (DMF), which was used to interact with the surface of h-BN and produced only milligram levels of pure nanosheets with thicknesses at the ranging of 2–10 nm [34]. The solvent selection can be also more optimized so that the energy of surface shall match with that of h-BN to overcome the vdW forces between layers [4].

2.4 *Chemical Vapor Deposition (CVD)*

The methods based on epitaxy could synthesis monolayer h-BN with high crystallinity structure, but they were highly sensitive to the substrates so that most of them with special orientations required very limited pretreatments, which resulted in long-term and complicated procedures. Consequently, the development of new simple processes seems necessary to produce 2D materials, especially h-BN, on the usual substrates with a large area and high quality. The process should adapt to silicon-based procedures and must meet the common demands. The chemical vapor deposition (CVD) method is a bottom-up process applied to fabricate high-quality nanoscale films. In reality, a substrate that acts as a catalyst or new catalysts could be deposited on a silica wafer for synthesizing h-BN film on top of it during the CVD method. The CVD of h-BN nanosheets is possible through several substrates and precursors. It should be noted that the reaction is mostly accompanied by toxic by-products. The deposition of thin h-BN on transition metals was formerly conducted by ultrahigh vacuum CVD (UHVCVD) [40], but CVD of the h-BN on metallic surfaces at high pressures has recently been prosperous [41]. Malcolm Basche [42] was carried out the thermal decomposition of ammonia (NH_3) with boron trichloride (BCl_3) at high temperatures to deposit the BN films on the surface. In the other work, ammonia and diborane (B_2H_6) as precursors were applied to deposit the BN coatings less than 600 nm thick on the surface of metals such as Mo, Si, Ge, and Ta at a temperature ranging from 600 to 1000 °C [43]. Furthermore, working with other precursors, like $H_3B_3Cl_3N_3$, BH_3, BF_3–NH_3, BCl_3–NH_3, and B_2H_6–NH_3 for getting uniform thin films has been also reported [44–47]. Some common CVD procedures for h-BN nanosheet growth were also performed, which used borazine and ammonia borane with substrates including Cu [48–50], Ni [51], and Pt [41, 52, 53]. Thermal CVD method was extended by heating B, MgO, FeO powders up to 1300 °C in a tube furnace under ammonia stream [54, 55]. Consequently, depending on synthesis temperature, vertically standing h-BNs were produced on Si/SiO_2 substrates with various morphologies and sizes. The nanosheets increased in lateral size by increasing temperature from 1000 to 1200 °C, so that nanosheets branching was done on the main nanosheet surface at temperature of 1300 °C. The produced h-BN nanosheets have regularly covered the surface of SiO_2 with less than 5 nm thickness. The microwave plasma CVD (MPCVD) based on catalyst-free method was carried out by Yu et al. [56] to grow h-BN nanosheets on Si substrates. The results indicated that triangular h-BN nanosheets were obtained with a size range of 0.8–2.5 mm and varied tilting angles to substrates. To this end, the growth of a few-layer h-BN with a wide area has been accomplished and even many works are currently in the planning steps. However, the production of wafer-scale h-BN with controlled layers and high crystalline quality remains a challenge.

3 Computer Simulation for Understanding the h-BN-Based Membrane Filtration and Adsorption

Simulation technique is one of the most potent tools available to decision-makers responsible for the design and evaluation of complex systems, which is defined as the operation of a model of the system. Usually, it is costly or impractical to carry out a range of experiments in the system. In other words, simulation makes possible the investigation, analysis, and evaluation of conditions which would not be otherwise practicable [57, 58]. The computer simulation methods are extended from macro [59] to nanoscale [60, 61]. In this regard, the approach for complicated nanoscale modeling at the molecular level has been provided by molecular simulations [62]. They have played a significant role in the past decades in our knowledge advance of the association between microscopic and macroscopic features of 2D nanomaterials. There are two types of molecular simulation, molecular dynamics (MD) simulation, and Monte Carlo (MC) simulation. In the MD simulation as a deterministic method, Newton equations of motion are solved for a system containing N particles, which interact through a force field, to explain the time evolution of the system. The macroscopic properties of the system can be calculated from its microscopic actions using the statistical mechanics and considering the trajectories of particles provided by the MD simulation at molecular level [63]. In the MC simulation, particles are moved randomly to represent a target probability distribution conforming to the desired state of the system. Therefore, MC calculates statistical thermodynamic probabilities of acceptance/rejection of moves [64].

Recently, MD simulations have been widely conducted to investigate the membrane separation processes [65]. This could play a considerable role in the prediction of 2D nanoporous materials capability in water desalination and adsorption performance of materials at the molecular level [66, 67]. Furthermore, due to the quantities computing connected to the system dynamics such as time-dependent fluctuations and transport coefficients, the MD method has benefits compared to other computational techniques [68].

3.1 Simulation of h-BN Membrane Filtration for Water Purification

Fnding of new ways to modify the performance of water desalination methods and diminish the capital and operating expenditures is a major issue in scientific researches [69]. The molecular simulations can prepare a novel insight, not available from experiments, which presents keys to the extension of new separation methods [70]. Garnier et al. [71] attempted to connect interfacial properties to water transfer across the BN and graphene membranes. They studied the surface tension of water on the graphene and BN surfaces using MD simulations. The local surface tension of water on the membranes was determined, which can be seen in Fig. 1. The results showed smaller surface tension on BN (7 mN.m^{-1}) in comparison with graphene (40 mN.m^{-1}), which leads to superior wetting of water on BN. They found that more desirable interactions between water molecules and atoms on BN membrane (B and N atoms) resulted in smaller surface tension on BN surface compared to graphene surface.

The pressure-driven MD simulations were carried out to comprise the pure water transfer across the nanoporous BN and graphene membranes. A nanopore with a diameter of 7 Å and a surface area of 67.2 Å^2 was carved in BN and graphene membranes. The results revealed that under the same condition, nanoporous BN possesses higher water permeability than graphene as shown in Fig. 2. The obvious gap in the surface tension influenced directly the water permeability across nanoporous membranes due to superior wetting, which enhances water molecule passage through the nanopore of BN membrane.

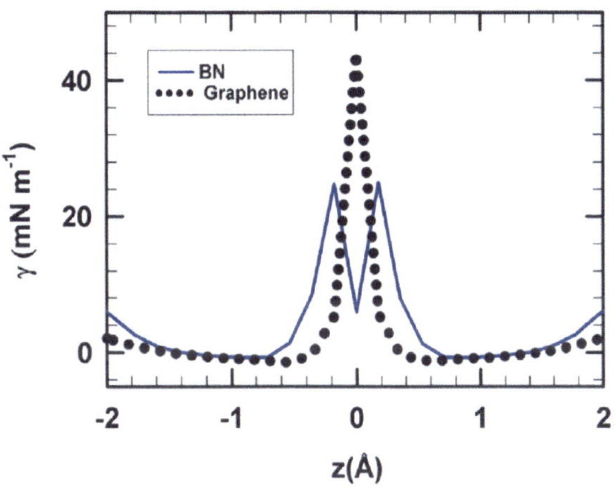

Fig. 1 Local surface tension of water on nanoporous graphene and BN membranes. The figure is adapted with permission from the American Chemical Society [71]

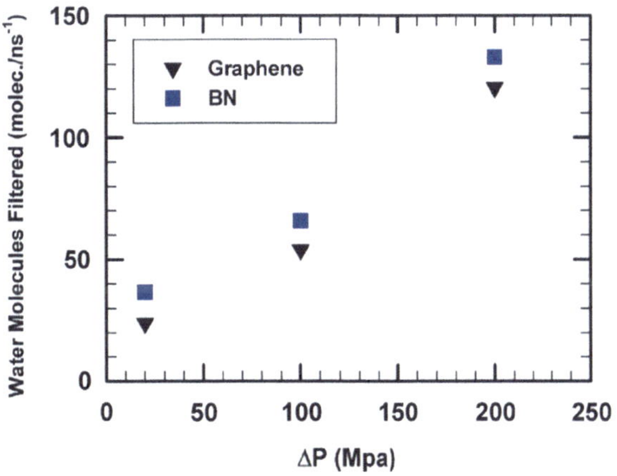

Fig. 2 The number of transferred water molecules through the nanoporous graphene and BN membranes. The figure is adapted with permission from the American Chemical Society [71]

In the other work, MD simulations were used by Gao et al. [72] to study the water desalination through the BNNS membrane. They created six equilaterals triangular nanopores on the membrane with two types of pore edges, as indicated in Fig. 3. The pore areas of the membranes are in the range of 42.1–97.7 $Å^2$ and have the sequence N3 ≈ B3 < N4 ≈ B4 < N5 ≈ B5. The external pressure as a driving force was employed on the system for water transport. The simulation box, containing 2170 water molecules and 18 Na^+ and 18 Cl^-, is also illustrated in Fig. 3, where the water and ions were uniformly placed on both sides of the membrane. The salt concentration was 27 g/L, whereas the salinity of seawater is about ~35 g/L. To study the efficiency of a membrane in water desalination, a well tradeoff between water permeability and salt rejection is an important criterion. Actually, flux passes membrane scales inversely related to the membrane thickness, and 2D nanoporous materials are very promising in this issue owing to their atomic thickness. Gromacs 4.5.5 software was used to carry out MD simulations and the TIP3P potential was employed for

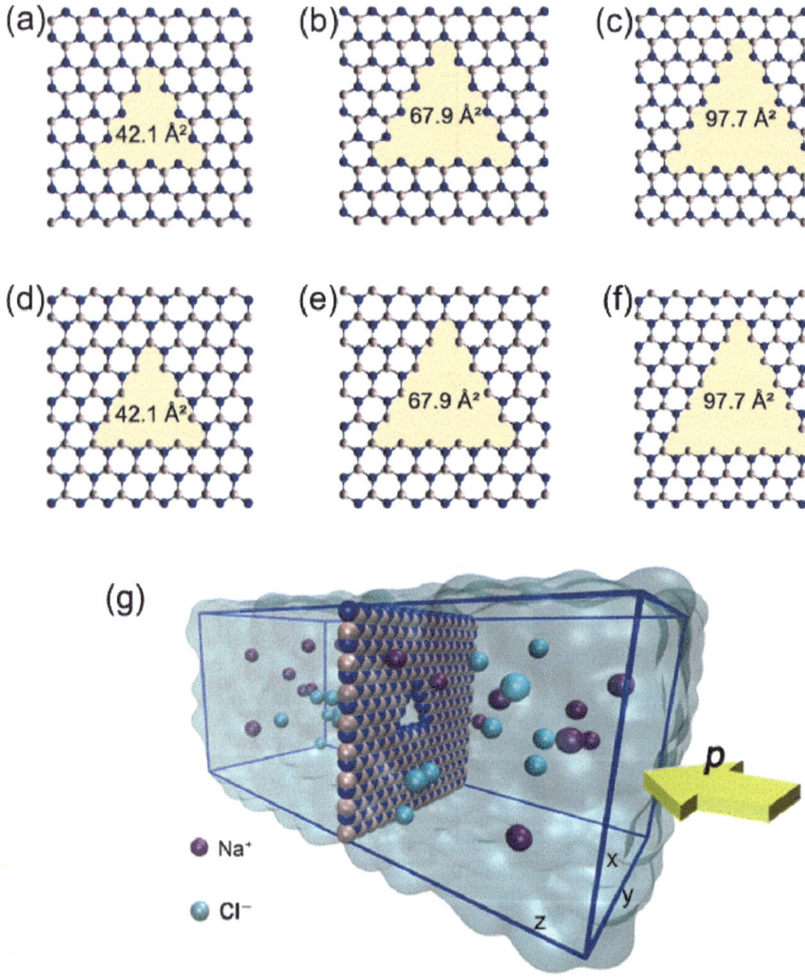

Fig. 3 Schematics and diameters of the **a** N3, **b** N4, **c** N5, **d** B3, **e** B4, and **f** B5 pores of the h-BN membranes. **g** Illustration of a simulation box for the N4 membrane, where the blue shade shows the water molecules. The figure is adapted with permission from the American Chemical Society [72]

water modeling. Excellent permeability, selectivity, and controllability were seen due to the rationally h-BN membrane's design. Indeed, the results demonstrated that the water flux and salt rejection can be regulated by pore size, edge chemistry, and applied pressure. Figure 4 exhibits that the net passed water molecules increases by increasing external pressure and the pore size. And the water permeability was higher for the system including only N atoms at the pore edge compared to the boron-edged system.

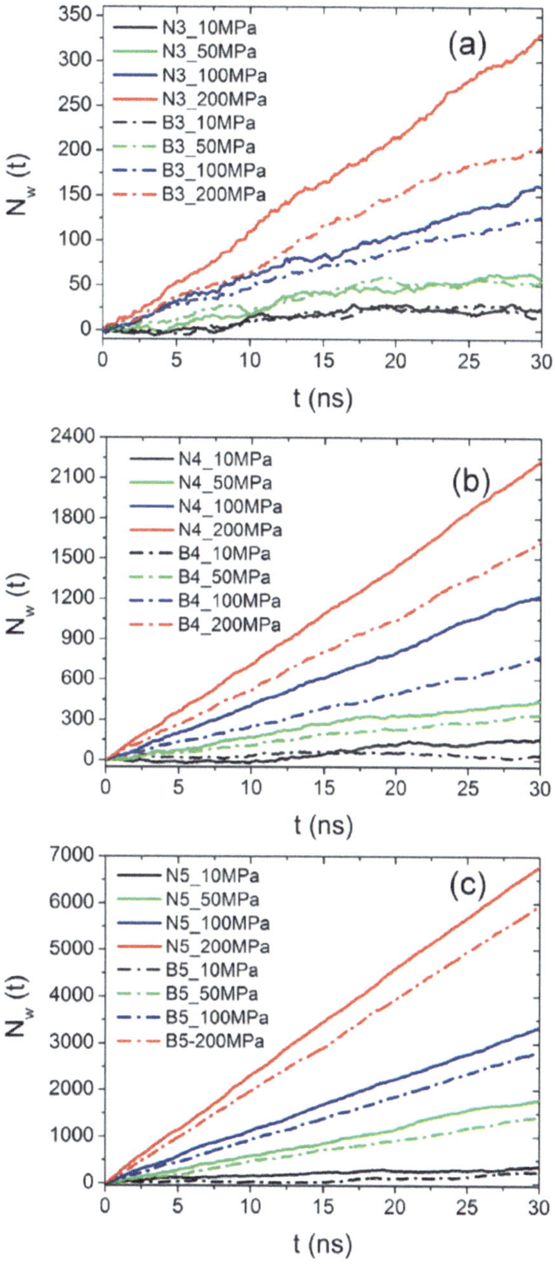

Fig. 4 The number of water molecules transferring across the **a** N3 and B3, **b** N4 and B4, and **c** N5 and B5 membranes at applied pressure ranging from 10 to 200 MPa. The figure is adapted with permission from the American Chemical Society [72]

Gu et al. [2] using MD simulation demonstrated rapid water permeability and effective salt rejection of nanoporous BN. Two types of systems including the triangular nanopores with N–H edges and those with B–H edges were investigated. Additionally, three different pore sizes were selected for each of the pore types as X–H–n (X = N, B; and n = 4, 5, 6). The X–H–n pore has n X–H pairs at each side of the triangular nanopore. The areas of N–H–4, B–H–4, N–H–5, B–H–5, N–H–6, and B–H–6 pores were calculated about 0.231, 0.234, 0.397, 0.401, 0.607, and 0.611 nm^2, respectively.

The system containing a total of 8300 water molecules and 0.599 M NaCl in saltwater was set and MD simulation was conducted by GROMACS software. Periodic boundary conditions were employed in all directions. The number of permeated water molecules through the nanoporous BN membrane was monitored during the simulation time. The external pressures at the range of 50–250 MPa were applied by the piston. Results revealed that the water flux enhances by increasing the applied pressure. Furthermore, the number of water molecules transferring across the B–H pores is higher than that of N–H pores at the same pressure, which is in accordance with pores area. Also, the results showed that when the pore size narrowed to N–H–4 pore, the permeability of water molecules starts to shut off. Besides, the salt rejection performance of pores was studied and results exhibited that B–H–4 pore has the highest salt rejection, which reaches 100%. They concluded that the orientation of water molecules close to the pores plays a prominent role, which affects water permeability and salt rejection.

In the other study by Davoy et al. [1], MD simulations were used to prepare a microscopic vision of the ions effect on water permeability across sub-nanometer boron nitride (sN-BN) monolayers. By elimination of few atoms at the center of BN monolayer, four sN-BN membranes with different geometries and pore sizes were made as shown in Fig. 5. Nonequilibrium MD (NEMD) simulations were performed using two rigid pistons (graphene walls), on which a given pressure was applied. The membrane was placed at the middle of the system, which was surrounded by

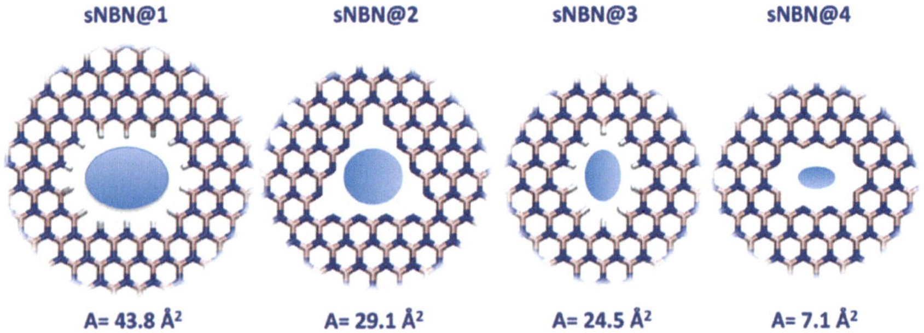

Fig. 5 View of the four sN-BN membranes (the pore area is presented for each membrane). sN-BN@1 and sN-BN@3 are hydrogenated pores, whereas sN-BN@2 and sN-BN@4 are dehydrogenated pores. B, N, and hydrogen atoms are shown in pink, blue and, white colors, respectively. The figure is adapted with permission from the American Chemical Society [1]

two water sections including 1800 water molecules. Additionally, 20 Na^+ and 20 Cl^- ions were initially set down on one side. Figure 6 demonstrates the illustration of the simulated system. All simulations were performed by DIPOLE package and periodic boundary conditions were used in three directions.

As can be seen in Fig. 7, sN-BN@4 as the thinnest pore was found to be more permeable compared to the wider sN-BN@3 pore. It revealed the difference among the water permeability of various sN-BN membranes cannot be described only by the varied pore size. Water flow through 2D materials is affected by the interfacial interactions and frictions between membrane and water molecules [73]. The results showed that the salt rejection rates by sN-BN@3 and sN-BN@4 are near to 100%.

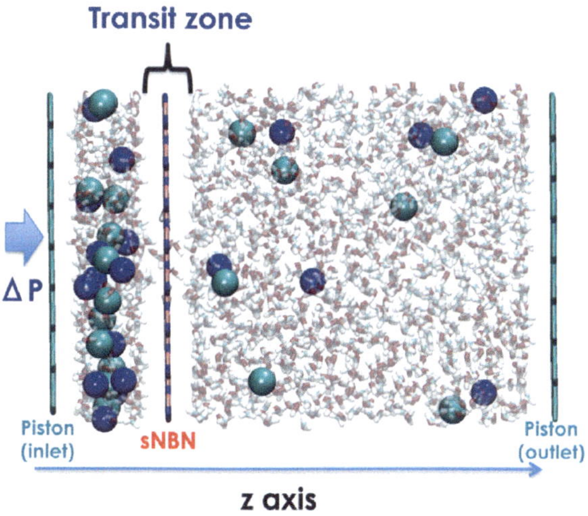

Fig. 6 Schematic of a simulated system. Cyan and blue colors correspond to the Cl^- and Na^+ ions. The figure is adapted with permission from the American Chemical Society [1, 2]

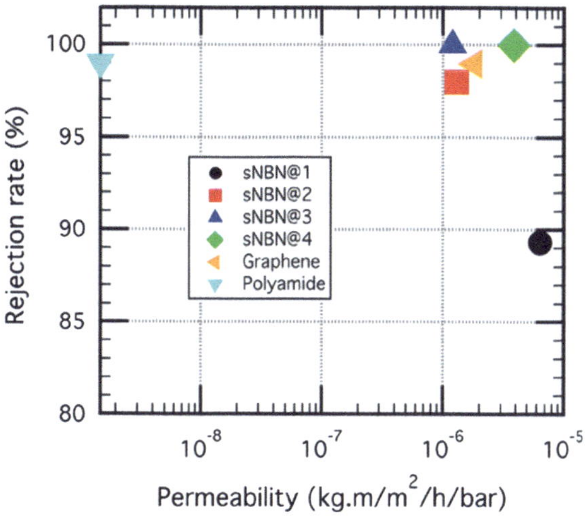

Fig. 7 Rejection rate of NaCl solution by various membranes versus water permeability. The figure is adapted with permission from the American Chemical Society [1, 2]

Therefore, the best efficiency was achieved by sN-BN@4 membrane. Water transport across sN-BN membranes was also found very high in comparison with that of common reverse osmosis (RO) polyamide membranes.

The water desalination performance of functionalized nanoporous BNNS membrane was evaluated using the MD simulation method [74]. The membrane was built in 30 × 30 Å^2 and a pore was developed in the BNNS membrane. Four chemically functionalized BNNS membranes were considered as shown in Fig. 8. The developed pores surface area was in the range of 6.2–15.8 Å^2. Functionalized membrane was located in the center of the system in an aqueous ionic solution consists of 26 Na^+ and 26 Cl^- ions to obtain the seawater salinity. An applied pressure was

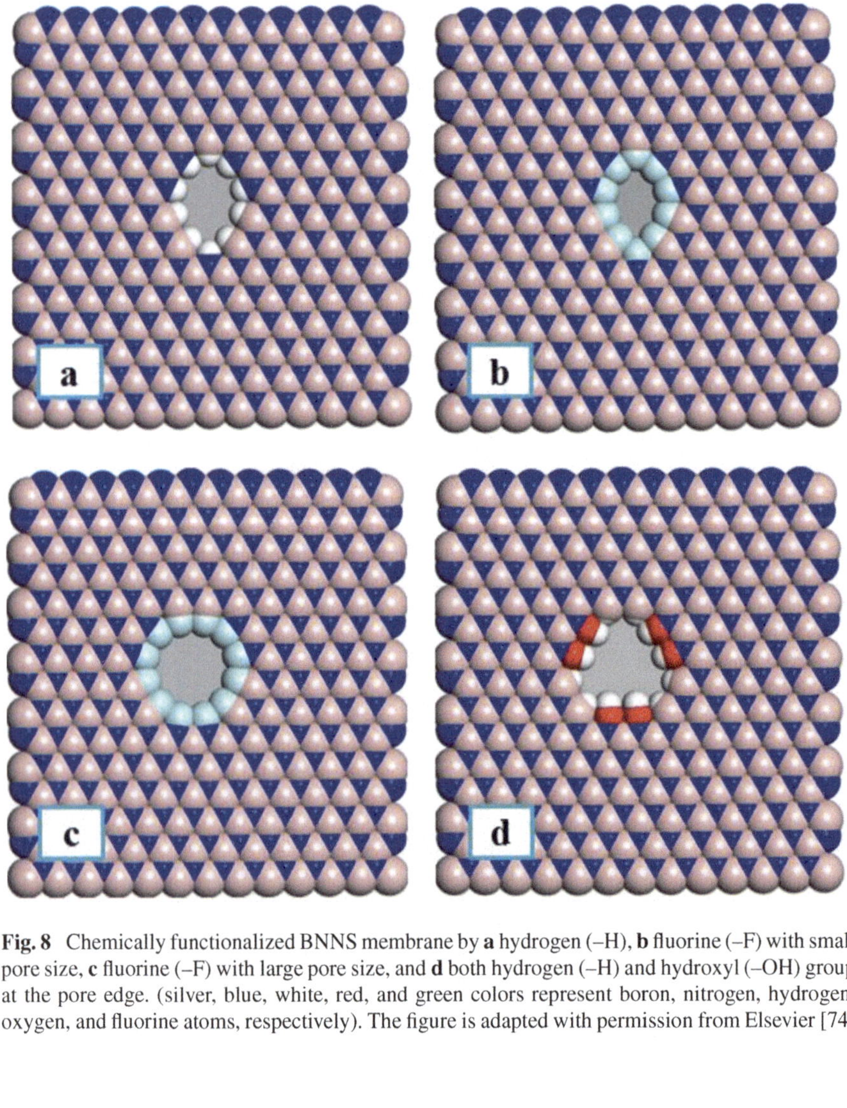

Fig. 8 Chemically functionalized BNNS membrane by **a** hydrogen (–H), **b** fluorine (–F) with small pore size, **c** fluorine (–F) with large pore size, and **d** both hydrogen (–H) and hydroxyl (–OH) group at the pore edge. (silver, blue, white, red, and green colors represent boron, nitrogen, hydrogen, oxygen, and fluorine atoms, respectively). The figure is adapted with permission from Elsevier [74]

Table 1 The permeability of the BNNS membranes [74]

Pore	System	Pore area (Å^2)	Permeability (L/m^2.h.bar)
a	System 1	14	663.16
b	System 2	6.2	224.78
c	System 3	15.8	3283.16
d	System 4	15.2	1219.53

exerted to the system for salt rejection in the range of 10–100 MPa. All MD simulations were carried out by NAMD 2.11 package with applying periodic boundary conditions in all directions. As illustrated in Table 1, the pore size showed a notable influence on the water permeability, which revealed that the chemical function types at the pore edge is a key factor.

The results revealed that the functionalities such as –OH and –F groups can enhance water permeability. Moreover, the salt rejection for studied systems at different applied pressures was indicated in Fig. 9, in which the fluorine modified BNNS membrane demonstrated the best performance for salt rejection. However, because of its small pore area (Table 1), the water permeability is also low. Furthermore, the large fluorine pore (pore c) with the highest water permeability is not proper for water desalination. The results demonstrated that pore d (hydrogen and hydroxyl pore) is more effective in the separation of salt from water compared with

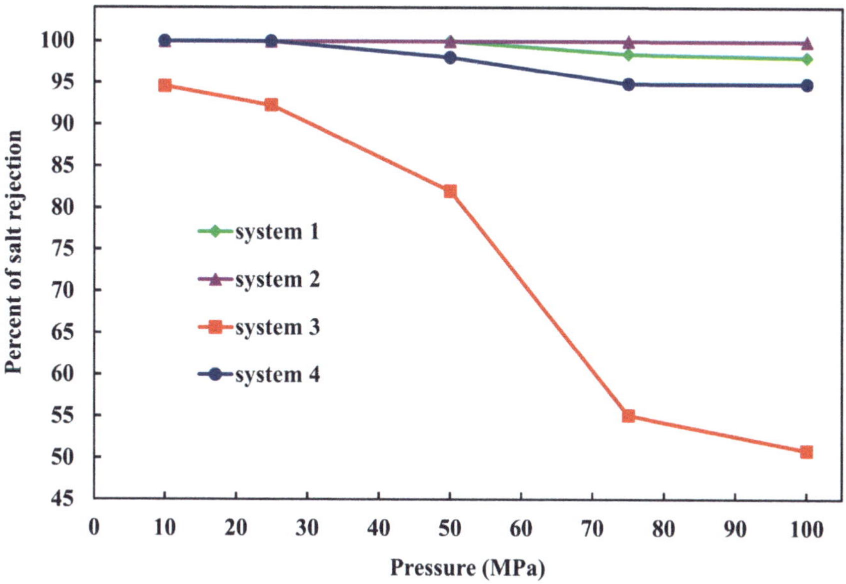

Fig. 9 The salt rejection percent at different applied pressure. The figure is adapted with permission from Elsevier [74]

other pore types at the applied pressures ranging up to 50 MPa. Actually, partial charges of membrane atoms as well as hydrophobicity/hydrophilicity properties of the membranes can affect the water permeability and salt rejection factors.

All of the above studies exhibited that water can permeate through the porous BNNSs, while ion transport can be blocked due to the larger size of the ion hydration shell compared to the molecular size of membrane pore.

3.2 Simulation of h-BN in Pollutants Adsorption Applications

Srivastava et al. [75] represented that the BNNSs are a promising material for removing arsenic ions from water. They performed MD simulations to study the adsorption capacity of As^{3+} on BN and graphene nanosheets. The simulation system was assumed as an aqueous solution of $As(NO_3)_3$ ionic salts, which comprising 3000 water molecules and the number of As^{3+} ions varying from 3 to 38 depending on ion concentration. All the MD simulations were performed by the LAMMPS package and the periodic boundary conditions were used in the three directions.

They found that the adsorption of arsenic ions follows the Langmuir isotherm. The results exhibited that the adsorption capacity of As^{3+} increases with rising ion concentration, which is greater on the BNNSs compared to graphene nanosheets. Additionally, the lower value of diffusion coefficient of As^{3+} ions on BNNSs compared to graphene indicated strong adsorption behavior of As^{3+} ions on BNNSs. The potential of mean force (PMF) calculations was also conducted, which depicted two distinct minima, the first one close the nanosheet is called the contact minima of ion and nanosheet, and the next one is called the solvent layer separated minima. The h-BN nanosheet showed lower contact minima for arsenic ion in comparison with graphene nanosheet, which demonstrated sturdy interaction between arsenic ion and BNNSs. These results disclosed BNNS is a preferable adsorbent relating to graphene nanosheet in the field of arsenic ions removing from water.

4 Experimental Work

4.1 Application of h-BN as Membrane Filtration in Water Treatment

Recently, considerable efforts have been given to investigate the improvement of water permeability by developing the new nanomaterial [76, 77]. This section highlights the performance of h-BN membrane in the field of water treatment and investigates the regeneration of h-BN for reusing. The composite membrane with extremely high separation performance based on porous BNNSs was developed by Liu et al. [78]. They examined Polyvinylidene fluoride/BNNS (PVDF/BNNS) composite on the treatment of synthetic wastewater contaminated with either oil, pharmaceuticals, or a dye. The morphology of the composite was evaluated by SEM and TEM in order to characterize the influence of porous BNNSs on the PVDF membrane. Oil/water emulsions with droplet size as feed were entered into a small system set-up with the composite membrane as indicated in Fig. 10. The droplets were not observed in the filtrate, which exhibited the high efficiency of the prepared composite membrane in de-emulsifying. The results showed the permeance value of oil/water through the composite membrane as high as 2 × 104 $L.\mu m.m^{-2}.h^{-1}.bar^{-1}$ with about 100% of separation performance (Fig. 11). The composite also demonstrated excellent recoverability as can be seen in Fig. 11.

The permeation of water solutions contaminated with an organic dye, methylene blue (MB), and pharmaceuticals through the composite membrane was also studied. The composite membrane demonstrated high permeability in comparison to PVDF

Fig. 10 The pressure-driven filtration system for separation of an oil/water emulsion, which only water passes through the PVDF/BNNSs composite membrane. The figure is adapted with permission from John Wiley & Sons, Inc. [78]

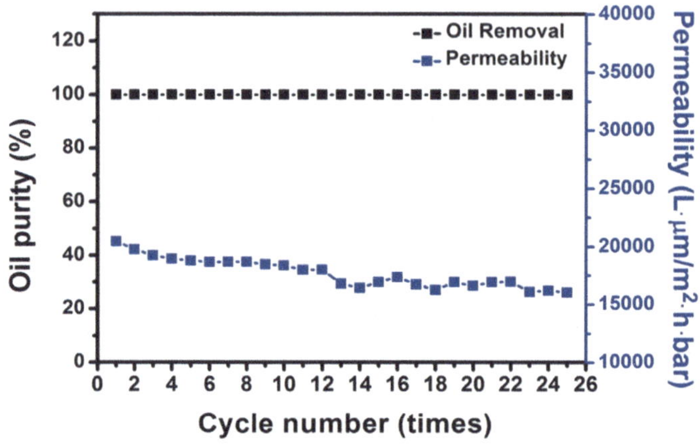

Fig. 11 Plots of the membrane permeability and oil removal of the filtrate. The figure is adapted with permission from John Wiley & Sons, Inc. [78]

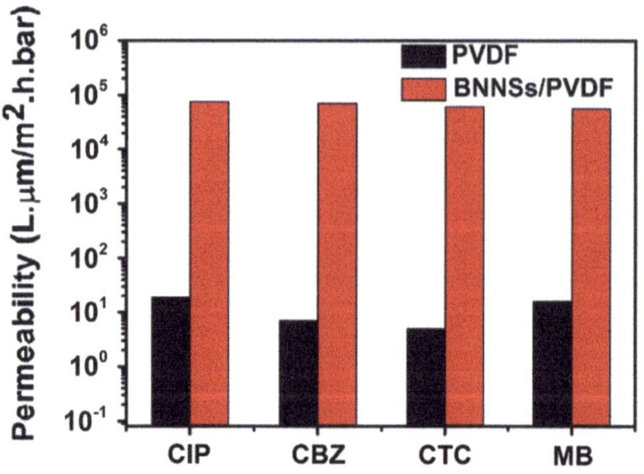

Fig. 12 Comparison of organic solvents permeability between PVDF and PVDF/BNNSs composite membranes. The figure is adapted with permission from John Wiley & Sons, Inc. [78]

membrane as shown in Fig. 12. Actually, the electrostatic attractions of the BN surface and dye/pharmaceuticals are the reason for excellent filter performance. The results indicated that the porous PVDF/BNNSs composite membrane has not only great potential in effective separation of oil, pharmaceuticals, and dyes but also has high recyclability.

4.2 Application of h-BN in Adsorption

A highly porous BNNS by a thermal treatment process without using of any catalyst was synthesized by Lei et al. [3], which showed high surface area (1,427 $m^2.g^{-1}$). Results indicated that both porous and non-porous BNNSs have a much higher absorbent performance for oils compared to other commercial bulk BN and activated carbon, which exhibits that the nanosheet structure influences the absorption capability and increases it. Due to the synergy of swelling and porosity features, the porous BNNSs showed a higher adsorption capacity compared with non-porous BNNSs. The porous BNNSs exhibited excellent adsorption of varied types of solvents, oils, and dyes, with mass uptakes values ranging from 2000 to 3300 wt%, which can be promising in the treatment of contaminated water. Furthermore, the saturated BNNSs can be reused several cycles for environmental restoration via regeneration. Considering the intense resistance of the BNNSs, the adsorbed materials can be simply eliminated from the collected BNNSs by heating in the furnace and directly burning in air. The BN adsorbents are very sturdy and can be reused many times without loss of activity. The simple recycling of BNNSs suggests the eminent potential it in wastewater treatment applications.

Li et al. [79] synthesized microporous and mesoporous BN materials through a simple two-step method, which exhibited ultrahigh specific surface area ($1687\ m^2.g^{-1}$) and a large pore volume ($0.99\ cm^3.g^{-1}$). The morphology and structural properties of the BN samples were checked applying XRD, SEM, FTIR, and SEM. They studied porous BN performance in water treatment by introducing 100 mg porous BN material into 250 ml of model dye solution with the initial concentration of $40\ mg.l^{-1}$. Results revealed that ~88 wt% of model dye is prominently eliminated from the solution after 5 min, and ~99 wt% of model dye is adsorbed after 2 h at room temperature. The obtained data were also fitted to the Langmuir isotherms, which has been extensively applied to characterize the adsorption of pollutants from solutions. The correlation coefficient and the maximum adsorption capacity of porous BN for model dye were calculated to be 0.991 and $298.3\ mg.g^{-1}$, respectively.

Moreover, the porous BN was used to remove copper ions from solutions. They investigated the adsorption capacity of copper ions by mixing 250-ml solution with copper ion concentration of $1.86\ mg.g^{-1}$ and 1 g porous BN, which results indicated $373\ mg.g^{-1}$ maximum removing capacity of copper ions. The BN exhibited outstanding potential for adsorption of model dye and copper ions, attributing to pore volume, superior surface area, and structural defects of porous BN. Additionally, regeneration of collected porous BN was easily performed after the treatment process by calcining at 350 °C for 2 h in air.

In the other work, an activated oxygen-rich porous BNNSs (OBNNSs) was synthesized by Li et al. [80] for improving the adsorption performance of BNNSs in water purification. They tested the morphology and structural properties of the samples using XRD, SEM, and TEM. In order to compare, the adsorption performances of commercial granular activated carbon (GAC) as well as bulk BN were also measured. In order to reach the contaminated water solution, each of $Pb(NO_3)_2$, $HgCl_2$, $CrCl_3$, and $CuCl_2$ was dissolved in water and then diluted to obtain the needed concentration. They performed adsorption kinetic tests to approximate the metallic ions adsorption rates in the OBNNSs, GAC, and bulk BN. As depicted in Fig. 13, the elimination of Pb(II) by the OBNNSs enhanced quickly by rising contact time.

GAC and bulk BN showed very low adsorption efficiency in contrast to the OBNNSs during the same time. Furthermore, the adsorption performance of various

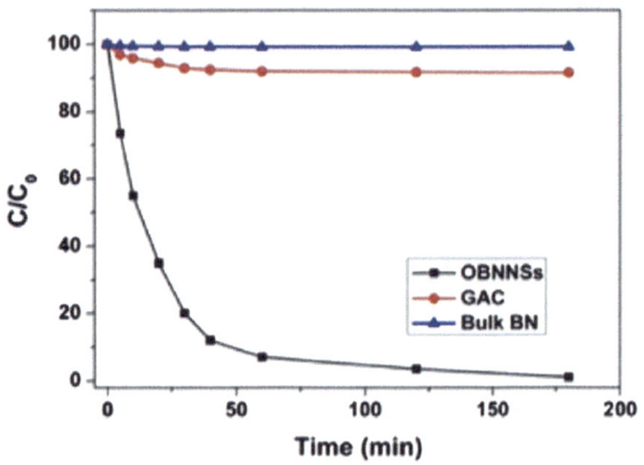

Fig. 13 Adsorption rates of Pb(II) on the OBNNSs, GAC and bulk BN (C_0 and C render the initial PB(II) concentration and those at various time distances during the adsorption, respectively). The figure is adapted with permission from Elsevier [80]

metallic ions (Pb^{2+}, Cr^{3+}, Hg^{2+}, and Cu^{2+}) was studied. The OBNNSs exhibited the same high efficiency as that of Pb(II) which can be seen in Fig. 14. The pivotal reason for the high removing capacity of the OBNNSs can be mainly considered by its illustrious polarity of B–O bonds and boron atoms positions.

As we know, to achieve the practical application of adsorbents, the possibility of several cycles' usage of the adsorbent is significant. In this regard, the regeneration of the Pb^{2+}-containing OBNNSs was also examined by simple acid elution. The results showed the removing efficiency of about 67% even after 10 cycles as shown in Fig. 15. This is due to its corrosion resistance, super anti-oxidation, and structural strength of OBNNSs. Therefore, this supreme recyclability indicates that OBNNSs are a promising nanomaterial for the elimination of the metallic ions from polluted water.

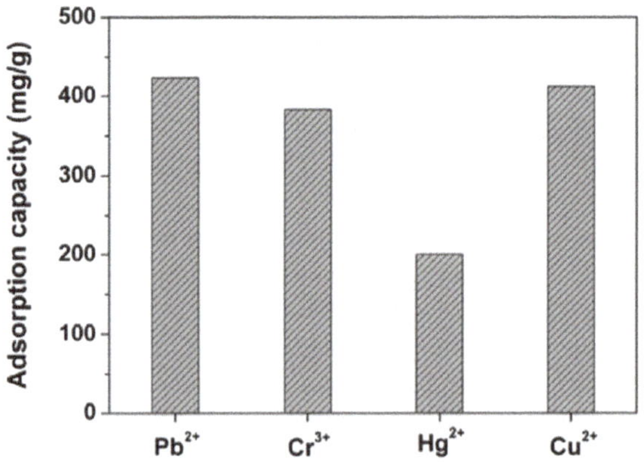

Fig. 14 Comparison of adsorption ability of OBNNSs for Pb^{2+}, Cr^{3+}, Hg^{2+}, and Cu^{2+}. The figure is adapted with permission from Elsevier [80]

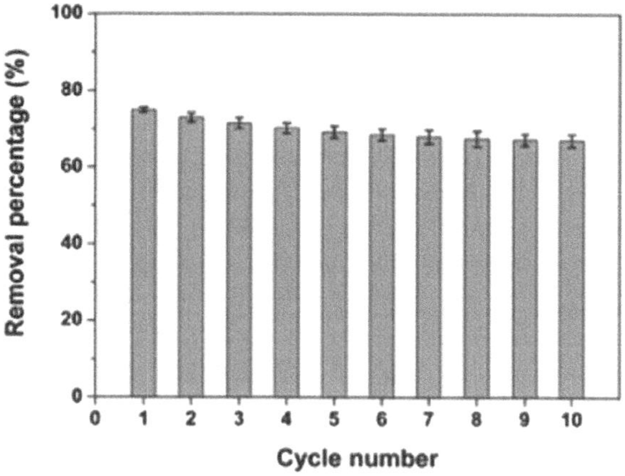

Fig. 15 Recyclable of the OBNNSs regenerated by simple acid elution. The figure is adapted with permission from Elsevier [80]

5 Comparative Study of Graphene Oxide and h-BN in Membrane Filtration

Recently, 2D nanomaterials, such as graphene oxide (GO) [81–83], boron nitride (BN) [84, 85], and transition metal dichalcogenide [86, 87] have been reported in membrane technology because of their particular chemical and physical properties. For instance, GO is suitable for synthesis filter membranes due to its atomic thin thickness, lamellar structure, and consequent particular features, such as high specific surface and chemical stability, its porous and rich oxygen-containing functional sites [88, 89]. In addition, GO-based membranes are most capable to separate ions and contaminants from water by regulation of interlayer space [90–92].

With advances in the development of 2D materials, chiefly graphene-based materials, 2D h-BN has been extremely attended as a type of various and outstanding 2D materials, because of its attractive features, which are distinct from bulk h-BN. BN, like graphene, has a hexagonal atomic structure, which is designed with intense covalent sp^2 bonds. On the contrary, BN shows a naturally inorganic character, while graphene exhibits a moderately organic one; hence, BN is commonly treated as chemically more inert in comparison to graphene [93].

Class 1 membranes, such as GO have demonstrated the efficient potency in the elimination of solutes from an aqueous solution at high fluxes and separation of the gas molecules with different kinetic diameters. This proposes that the external micro porosity covers ranging from ultra-micropore to supermicropore (0.7–2.0 nm) [94] based on different nanosheets and stacking modes. Nair et al. [95] investigated the passage of water molecules across the GO membranes with submicron thickness. The non-oxidized section of the GO laminate can cause a nanoscale capillary network for increasing water transfer through GO sheets, while the oxide region supplies mechanical support to maintain the laminate structure with particular interlayer spacing [93].

Regarding the exclusive and attractive features of the h-BN, such as high strength and chemically inert, conductivity, and thermal stability, it is a promising 2D nanomaterial in membrane technology [96–98]. The performance of BN laminates for adsorptive separation of oils, organic solvents, and water was studied by Lei et al. [3]. They performed a dynamic templating method to obtain the layered structure of BN. The obtained BN with a remarkable total surface area (1427 m^2/g) and pore volume demonstrated the possibility of the BN nanosheet in membrane-based applications. Very recently, nanostructural h-BN with the excellent surface area has been confirmed to be an original and appealing sorbent material for efficient elimination of organic pollutants from water [99, 100]. Actually, this is generally due to the eminent polarity of B–N bonds, lightweight, thermal stability, exceptional chemical, and ultrahigh surface area of h-BN, which suggests great adsorption performance for varied organic pollutants [101, 102].

On the other hand, GO has a surface area of around 2630 m^2/g [103], which is higher than that of h-BN, and the hydrophobicity of the un-oxidized region of the GO carbon basal plane may offer high adsorption capability towards organic foulants. In addition to the high surface area, the GO membrane surface has a water contact angle of less than 20° [104] and thus is much more hydrophilic than h-BN whose contact angle values reported in experiments are in the range of 40°–55° [105]. Such great hydrophilicity of the GO membrane causes reduced interactions between the membrane surface and foulants, which results in the decrement of the membrane fouling tendency [104, 106].

Besides the nanostructural similarity of h-BN to the graphene-based materials, the functionalization of h-BN materials can be discovered. Unlike the functionalization of graphene-based materials, which can refer to many organic chemicals, the functionalization of h-BN is not curreent with chemically active sites in organic chemistry [107]. Functionalization of h-BN nanosheets is essential to avoid them from the robust propensity to aggregation and to modify their interfacial properties and dispersity in solvents and polymer composites [108].

6 New Fabrication Methods

6.1 Physical Vapor Deposition

In addition to synthesis methods mentioned above, 2D substances have also been exhibited to nucleate and rise crystalline, highly consecutive films through physical vapor deposition (PVD) methods. In this method, the plasma produced inside a high vacuum chamber can display ion and electron energies in the order of 1–100 eV. The metastable and complicated structures, like 2D substance planes, were created by these processes with controlling the rate and concentration of species inside the plasma [109]. In sputtering, a thin film on a wafer facing the target is prepared by bombarding the solid target of the chosen film material with energetic particles. To achieve high-quality and few-layer films, Ar ion beams are employed to sputter the h-BN on Cu foils. Moreover, magnetron sputtering of B in N_2/Ar is used to produce high-quality h-BN on Ru. This procedure leads to two atomic film layers thick, which thicker films can be grown by adjustment of ambient temperature deposition and annealing cycles [110, 111].

High area coverage, compatibility in device manufacturing, and decrease in a required temperature for h-BN formation are some advantages of the high-energy plasmas usage compared to conventional CVD techniques [112]. Radio-frequency (RF) magnetron sputtering of a B target is used by Sutter et al. [113] to commit mono- and few-layer h-BN in an ultrahigh-vacuum N_2/Ar atmosphere. Applying Ru(0001) substance causes N and B species orientation to take into an ordered h-BN layered film. Undoubtedly, PVD method offers a distinct way to obtain the high-quality and wafer-scale manufacturing of 2D layered h-BN. However, it is needed to optimize the crystallinity of the obtained nanosheets and the type of substrates.

6.2 Surface Segregation-Based Method

Surface segregation is well known as a practicable method in manufacturing high surface area 2D layered substances, which was first planned to fabricate 2D h-BN [114]. Using heat treatment in a high-vacuum condition, few-layer h-BN nanosheets were grown by suppling the electropolished Fe–Cl–Ni alloy doped with N and B. Triangular h-BN fields were seen on the surface of alloy with a layered form, which offers the controllability of grain morphologies, structural orientations and number of layers. The structure of Ni(C)/(B, N)/Ni squeezed growth material and the annealing temperature are the two key controlling parameters in this method. The diffusing amount of N and B atoms to the metal surface enhances with increasing the temperature, which uses to control the h-BN thickness.

Surface segregation of the different layers, while the manufacturing of in-plane heterostructures requires newly developed synthesis methods. Several methods including direct chemical conversion, templated growth, and patterned regrowth have been studied for the synthesis of in-plane Gr-h-BN heterostructures. Considering the segregation growth of individual h-BN and graphene thin films, direct growth of h-BN/Gr vertical heterostructures by a co-segregation technique was reported [115].

6.3 Pulsed Laser Deposition

The pulsed laser deposition (PLD) is a substituted method for PVD procedure to reactive sputtering and molecular beam epitaxial. In the present method, fast heating on the targeted surface is obtained by a focused laser beam pulse with high energy. The ionized and neutral components, in the form of a normal plasma plume, accelerate to the target surface. The plasma plume spread through the substrate and, regarding the pressure and distance from target to surface, can be thermalized by collisions with the background gas [112].

Glavin et al. [116] carried out the PLD method to synthesize very thin and crystalline h-BN from an amorphous BN target. The study was conducted with the aim of the growth temperature decreasing and well-stoichiometric h-BN nanosheets achievement. They deposited the thin h-BN films on the sapphire substances and highly ordered pyrolytic graphite (HOPG). Then, the high area film was covered without the separation of the micro-sized domains or any pinholes.

Alternatively, Velazquez et al. [117] fabricated mono and few-layer h-BN by Ag films, which performed as a layer on $SiTiO_3(001)$ with the laser parameters including 266 nm, 100 mJ per pulse, and 5 ns. The neutrally weak interaction of the Ag substrates and few-layer h-BN led to the exfoliation of the as-grown h-BN layers by a sticky tape. Although the h-BN sheets synthesized by the PLD method could carry out the large-area production without grain boundaries or deformations, the obtained crystallinity from this method is fairly low compared to that derived from liquid exfoliation or CVD.

7 Research Gap

Notwithstanding the excellent developments up to date, there are also notable ambiguities to be explored in the field of using h-BN for water purification. For instance, h-BN as a foundational substance plays an important role in the heterostructures investigation, however, the ideal quality and size of the h-BN film are still not achieved and the greatest single-crystal domain size is only less than 1 mm. Hence, extending the size of h-BN domain is strongly required, especially for freestanding nanoporous h-BN membrane fabrication; but has remained as a great challenge.

The quality and scale-up possibility of 2D h-BN membranes are also serious requisitions for industrial applications. Furthermore, another important issue about this type of membranes is the suffering of CVD-grown substances from inherent deficiencies, wrinkles, and grain boundaries. Therefore, more studies are needed on this subject [83].

The 2D membranes like GO and graphene have indicated successful performance in the separation of various liquid and gas mixtures. In this regard, there is a quick development of data organization available for such materials, which helps to guide new 2D materials fabrication e.g., h-BN. In fact, most of these 2D materials have not yet been investigated in several separation fields so that further study information is required in this field.

8 Conclusions

The 2D h-BN nanosheets have been a topic of remarkable awareness to consider the typical 2D materials and investigate the practical usage, because of their exclusive structure and characteristics including low density, high mechanical strength, great thermal conductivity, low friction coefficient, electrical insulation, and excellent chemical inertness. The structural and functional features of h-BN can well be operated by merging or embedding carbon and graphene due to their atomic similarities. Thus, various classes of 2D heterostructures and incorporating based on graphene and h-BN nanosheets have been created with adjustable permeation, chemical, and mechanical characteristics.

Recently, considerable molecular simulations were carried out to investigate the water purification performance of h-BN membrane. Using molecular simulations as an ideal experimental system, allows us to access the microstructure of a system and to provide a microscopic picture of the water permeation through the membrane. A tradeoff between water permeability and salt rejection is an important key to assess the membrane performance in water purification applications. All performed simulations demonstrated that BNNS possesses outstanding water desalination capacity with efficient water permeability and high salt rejection. Adsorption of arsenic ions on the h-BN membrane has been also studied using molecular simulations, which revealed that the arsenic ions adsorption is more favorable on the BNNSs, due to their strong interaction. The results suggested that h-BN membrane has overall better desalination performance in comparison to existing commercial techniques, which make it a promising candidate for future desalination membrane filters and also, removing arsenic ions from aqueous media. The thorough review represented that the types of 2D h-BN made by different fabrication methods, which the conventional ones comprise exfoliation, CVD, and gas-phase epitaxy, with numerous other new approaches, have been prosperously extended over the past years. Until now, both in-plane and vertical stacked h-BN have been successfully synthesized by exfoliation transfer processes based on top-down methods or bottom-up strategies associated with CVD approaches. Several other procedures, like mechanical and liquid exfoliation, PVD, surface segregation, and PLD, have been developed to produce various quality and configurations of 2D h-BN.

Moreover, some experimental works were successfully developed to investigate the adsorption capacity of h-BN in water treatment. The h-BN indicated great adsorption potential for a wide range of oils, model dyes, and various metallic ions. As we know, the regeneration of adsorbent is a key factor, which was also examined by simple methods. The results exhibited that the porous BNNSs membrane possesses not only excellent potential in the effective adsorption of oils, metallic ions, and dyes, but also has high recoverability. Actually, these adsorbed materials can be easily removed from the BNNSs. It demonstrated that h-BN can be used as hopeful material in water treatment.

References and Future Readings

1. Davoy X, Gellé A, Lebreton J-C, Tabuteau H, Soldera A, Szymczyk A, Ghoufi A (2018) High water flux with ions sieving in a desalination 2D sub-nanoporous boron nitride material. ACS Omega 3:6305–6310
2. Gu Z, Liu S, Dai X, Chen S, Yang Z, Zhou R (2018) Nanoporous Boron nitride for high efficient water desalination, bioRxiv 500876
3. Lei W, Portehault D, Liu D, Qin S, Chen Y (2013) Porous boron nitride nanosheets for effective water cleaning. Nature Commun 4:1777
4. Zhang K, Feng Y, Wang F, Yang Z, Wang J (2017) Two dimensional hexagonal boron nitride (2D-hBN): synthesis, properties and applications. J Mater Chem C 5:11992–12022
5. Li X, Wu X, Zeng XC, Yang J (2012) Band-gap engineering via tailored line defects in boron-nitride nanoribbons, sheets, and nanotubes. ACS Nano 6:4104–4112
6. Azamat J, Sattary BS, Khataee A, Joo SW (2015) Removal of a hazardous heavy metal from aqueous solution using functionalized graphene and boron nitride nanosheets: Insights from simulations. J Mol Graph Model 61:13–20
7. Giovannetti G, Khomyakov PA, Brocks G, Kelly PJ, Van Den Brink J (2007) Substrate-induced band gap in graphene on hexagonal boron nitride: Ab initio density functional calculations. Phys Rev B 76:
8. Meric I, Dean CR, Petrone N, Wang L, Hone J, Kim P, Shepard KL (2013) Graphene field-effect transistors based on boron–nitride dielectrics. Proc IEEE 101:1609–1619
9. Gao R, Yin L, Wang C, Qi Y, Lun N, Zhang L, Liu Y-X, Kang L, Wang X (2009) High-yield synthesis of boron nitride nanosheets with strong ultraviolet cathodoluminescence emission. J Phys Chem C 113:15160–15165
10. Zhi C, Bando Y, Tang C, Huang Q, Golberg D (2008) Boron nitride nanotubes: functionalization and composites. J Mater Chem 18:3900–3908
11. Wang Y, Shi Z, Yin J (2011) Boron nitride nanosheets: large-scale exfoliation in methanesulfonic acid and their composites with polybenzimidazole. J Mater Chem 21:11371–11377
12. Tang C, Bando Y, Liu C, Fan S, Zhang J, Ding X, Golberg D (2006) Thermal conductivity of nanostructured boron nitride materials. J Phys Chem B 110:10354–10357
13. Azamat J, Khataee A, Sadikoglu F (2016) Separation of carbon dioxide and nitrogen gases through modified boron nitride nanosheets as a membrane: insights from molecular dynamics simulations. RSC Advances 6:94911–94920
14. Watanabe K, Taniguchi T, Kanda H (2004) Direct-bandgap properties and evidence for ultraviolet lasing of hexagonal boron nitride single crystal. Nat Mater 3:404
15. Pacile D, Meyer J, Girit Ç, Zettl A (2008) The two-dimensional phase of boron nitride: few-atomic-layer sheets and suspended membranes. Appl Phys Lett 92:
16. Britnell L, Gorbachev RV, Jalil R, Belle BD, Schedin F, Katsnelson MI, Eaves L, Morozov SV, Mayorov AS, Peres NM (2012) Electron tunneling through ultrathin boron nitride crystalline barriers. Nano Lett 12:1707–1710
17. Neek-Amal M, Beheshtian J, Sadeghi A, Michel K, Peeters F (2013) Boron nitride monolayer: a strain-tunable nanosensor. J Phys Chem C 117:13261–13267
18. Pakdel A, Bando Y, Golberg D (2014) Nano boron nitride flatland. Chem Soc Rev 43:934–959
19. Yin J, Li J, Hang Y, Yu J, Tai G, Li X, Zhang Z, Guo W (2016) Boron nitride nanostructures: fabrication, functionalization and applications. Small 12:2942–2968
20. Wang H, Zhao Y, Xie Y, Ma X, Zhang X (2017) Recent progress in synthesis of two-dimensional hexagonal boron nitride. J Semicond 38:
21. Xu M, Liang T, Shi M, Chen H (2013) Graphene-like two-dimensional materials. Chem Rev 113:3766–3798
22. Sperber JL (2016) Investigations of hexagonal boron nitride: bulk crystals and atomically-thin two dimensional layers, Kansas State University
23. Novoselov KS, Jiang D, Schedin F, Booth T, Khotkevich V, Morozov S, Geim AK (2005) Two-dimensional atomic crystals. Proc Natl Acad Sci 102:10451–10453

24. Zhang Y, Small JP, Pontius WV, Kim P (2005) Fabrication and electric-field-dependent transport measurements of mesoscopic graphite devices. Appl Phys Lett 86:
25. Bunch JS, Yaish Y, Brink M, Bolotin K, McEuen PL (2005) Coulomb oscillations and Hall effect in quasi-2D graphite quantum dots. Nano Lett 5:287–290
26. Berger C, Song Z, Li T, Li X, Ogbazghi AY, Feng R, Dai Z, Marchenkov AN, Conrad EH, First PN (2004) Ultrathin epitaxial graphite: 2D electron gas properties and a route toward graphene-based nanoelectronics. J Phys Chem B 108:19912–19916
27. Island JO, Steele GA, der Zant HS, Castellanos-Gomez A (2015) Mechanical manipulation and exfoliation of boron nitride flakes by micro-plowing with an AFM tip, arXiv preprint arXiv:1501.06437
28. Novoselov KS, Geim AK, Morozov SV, Jiang D, Zhang Y, Dubonos SV, Grigorieva IV, Firsov AA (2004) Electric field effect in atomically thin carbon films. Science 306:666–669
29. Pakdel A, Zhi C, Bando Y, Golberg D (2012) Low-dimensional boron nitride nanomaterials. Mater Today 15:256–265
30. Castellanos-Gomez A, Agraït N, Rubio-Bollinger G (2010) Optical identification of atomically thin dichalcogenide crystals. Appl Phys Lett 96:213116
31. Wang YY, Gao RX, Ni ZH, He H, Guo SP, Yang HP, Cong CX, Yu T (2012) Thickness identification of two-dimensional materials by optical imaging. Nanotechnology 23:495713
32. Venkatachalam DK, Parkinson P, Ruffell S, Elliman RG (2011) Rapid, substrate-independent thickness determination of large area graphene layers. Appl Phys Lett 99:
33. Gorbachev RV, Riaz I, Nair RR, Jalil R, Britnell L, Belle BD, Hill EW, Novoselov KS, Watanabe K, Taniguchi T (2011) Hunting for monolayer boron nitride: optical and Raman signatures. Small 7:465–468
34. Warner JH, Rummeli MH, Bachmatiuk A, Büchner B (2010) Atomic resolution imaging and topography of boron nitride sheets produced by chemical exfoliation. ACS Nano 4:1299–1304
35. Zhi C, Bando Y, Tang C, Kuwahara H, Golberg D (2009) Large-scale fabrication of boron nitride nanosheets and their utilization in polymeric composites with improved thermal and mechanical properties. Adv Mater 21:2889–2893
36. Li X, Hao X, Zhao M, Wu Y, Yang J, Tian Y, Qian G (2013) Exfoliation of hexagonal boron nitride by molten hydroxides. Adv Mater 25:2200–2204
37. Du M, Wu Y, Hao X (2013) A facile chemical exfoliation method to obtain large size boron nitride nanosheets. CrystEngComm 15:1782–1786
38. Hummers WS Jr, Offeman RE (1958) Preparation of graphitic oxide. J Am Chem Soc 80:1339
39. Han W-Q, Wu L, Zhu Y, Watanabe K, Taniguchi T (2008) Structure of chemically derived mono-and few-atomic-layer boron nitride sheets. Appl Phys Lett 93:
40. Nagashima A, Tejima N, Gamou Y, Kawai T, Oshima C (1995) Electronic structure of monolayer hexagonal boron nitride physisorbed on metal surfaces. Phys Rev Lett 75:3918
41. Kim G, Jang A-R, Jeong HY, Lee Z, Kang DJ, Shin HS (2013) Growth of high-crystalline, single-layer hexagonal boron nitride on recyclable platinum foil. Nano Lett 13:1834–1839
42. Basche M, Schiff D (1964) New pyrolytic boron nitride. Mater Design Eng 59:78–81
43. Rand MJ, Roberts JF (1968) Preparation and properties of thin film boron nitride. J Electrochem Soc 115:423–429
44. Pierson HO (1975) Boron nitride composites by chemical vapor deposition. J Compos Mater 9:228–240
45. Rozenberg A, Sinenko YA, Chukanov N (1993) Regularities of pyrolytic boron nitride coating formation on a graphite matrix. J Mate Sci 28:5528–5533
46. Auwärter W, Suter HU, Sachdev H, Greber T (2004) Synthesis of one monolayer of hexagonal boron nitride on Ni (111) from B-trichloroborazine (ClBNH) 3. Chem Mater 16:343–345
47. Müller F, Stöwe K, Sachdev H (2005) Symmetry versus commensurability: epitaxial growth of hexagonal boron nitride on Pt (111) from B-trichloroborazine (ClBNH) 3. Chem Mater 17:3464–3467
48. Lee KH, Shin H-J, Lee J, Lee I-Y, Kim G-H, Choi J-Y, Kim S-W (2012) Large-scale synthesis of high-quality hexagonal boron nitride nanosheets for large-area graphene electronics. Nano Lett 12:714–718

49. Tay RY, Park HJ, Ryu GH, Tan D, Tsang SH, Li H, Liu W, Teo EHT, Lee Z, Lifshitz Y (2016) Synthesis of aligned symmetrical multifaceted monolayer hexagonal boron nitride single crystals on resolidified copper. Nanoscale 8:2434–2444
50. Song X, Gao J, Nie Y, Gao T, Sun J, Ma D, Li Q, Chen Y, Jin C, Bachmatiuk A (2015) Chemical vapor deposition growth of large-scale hexagonal boron nitride with controllable orientation. Nano Research 8:3164–3176
51. Shi Y, Hamsen C, Jia X, Kim KK, Reina A, Hofmann M, Hsu AL, Zhang K, Li H, Juang Z-Y (2010) Synthesis of few-layer hexagonal boron nitride thin film by chemical vapor deposition. Nano Lett 10:4134–4139
52. Gao Y, Ren W, Ma T, Liu Z, Zhang Y, Liu W-B, Ma L-P, Ma X, Cheng H-M (2013) Repeated and controlled growth of monolayer, bilayer and few-layer hexagonal boron nitride on Pt foils. ACS Nano 7:5199–5206
53. Park J-H, Park JC, Yun SJ, Kim H, Luong DH, Kim SM, Choi SH, Yang W, Kong J, Kim KK (2014) Large-area monolayer hexagonal boron nitride on Pt foil. ACS Nano 8:8520–8528
54. Pakdel A, Bando Y, Golberg D (2013) Morphology-driven nonwettability of nanostructured BN surfaces. Langmuir 29:7529–7533
55. Pakdel A, Zhi C, Bando Y, Nakayama T, Golberg D (2011) Boron nitride nanosheet coatings with controllable water repellency. ACS Nano 5:6507–6515
56. Yu J, Qin L, Hao Y, Kuang S, Bai X, Chong Y-M, Zhang W, Wang E (2010) Vertically aligned boron nitride nanosheets: chemical vapor synthesis, ultraviolet light emission, and superhydrophobicity. ACS Nano 4:414–422
57. Shannon RE (1998) Introduction to the art and science of simulation. In: 1998 winter simulation conference. Proceedings (Cat. No. 98CH36274), IEEE, pp 7–14
58. Maria A (1997) Introduction to modeling and simulation, Winter simulation conference, pp 7–13
59. Saghatchi R, Ghazanfarian J, Gorji-Bandpy M (2014) Numerical simulation of water-entry and sedimentation of an elliptic cylinder using smoothed-particle hydrodynamics method. J Offshore Mech Arct Eng 136:
60. Zen A, Luo Y, Mazzola G, Guidoni L, Sorella S (2015) Ab initio molecular dynamics simulation of liquid water by quantum Monte Carlo. J Chem Phys 142:
61. Hasanzadeh A, Pakdel S, Azamat J, Erfan-Niya H, Khataee A (2020) Atomistic understanding of gas separation through nanoporous DDR-type zeolite membrane. Chem Phys, 110985
62. Rapaport DC, Rapaport DCR (2004) The art of molecular dynamics simulation. Cambridge University Press
63. Ebro H, Kim YM, Kim JH (2013) Molecular dynamics simulations in membrane-based water treatment processes: a systematic overview. J Membr Sci 438:112–125
64. Metropolis N, Rosenbluth AW, Rosenbluth MN, Teller AH, Teller E (1953) Equation of state calculations by fast computing machines. J Chem Phys 21:1087–1092
65. Hasanzadeh A, Azamat J, Pakdel S, Erfan-Niya H, Khataee A (2020) Separation of noble gases using CHA-type zeolite membrane: insights from molecular dynamics simulation. Chemical Papers, pp 1–9
66. Azamat J, Khataee A, Sadikoglu F (2018) Computational study on the efficiency of MoS2 membrane for removing arsenic from contaminated water. J Mol Liq 249:110–116
67. Jafarzadeh R, Azamat J, Erfan-Niya H (2018) Fluorine-functionalized nanoporous graphene as an effective membrane for water desalination. Struct Chem 29:1845–1852
68. Andersen HC (1980) Molecular dynamics simulations at constant pressure and/or temperature. J Chem Phys 72:2384–2393
69. Werber JR, Osuji CO, Elimelech M (2016) Materials for next-generation desalination and water purification membranes. Nature Reviews Materials 1:16018
70. Azamat J, Khataee A, Joo SW (2014) Functionalized graphene as a nanostructured membrane for removal of copper and mercury from aqueous solution: a molecular dynamics simulation study. J Mol Graph Model 53:112–117
71. Garnier L, Szymczyk A, Malfreyt P, Ghoufi A (2016) Physics behind water transport through nanoporous boron nitride and graphene. J Phys Chem Lett 7:3371–3376

72. Gao H, Shi Q, Rao D, Zhang Y, Su J, Liu Y, Wang Y, Deng K, Lu R (2017) Rational design and strain engineering of nanoporous boron nitride nanosheet membranes for water desalination. J Phys Chem C 121:22105–22113
73. Michaelides A (2016) Nanoscience: slippery when narrow. Nature 537:171
74. Jafarzadeh R, Azamat J, Erfan-Niya H, Hosseini M (2019) Molecular insights into effective water desalination through functionalized nanoporous boron nitride nanosheet membranes. Appl Surf Sci 471:921–928
75. Srivastava R, Kommu A, Sinha N, Singh J (2017) Removal of arsenic ions using hexagonal boron nitride and graphene nanosheets: a molecular dynamics study. Mol Simul 43:985–996
76. Das R, Vecitis CD, Schulze A, Cao B, Ismail AF, Lu X, Chen J, Ramakrishna S (2017) Recent advances in nanomaterials for water protection and monitoring. Chem Soc Rev 46:6946–7020
77. Xu G-R, Xu J-M, Su H-C, Liu X-Y, Zhao H-L, Feng H-J, Das R (2019) Two-dimensional (2D) nanoporous membranes with sub-nanopores in reverse osmosis desalination: Latest developments and future directions. Desalination 451:18–34
78. Liu D, He L, Lei W, Klika KD, Kong L, Chen Y (2015) Multifunctional polymer/porous boron nitride nanosheet membranes for superior trapping emulsified oils and organic molecules. Adv Mater Interf 2:1500228
79. Li J, Lin J, Xu X, Zhang X, Xue Y, Mi J, Mo Z, Fan Y, Hu L, Yang X (2013) Porous boron nitride with a high surface area: hydrogen storage and water treatment. Nanotechnology 24:155603
80. Li J, Jin P, Dai W, Wang C, Li R, Wu T, Tang C (2017) Excellent performance for water purification achieved by activated porous boron nitride nanosheets. Mater Chem Phys 196:186–193
81. Huang K, Liu G, Lou Y, Dong Z, Shen J, Jin W (2014) A graphene oxide membrane with highly selective molecular separation of aqueous organic solution. Angew Chem Int Ed 53:6929–6932
82. Huang L, Huang S, Venna SR, Lin H (2018) Rightsizing Nanochannels in reduced graphene oxide membranes by solvating for dye desalination. Environ Sci Technol 52:12649–12655
83. Mohammad A, Asiri AM (2017) Inorganic pollutants in wastewater: methods of analysis, removal and treatment. Mater Res Forum LLC
84. Fan Y, Yang Z, Hua W, Liu D, Tao T, Rahman MM, Lei W, Huang S, Chen Y (2017) Functionalized Boron Nitride Nanosheets/Graphene Interlayer for Fast and Long-Life Lithium-Sulfur Batteries. Adv Energy Mater 7:1602380
85. Falin A, Cai Q, Santos EJ, Scullion D, Qian D, Zhang R, Yang Z, Huang S, Watanabe K, Taniguchi T (2017) Mechanical properties of atomically thin boron nitride and the role of interlayer interactions. Nature Commun 8:15815
86. Hirunpinyopas W, Prestat E, Worrall SD, Haigh SJ, Dryfe RA, Bissett MA (2017) Desalination and nanofiltration through functionalized laminar MoS_2 membranes. ACS Nano 11:11082–11090
87. Pan L, Liu YT, Xie XM, Ye XY (2016) Facile and Green Production of Impurity-Free Aqueous Solutions of WS2 Nanosheets by Direct Exfoliation in Water. Small 12:6703–6713
88. Fathizadeh M, Xu WL, Zhou F, Yoon Y, Yu M (2017) Graphene oxide: a novel 2-dimensional material in membrane separation for water purification. Adv Mater Interf 4:1600918
89. Hosseini M, Azamat J, Erfan-Niya H (2019) Water desalination through fluorine-functionalized nanoporous graphene oxide membranes. Mater Chem Phys 223:277–286
90. Mi B (2014) Graphene oxide membranes for ionic and molecular sieving. Science 343:740–742
91. Jiao S, Xu Z (2017) Non-continuum intercalated water diffusion explains fast permeation through graphene oxide membranes. ACS Nano 11:11152–11161
92. Cheng C, Jiang G, Garvey CJ, Wang Y, Simon GP, Liu JZ, Li D (2016) Ion transport in complex layered graphene-based membranes with tuneable interlayer spacing. Sci Adv 2:
93. Hyun T, Jeong J, Chae A, Kim YK, Koh D-Y (2019) 2D-enabled membranes: materials and beyond. BMC Chem Eng 1:1–26
94. Thommes M, Kaneko K, Neimark AV, Olivier JP, Rodriguez-Reinoso F, Rouquerol J, Sing KS (2015) Physisorption of gases, with special reference to the evaluation of surface area and pore size distribution (IUPAC Technical Report). Pure Appl Chem 87:1051–1069

95. Nair R, Wu H, Jayaram P, Grigorieva I, Geim A (2012) Unimpeded permeation of water through helium-leak–tight graphene-based membranes. Science 335:442–444
96. Qin S, Liu D, Chen Y, Chen C, Wang G, Wang J, Razal JM, Lei W (2018) Nanofluidic electric generators constructed from boron nitride nanosheet membranes. Nano Energy 47:368–373
97. Li Z, Zhang Y, Chan C, Zhi C, Cheng X, Fan J (2018) Temperature-dependent lipid extraction from membranes by boron nitride Nanosheets. ACS Nano 12:2764–2772
98. Weber M, Koonkaew B, Balme S, Utke I, Picaud F, Iatsunskyi I, Coy E, Miele P, Bechelany M (2017) Boron nitride nanoporous membranes with high surface charge by atomic layer deposition. ACS Appl Mater Interf 9:16669–16678
99. Han W-Q, Zettl A (2003) Functionalized boron nitride nanotubes with a stannic oxide coating: a novel chemical route to full coverage. J Am Chem Soc 125:2062–2063
100. Yin J, Li X, Zhou J, Guo W (2013) Ultralight three-dimensional boron nitride foam with ultralow permittivity and superelasticity. Nano Lett 13:3232–3236
101. Zhang X, Lian G, Zhang S, Cui D, Wang Q (2012) Boron nitride nanocarpets: controllable synthesis and their adsorption performance to organic pollutants. CrystEngComm 14:4670–4676
102. Portehault D, Giordano C, Gervais C, Senkovska I, Kaskel S, Sanchez C, Antonietti M (2010) High-surface-area nanoporous boron carbon nitrides for hydrogen storage. Adv Func Mater 20:1827–1833
103. Lambert TN, Chavez CA, Hernandez-Sanchez B, Lu P, Bell NS, Ambrosini A, Friedman T, Boyle TJ, Wheeler DR, Huber DL (2009) Synthesis and characterization of titania—graphene nanocomposites. J Phys Chem C 113:19812–19823
104. Hu M, Zheng S, Mi B (2016) Organic fouling of graphene oxide membranes and its implications for membrane fouling control in engineered osmosis. Environ Sci Technol 50:685–693
105. Essafri I, Le Breton J-C, Saint-Jalmes A, Soldera A, Szymczyk A, Malfreyt P, Ghoufi A (2019) Contact angle and surface tension of water on a hexagonal boron nitride monolayer: a methodological investigation. Mol Simul 45:454–461
106. Liu Q, Xu G-R, Das R (2019) Inorganic scaling in reverse osmosis (RO) desalination: mechanisms, monitoring, and inhibition strategies. Desalination 468:
107. Weng Q, Wang X, Wang X, Bando Y, Golberg D (2016) Functionalized hexagonal boron nitride nanomaterials: emerging properties and applications. Chem Soc Rev 45:3989–4012
108. Liu S, Ji J, Zeng H, Xie Z, Song X, Zhou S, Chen P (2018) Functionalization of hexagonal boron nitride nanosheets and their copolymerized solid glasses. 2D Mater 5:035036
109. Muratore C, Varshney V, Gengler JJ, Hu J, Bultman JE, Roy AK, Farmer BL, Voevodin AA (2014) Thermal anisotropy in nano-crystalline MoS 2 thin films. Phys Chemstr Chem Phys 16:1008–1014
110. Wang H, Zhang X, Meng J, Yin Z, Liu X, Zhao Y, Zhang L (2015) Controlled growth of few-layer hexagonal boron nitride on copper foils using ion beam sputtering deposition. Small 11:1542–1547
111. Sutter P, Lahiri J, Zahl P, Wang B, Sutter E (2013) Scalable synthesis of uniform few-layer hexagonal boron nitride dielectric films. Nano Lett 13:276–281
112. Glavin NR (2016) Ultra-thin boron nitride films by pulsed laser deposition: Plasma diagnostics, synthesis, and device transport
113. Sutter P, Lahiri J, Zahl P, Wang B, Sutter E (2012) Scalable synthesis of uniform few-layer hexagonal boron nitride dielectric films. Nano Lett 13:276–281
114. Xu M, Fujita D, Chen H, Hanagata N (2011) Formation of monolayer and few-layer hexagonal boron nitride nanosheets via surface segregation. Nanoscale 3:2854–2858
115. Zhang C, Zhao S, Jin C, Koh AL, Zhou Y, Xu W, Li Q, Xiong Q, Peng H, Liu Z (2015) Direct growth of large-area graphene and boron nitride heterostructures by a co-segregation method. Nature Commun 6:6519

116. Glavin NR, Jespersen ML, Check MH, Hu J, Hilton AM, Fisher TS, Voevodin AA (2014) Synthesis of few-layer, large area hexagonal-boron nitride by pulsed laser deposition. Thin Solid Films 572:245–250
117. Velázquez D, Seibert R, Man H, Spentzouris L, Terry J (2016) Pulsed laser deposition of single layer, hexagonal boron nitride (white graphene, h-BN) on fiber-oriented Ag (111)/SrTiO3 (001). J Appl Phys 119:

Molybdenum Disulfide and Tungsten Disulfide as Novel Two-Dimensional Nanomaterials in Separation Science

Mateus H. Köhler, João P. K. Abal, Gabriel V. Soares, and Marcia C. Barbosa

Abstract Beyond graphene-based membranes, for water desalination, a vast horizon of new materials has been discovered for solutes separation from water. In this realm, the transition-metal dichalcogenides (TMDs) molybdenum and tungsten disulfide (MoS_2 and WS_2, respectively) stand as promising two-dimensional (2D) materials. Their tailoring for nanofluidics as well as the emerging synthesis and production methods unfold the possibility of applying MoS_2 and WS_2 in modern desalination processes based on 2D membranes. We present here an overview from their theoretical conception to their state-of-the-art applications, highlighting the challenges and opportunities associated with measuring water flow and ionic rejection rates at nanoscale. In a world full of environmental concerns, both the theoretical gaps and experimental perspectives point toward a promising use of MoS_2 and WS_2 as green components in separation technologies, contributing to increase the availability of clean, potable water.

Keywords Transition metal dichalcogenides · Molybdenum disulfide · Tungsten disulfide · Desalination

1 Introduction

Since the very beginning of this century, we have witnessed a race to shrink the dimensions of fluidic devices to the nanometer scale. The use of nanotubes or nanopores made new discoveries on fluid transport possible, putting even classical hydrodynamics in check [1]. In fact, two-dimensional (2D) materials were not even considered beyond the desks of physicists and chemists until quite recently with the rise

M. H. Köhler (✉)
Department of Physics, Federal University of Santa Maria, 97105-900 Santa Maria, Brazil
e-mail: mateus.kohler@ufsm.br

J. P. K. Abal · G. V. Soares · M. C. Barbosa
Institute of Physics, Federal University of Rio Grande do Sul, 91501-970 Porto Alegre, Brazil

R. Das (ed.), *Two-Dimensional (2D) Nanomaterials in Separation Science*,
Springer Series on Polymer and Composite Materials,
https://doi.org/10.1007/978-3-030-72457-3_8

of graphene, a one-carbon-atom-thick structure. The discovery of this fascinating material and its exceptional properties [2, 3] completely changed the game.

While much research is still focused on carbon nanotubes (CNTs) [4], zeolites [5], polymers [6], or ceramics [7] to act as membranes for ion separation, the fact is that graphene and graphene-related materials have taken over the news with promising nanofiltration results. More recently, van der Waals (vdW) assembly of 2D materials has been used to create artificial channels with sub-nanometer-scale precision [8]. Two of these vdW structures, molybdenum disulfide (MoS_2) and tungsten disulfide (WS_2), stand as prominent alternatives to graphene, exhibiting many similar characteristics as atomic thickness, large surface area, mechanical strength, extreme durability, and also the most important: an exotic love-hate relationship with water that leads to high permeation rates. MoS_2 is the most widely employed TMD in a range of applications and has recently been investigated for its potential in separation techniques. This prototypical TMD is made up of a middle layer of molybdenum sandwiched between two sulfur layers with a thickness of ~1 nm and a robust Young's modulus of ~300 GPa [9] (comparable to Young's modulus of steel). TMDs, a family of over 40 materials, are represented by the generalized formula MX_2 and consist of a transition metal (M), for example, Mo, W, or Ni packed between two chalcogens (X) such as S, Se, or Te. The coordination of a transition metal by chalcogens in a TMD structure opens up the possibility for multiple stacking sequences. As illustrated in Fig. 1, a single-layered TMD generally presents either an octahedral or a trigonal

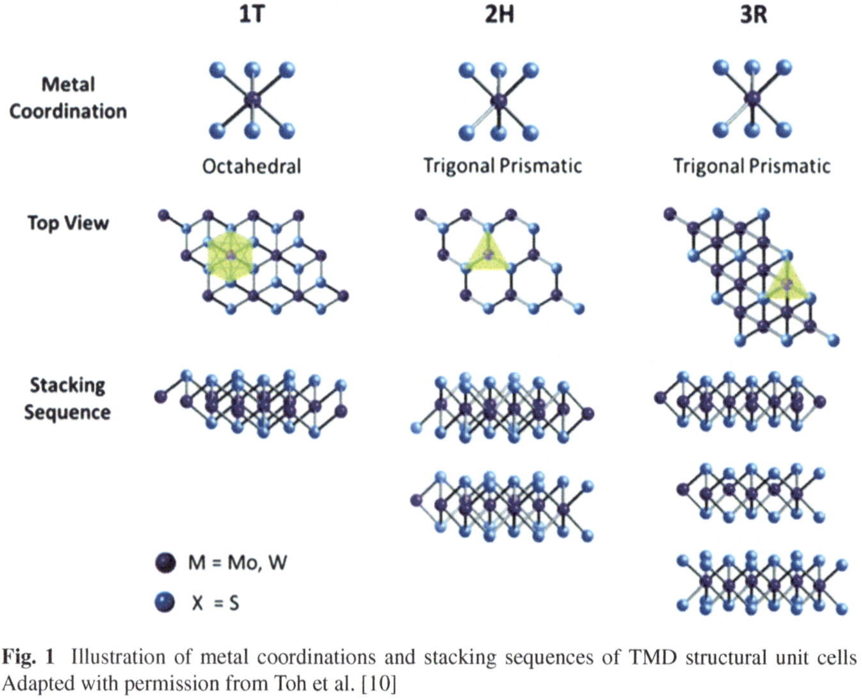

Fig. 1 Illustration of metal coordinations and stacking sequences of TMD structural unit cells Adapted with permission from Toh et al. [10]

prismatic coordination phase. In multi-layered TMDs, a large variety of polymorphic structures arise as each individual layer can possess any of the two coordination phases. The three commonly found polymorphs are defined as 1T, 2H, and 3R, where the digit is indicative of the number of layers in the crystallographic unit cell, and the letter designates the symmetry: T for tetragonal (D_{3d} group), H for hexagonal (D_{3h} group), and R for rhombohedral (C_{3V}^{5} group). The 1T form displays metallic behavior, while both 2H and 3R forms exhibit semiconducting behavior. Each TMD polymorph possesses unique structural and electronic properties, which can be further explored to build efficient adsorbent devices and desalination membranes.

Recent works have reported high water permeability and selectivity of MoS_2 nanosheets, both desired features in desalination membranes. A flexible laminar separation membrane prepared from MoS_2 sheets exhibited a water flux from 3 to 5 times higher than that reported for graphene oxide (GO) and rejected 89 and 98% of Evans blue and cytochrome C molecules, respectively [11]. The possibility to craft the pore edge with Mo, S, or both provides flexibility to design nanopores within the membrane with the desired functionality. Another option for creating a desalination membrane is to use stacking layers of MoS_2 instead of crafting a pore. A few-layer MoS_2 membrane of only 7 nm thick, grown by Chemical Vapor Deposition (CVD) technique, has allowed for an excellent combination of high water permeability (>322 $L \cdot m^{-2} \cdot h^{-1} \cdot bar^{-1}$) and high ionic sieving capability (>99%) for various seawater salts including Na^+, K^+, Ca^{2+}, and Mg^{2+} with a range of concentrations [12]. Near 100% of salt ion rejection rates for actual seawater obtained from the Atlantic coast was also reported, significantly outperforming the previously developed 2D MoS_2 layer membranes of micrometer thickness as well as conventional reverse osmosis (RO) membranes. These results confirmed previous theoretical and computational predictions about desalination capacity and water permeation of 2D nanoporous MoS_2 [13, 14].

Advancing the search for high-performance lamellar separation membranes, Sun et al. [15] investigated the potential of WS_2, a semiconductor material similar to MoS_2, with a Young's modulus of ~270 GPa [16]. The bulk structure was exfoliated and a thin film was constructed via filtration. The lamellar WS_2 membrane exhibited water flux five times greater than GO membranes and two times greater than MoS_2 laminar membranes, rejecting 90% of Evans blue molecules. This impressive water permeance further increased from 450 to 930 $L \cdot m^{-2} \cdot h^{-1} \cdot bar^{-1}$ with the addition of metal hydroxide nanostrands. They created additional channels between the WS_2 layers allowing for increased water transport without degrading the membrane's salt rejection properties.

Understanding the mechanisms and peculiarities of 2D membrane-based desalination is the ultimate frontier to reach industrial scale. So far, cutting-edge theoretical work has been driving advances and pointing directions for experimental work with some success. But we still need more. We need to merge theoretical advances with new experimental approaches, such as the scalable method to controllably make nanopores in single-layer TMDs or nanoscale water velocity profile mapping introduced by Secchi and collaborators [17], always sharing the goal of making large-scale application of MoS_2 and WS_2 membranes in water purification possible. It is important to note that in terms of using nanostructured membranes in desalination technology, such as RO systems, these advances open the door to work with a new paradigm of membrane permeability and selectivity.

2 Computer Simulations

Before computers took over every laboratory—and even our daily life—science was based on the assumption that we could only model the natural world through the lenses of experiments and purely theoretical works. Take the example of the Dutch: they built the Netherlands as we know it today with the help of the Delta Works, a set of megastructures that hold back the ocean. To put together this enormous project they had to build the Waterloopkundig Laboratorium, a massive concrete-based hydrological laboratory conceived after World War II, in the pre-computer age, where water could be guided into and out of large-scale trial models. Contemporarily it is almost impossible to think of a world where things, from a small pen to a huge hydroelectric plant, are not designed on a computer.

During an experiment, a set of results can be obtained directly from measuring the properties of a system. Alternatively, a mathematical description could be used to create a model, which in turn can be validated by its ability to describe some physical behavior. Today, we have another tool to probe a physical system: computer simulations. A model is provided by theorists but the calculations can be carried by machines following some recipe. In this way, computer simulations unlocked the possibility to study more complex and realistic systems, becoming a bridge between theoretical models and real-world experiments [18].

There is no doubt that computer simulations play an important role in contemporary science development. Several different computational approaches can be used to study physical systems. For example, when we aim to investigate nanoscaled structures the interactions between atoms are the core of the simulation. Often, empirical interatomic potentials (as Lennard-Jones potentials) are fitted to reproduce a given experimental property using van der Waals systems as a basis. Thereby, molecular dynamics (MD) simulations are used to obtain the temporal evolution of different systems. But if the interest resides on the electronic structure of a strongly covalent material, electronic correlations are very important and we need to use ab initio methods such as density functional theory (DFT). We can even merge both approaches and use DFT to parameterize Lennard-Jones potentials that will be further used in MD simulations. Method suitability will depend on which kind of properties we are interested in, from infrared spectra to dynamical and thermodynamical information.

Just like a puzzle that fits piece by piece, MD simulations adapted perfectly to the study of physical and chemical properties of nanofluidic systems. They have allowed us to probe a wide range of microscopic behaviors that otherwise would be tremendously difficult to access at nanoscale. Simulations within MD machinery are usually performed in a feasible timescale with high accuracy. In other words, the MD approach is the mechanism in which we can perform computational simulations that take into account thermodynamic and dynamic behaviors of nanofluidic systems. It means that this approach can be viewed as a bridge between the quantum realm (hidden in atomic interactions and sizes) and classical hydrodynamics.

While there are clear benefits to using MD simulations, there are also major challenges. Computational materials scientists have worked hard to design general, accurate, and reliable molecular and atomistic models. For example: in the case of water, the model chosen to represent atomic geometries and interaction parameters is the seed in which the whole dynamic relies on so that the physics can emerge following the classical equations of motion. The mainstream model in MD simulations is based on the Lennard–Jones potential plus a Coulombic term. The task of finding the right model—which is generally suitable only for a handful of specific systems—has haunted theoretical physicists and chemists since the first computer simulation of liquid water with the Bernal–Fowler model in 1969. Even with the help of new experimental data and theories being developed, water is still notoriously hard to model and remains relatively poorly understood, with several anomalous properties—behaviors which contradict general theories on the liquid state of matter.

The interest in nanometric desalination systems has led us to question how the key properties of saline solutions can be captured in a computer simulation (e.g., mixing different interaction parameters). One of the most widely used methodologies to classically simulate saltwater desalination consists of creating a box with the membrane located between two reservoirs, each one pressed by a piston—usually graphene-made. This imposes a controlled pressure gradient in water as illustrated in Fig. 2a. Here we can highlight that (*i*) as we apply different pressures in each reservoir and (*ii*) the solutions are at different concentrations, the system is not under thermodynamic equilibrium. In fact, these simulations aim to find water transport and salt rejection rates acquired at a steady-state flow, which is only possible in non-equilibrium states. Of course, that as the simulation runs the system will eventually reach equilibrium. But to mimic an actual nanoscaled RO scenario the simulation should be far from this point.

In order to filter water using 2D membranes, we can use either the interlayer spacing, forcing water to flow through the structure gaps, or drill holes (nanopores) in the membrane. In any case, it is important to ensure that the nanopore size allows only for the flow of water molecules. Interestingly, nanopores in 2D materials such as TMDs will naturally appear during the growth process. Point defects, grain boundaries, and van der Waals (vdW) gaps, among other structural deformations, have been observed in CVD grown of MoS_2 and WS_2 monolayers [19–21]. At first glance, this might seem like a disadvantage, but the truth is that these "flaws" lead to the emergence of nanopores and nanogaps that can be used to desalinate water and remove heavy metals, biomolecules, and other pollutants. In the case of MoS_2, for instance, these intrinsic defects appear in high concentrations ($\sim 10^{13}$ cm^{-2} for sulfur vacancies [22]). However, these "natural" pores are randomly distributed with varied sizes, which pose a challenge for scalability—a very important aspect in large-scale production.

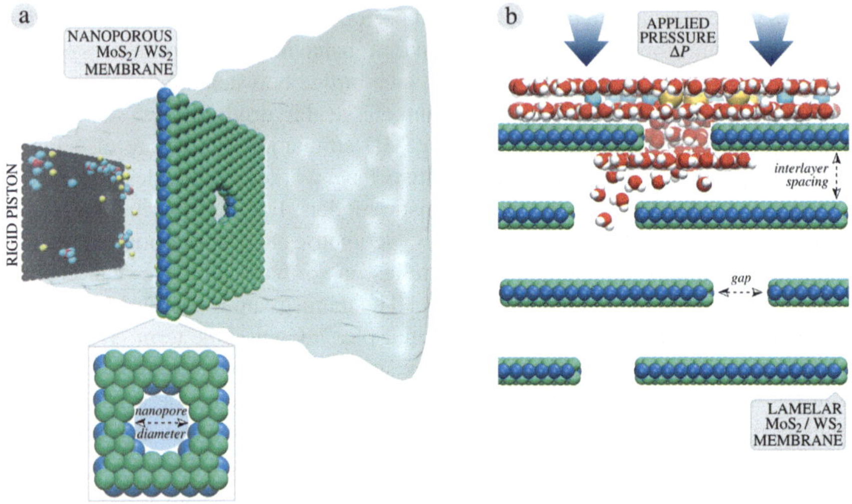

Fig. 2 Archetypal systems used in computer simulations of water desalination with **a** nanoporous and **b** lamellar MoS_2 and WS_2 membranes

MoS_2 and WS_2 nanoporous membranes were idealized by theoretical calculations based on MD simulations, which predicted superior desalination performances. While simulations involving MoS_2 interaction potentials have been more frequent, results for WS_2 are more scarce due to the lack of proper force fields. There are frictional losses in these systems that dominate fluid transport, and this phenomenon is mostly related to the nanopore's entrance. MD simulations have shown that in this region, water flux is intrinsically connected with a complex hydrogen bonding (HB) network and the ability to enter the nanopore is governed by a combination of favorable geometric orientations and HB configurations [23]. Interestingly, water flux scales linearly with pore area when there is a big enough pore size [24]. For smaller pores, a nonlinear relationship between water transport and pore area appears, which implies that the phenomenon cannot be explained based on classical hydrodynamics (continuum fluid model). This happens when the water membrane terms acquire higher importance than the HB network interaction, which is a result of a reasonable fraction of total confined water interacting with the wall—something that does not occur with bigger pores.

We could think of the desalination process through 2D membranes as a collection of small events, all happening at the same time. We could thus highlight the main mechanisms as size exclusion, steric exclusion of the hydration shell, charge repulsion (pore chemistry), nanopore morphology, complex solute-pore interactions, and entropy gradients. The first is directly related to the ion/nanopore size ratio, and the second comes from the fact that these ions in water are surrounded by a hydration shell, which means that in order to enter the nanopore the ions need to bare themselves from the water shell at some energy penalty. Both the pore's chemistry and morphology are going to affect possible HB configurations, that results in an impact on salt permeation. The fifth mechanism can be viewed as the possibility of mixing different ions and observing different salts rejection. At last, entropic differences can be expressed as a mix of nanopore morphology, charge distribution, and so on as a result of the possible configurational restrictions, which in turn affects the free energy barrier through the membrane.

When it comes to cleaning aqueous solutions using membranes, we need to face a range of ions with different valences. Fortunately, MD simulations on solutions of water and three cations with different valences (Na^+, Zn^{2+}, and Fe^{3+}) indicate that the higher the valence, the greater the ion rejection through 2D nanoporous MoS_2 and graphene [13]. The desalination rates were confirmed even under a wide range of test conditions, including high pressures and different nanopore sizes.

The cation charge dependence leads to another question: what if we add more cations with higher valence to the solution? It would seem very inappropriate at first, but surprisingly, simulations have shown that this procedure can lead to higher desalination rates [25]. The possible reason is that the whole membrane-mediated desalination process is based on the assumption that size matters. Furthermore, the combination of larger ions and smaller nanopores (large enough to let water pass) is crucial for efficient selectivity. When high valence ions are added, they aggregate with counterions to form clusters that in time will be rejected by the membrane.

There are physical and chemical aspects that can be decisive either to improve or to hinder 2D membrane-based desalination. Unfortunately, very few experimental studies of ionic conductance through MoS_2 and WS_2 nanoporous membranes with diameters lower than 2.0 nm have been reported. Almost all of our knowledge is based on computational simulations and theoretical models. For instance, one important parameter to control the flow through the nanopore is ionic conductance. Perez et al. [26] developed a continuum model of ionic conductivity for a KCl electrolyte through a sub-5-nm single-layer MoS_2 nanopore using all-atom MD simulations. They showed that electrolyte behavior deviates by 50% from bulk properties for diameters below 2.0 nm: ion pore conductivity is about half of the bulk value for 2.0 nm and only a third when the diameter approaches 1.0 nm. Their results corroborate the idea that the nanopore's size plays a fundamental role in the desalination process.

Usually, layer-stacked membranes made of 2D MoS_2 and WS_2 are synthesized with hundreds to thousands of nanosheets. But this thickness is unlikely to be modeled in a traditional MD simulation due to computational limitations. Alternatively, we can take advantage of the fact that there is a relation between measured flux and membrane thickness. Indeed, if we fit the water flux versus the membrane thickness, we can get a parabolic dependence in a way that we can estimate the experimental value. For instance, Wang et al. [27] noticed that the predicted flux decreases as membrane thickness increases, consistent with their experimental data. They found that as their model membrane thickness increased to ~500 nm, water flux would decrease to around 50 $L \cdot m^{-2} \cdot h^{-1} \cdot bar^{-1}$, matching the experimental value.

Whether measuring mass transport properties or the membrane's mechanical strength, computer simulations have constantly contributed to the prediction and confirmation of 2D membranes' use for desalination. There are many aspects which can make a certain material a good candidate to be used in desalination membranes. High structural strength when subjected to high pressures is certainly one of them. Therefore, as important as the ability to retain ions is the membrane's ability to remain stable under strong external pressures. Unfortunately, an experimental work by Sun et al. [15] showed that nanochannels created in lamellar thin films of WS_2 with the addition of metal hydroxide nanostrands can crack under high pressure. Although first-principles calculations revealed Young's modulus $Y = 200$ GPa for 2D WS_2, comparable to stainless steel, finite-element-based mechanical simulations confirmed the collapse of cylindrical channels at a critical pressure of ~ 1.6 GPa. All of these structural fractures could pose a significant problem regarding the membrane's desalination capacity. Remarkably, the formation of nanocracks within the WS_2 membrane was found to increase water permeance as a consequence of increased porosity. In other words, the crack produces new fluidic nanochannels that further results in water flux four times higher than that of the as-prepared WS_2 membrane, without rejection performance degradation. According to Sun's group, these membranes have separation performances 2 orders of magnitude higher than that of commercial membranes with similar rejection rates.

3 MoS_2 and WS_2 Syntheses and New Fabrication Methods

Usually, 2D materials are the basic building blocks of bulk materials when they are stacked together layer by layer. The possibility to obtain a single layer of carbon (graphene) from bulk graphite has led to exciting new physics and material properties, for instance: atomically thin electrical and thermal conductors, high transparency to visible light, and high carrier mobility [28]. After the first works on graphene isolation by micromechanical exfoliation, several other methods to obtain graphene were investigated, such as molecular beam epitaxy [29], CVD on metal surfaces [30], and graphene films on silicon carbide (SiC) single crystals formed by silicon evaporation [31], among others. Yet, none of these growth methods completely fulfills every basic requirement for applications where suitable size and quality are mandatory, such as in high-performance electronic devices. Still, some of these growth methods such as CVD and silicon evaporation from SiC crystals can be used in applications with less restricted requirements, namely: anti-corrosion coatings and paints [32], nanoporous membranes for desalination [33], and pesticide biosensing [34]. Moreover, the electronic band structure of graphene has a linear dispersion near the K point, where the bottom of the conduction band touches the top of the valence band, making it a zero-gap semiconductor. This provides a great challenge to produce electronic devices for many applications where high "on and off" electrical current ratios are required.

In the past decade, many other 2D materials have been isolated and studied, such as hexagonal boron nitride (h-BN), silicene, phosphorene, and TMDs. TMDs are a class of materials composed of a transition metal from groups IV to VI (such as Ti, Nb, W, Mo, etc.) and a chalcogen (S, Se, or Te). Different from graphene, these 2D materials are composed of three layers of atoms, where a metal plane is bonded to a chalcogen plane on both sides. They usually present a hexagonal crystal structure, where the wide range of possible compositions leads to several distinct optical, electrical, and mechanical properties. For instance, TMDs can present metallic ($NbTe_2$, $TaTe_2$) and superconducting [35] (NbS_2, TaS_2) or semiconducting [36] (MoS_2, WS_2) electronic properties. Interestingly, TMD properties have some degree of reliance upon the number of layers, where the chalcogen–metal–chalcogen structure is considered a monolayer. One of the most common properties dependent on the number of layers is the TMDs' band structure. Many of them present a transition from an indirect (bulk material) to a direct (monolayer) band gap, allowing for optoelectronic device application [37]. One example is the MoS_2 transition from bulk with a 1.3 eV indirect band gap to a 1.8 eV direct band gap in monolayer form [38]. A similar effect is also observed in graphene when it presents a band gap different from zero [39] or superconductivity [40] in a bilayer structure.

The reproducible synthesis of 2D materials is mandatory for characterizing these layer properties, as well as providing a path for their integration into a variety of applications. It is possible to distinguish between the two most widely used methods to synthesize single- and few-layer 2D materials: the top-down and bottom-up methods. The first is based on the exfoliation of layered bulk van der Walls crystals, while the latter is obtained by CVD growth on different substrates. The micromechanical exfoliation has been used routinely to obtain 2D flakes, which may contain several crystal layers to a single monolayer. Van der Walls crystalline solids are usually employed as the base material for exfoliation. The crystal structure is composed of atom layers strongly bonded covalently or ionically in-plane, and these planes are stacked together by weak van der Walls forces. These van der Walls energies are in the range of a few of meV, which enable the easy exfoliation of the atomic layers. This was the approach used to obtain the first graphene layers [2] and it is still used for a wide range of materials, predominantly using a tape as the (dry) exfoliation method. 2D materials such as MoS_2, black phosphorus [41], and h-BN [42] have been obtained using this procedure. The obtained 2D monolayers present the same crystal structure as their bulk counterparts and can remain thermodynamically stable under ambient conditions up until weeks. This method of isolation of individual and few layers still is one of the most widely used to study their properties and for the fabrication of electronic devices, since it can produce monolayers with high quality. Nevertheless, it is mostly limited to crystal sizes in the order of some μm, making it unsuitable for large-area applications.

Conversely, CVD growth is able to produce high-quality 2D materials with controlled size, number of layers, and superior electronic properties. Specifically, this method has been used to produce large-area MoS_2, WS_2, graphene, and h-BN, which can be integrated into the fabrication of nanoelectronic devices.

MoS_2 growth can be obtained using two different precursors, one that contains Mo and another that contains sulfur. The process involves typically MoO_3 and S powders, where the substrate can be a dielectric (such as SiO_2) or conductor (graphene), among others. It consists of heating MoO_3 to become volatile as some sub-oxide (MoO_x) and then reacts with the sulfur vapor to produce MoS_2 of the desired substrate. There are many variations of this process that can lead to MoS_2 layers with sizes up to a few mm. In some cases, the MoO_3 is first deposited on the substrate and then it is submitted to the sulfur vapor annealing. Some works have reported on the deposition of Mo films that were later sulfurized to produce MoS_2 [43]. Modulating the homogeneity and thickness of Mo and/or MoO_3 film determines the quality and thickness of the MoS_2 film. In any case, this CVD approach has been shown to be highly scalable.

In some cases, a single precursor which contains both elements has been used [44]. $(NH_4)_2MoS_4$ can be thermally decomposed on insulating substrates and, similarly to the process described above, can also yield large-area MoS_2 in the presence of sulfur vapor. Moreover, transistors based on this CVD-prepared MoS_2 showed good electronic properties with large on/off ratios [45]. Nevertheless, these values are lower than those obtained from the mechanically exfoliated MoS_2 because the crystal defects shown in the CVD-grown films can be deleterious to carrier mobility [46]. Finally, recent works on a variation of CVD, the metal-organic (MO)CVD using bis(tert-butylimido)-bis(dimenthylamido)molybdenum and dietyl disulfide have produced MoS_2 layers from 1 to 25 nm thick at short deposition times (90 s) and with great uniformity on 50 mm SiO_2/Si wafers [47], making it very promising for large-scale applications.

WS_2 posses similar electrical, mechanical, and optical properties as MoS_2 and can be synthesized in a similar fashion. For instance, the band structure of WS_2 is also dependent on the number of layers since bulk WS_2 is a semiconductor with an indirect band gap of 1.4 eV, while monolayer WS_2 presents a direct band gap of 2.1 eV [48]. It is also possible to obtain WS_2 through micromechanical exfoliation and CVD. Dry exfoliation is very similar as in the case of graphene and MoS_2: a tape is applied to a bulk WS_2 crystal removing from single to few layers of WS_2 which can be transferred on the desirable substrate. Superior crystal and electronic properties are obtained using this method, but it lacks scalability. Chemical (wet) exfoliation has also been reported [49].

The W-based precursors for CVD growth of WS_2 present higher melting points than the ones used in the MoS_2 growth. For instance, W has a melting point of 3422 °C and WO_3 of 1473 °C [43]. This makes WS_2 deposition more challenging than MoS_2 since the Mo-precursors present a lower melting point. Still, the basic approach is similar for both materials: reaction of sulfur vapor with a W-precursor at high temperatures, which is deposited on the substrate.

The most common precursors are WO_3 and sulfur powder. They are heated inside a reactor under an inert gas flow, such as argon. The reactor temperature is usually in the 850–950 °C range. Among the challenges also observed in the growth of MoS_2 (homogeneity, crystal quality, mechanical and electrical properties, layer number control, etc.), one of the main issues for WS_2 growth is the high temperatures employed in its process. In this manner, there is a limitation on the type of substrate that can be used to grow WS_2. Some works have reported on the use of alternative W-precursors in order to lower growth temperature, such as the use of WCl 6 [50]. A great effort must be made in order to grow WS_2 directly on several types of substrates and materials.

4 Feasibility of Free-Standing MoS_2 and WS_2 Membranes in Filtration

Recently, free-standing membranes of 2D materials have attracted attention due to their wide range of applications, including piezoelectricity in MoS_2 [51], single-molecule DNA sensing [52], high Young modulus [9], and water desalination [11, 53]. They can also be used to investigate fundamental properties of 2D materials since free-standing membranes are not influenced by underlying substrates [54–56].

Free-standing WS_2 membranes have also been produced using similar methods. The most used approach to produce these membranes is to use a substrate containing an array of circular or square holes, usually produced by chemical or e-beam lithography. After that, the desirable 2D material (MoS_2 or WS_2) is transferred to this substrate's surface. Since the transference process of CVD 2D materials can be difficult, the use of exfoliated materials is preferred.

Eda et al. [57] have prepared MoS_2 nanosheets by chemical exfoliation using organolithium intercalation and forced hydration. Li intercalates between MoS_2 layers, and when it reacts with water, it increases the plane spacing with hydrogen gas. The resulting MoS_2 is then used to produce layer-stacked membranes. Using a CVD approach, Waduge et al. [56] prepared micrometer-scale apertures in silicon nitride membranes, which were placed above a boat containing MoO_2. A second boat containing sulfur was heated and the carrier gas (Ar) transported the sulfur vapor to the silicon nitride membranes. The growth process takes place at 750 °C leading to selective MoS_2 growth near the apertures.

4.1 MoS_2 and WS_2 Membranes Compared to Other 1D/2D Materials

Figure 3 compares MoS_2/WS_2, graphene, and CNT membranes. One of the most interesting discoveries of our time and hypothesized as fruitful for a large range of future technological applications is the role of dimensionality in determining the properties of a material. Ultrathin 2D nanosheets of layered TMDs are fundamentally intriguing. Aside from presenting electronic properties diverse from the bulk, this group of materials exhibits versatile chemistry in contrast with graphene's chemically inert behavior [35]. Furthermore, 1D materials have attracted a lot of attention once theoretical and experimental observations led to the discovery of anomalous water transport under certain conditions, enabling nanofluidic flow enhancement in CNTs [58]. In this section, the MoS_2 and WS_2 membranes will be compared to other 1D/2D materials from the computational and experimental perspective.

A general 2D material can be used to transform a very thick membrane into two kinds of construction design: a nanoporous membrane or a layer-stacked membrane [59]. In terms of desalination purposes, nanoporous membranes of MoS_2 and WS_2 are very peculiar for they present a mix of hydrophobic and hydrophilic edges, which

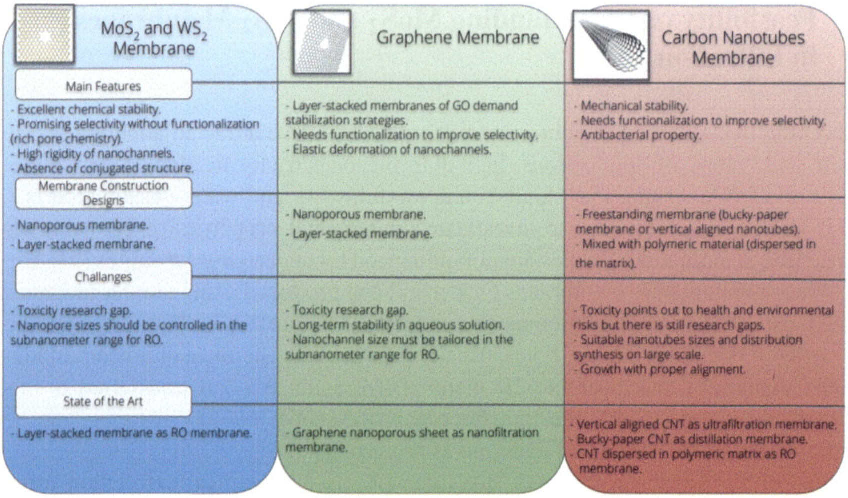

Fig. 3 Comparison between MoS_2/WS_2 [59], graphene [66–68] and CNT membranes [69–71]

in turn can be tuned to enhance water permeability and salt rejection. Heiranian et al. [14] constructed nanoporous membranes with three pore edge types: the first labeled as mixed (a combination of molybdenum and sulfur atoms), and the other two labeled as Mo only and S only (terminated only by molybdenum or sulfur atoms, respectively). MD simulation analysis of water permeation through each membrane allowed them to conclude that Mo only pores and mixed pores perform better than S only pores regarding water flux. The reason is related to the fact that Mo-only regions achieve higher local water density, hence attracting more water throughout all possible MoS_2 pore architectures. They applied the same methodology for different force-field parameters to mimic different membrane compounds, such as $MoSe_2$, $MoTe_2$, WS_2, and WSe_2, and reported that the transition metal atom plays a more important role than the chalcogen atom when it comes to choosing the best TMD material for desalination. Water permeability was found to be two to five orders of magnitude greater than in current technology and 70% better than graphene nanopore with similar sizes. These results demonstrate how the material's chemistry (especially in nanopores) leads to exotic relationships with water, which is attracted to the inner pore, enhancing both water permeation and rejection of unwanted substances.

Taking advantage of MoS_2 nanosheets imperfections, Li et al. [60] used MD simulations to propose a desalination process regulated by "open" and "closed" states induced by a mechanical strain. By applying lateral strain to the MoS_2 membrane, they observed a high water transport rate ($355.3\ L{\cdot}m^{-2}{\cdot}h^{-1}{\cdot}bar^{-1}$) and excellent salt rejection capability when the membrane reached the open state, which corresponds to a strain of about 6% (~6% cross-sectional expansion in the membrane). The membrane demonstrated high water transparency and strong salt filtering capability even under a 12% strain. In this case, the mechanical strength associated with TMDs is a critical parameter, paving the way for a large-scale industrial application.

When it comes to the experimental realization of 2D MoS_2 membranes, the layer-stacked scheme is the current feasible option [12, 59, 61] due to the challenges concerning scalability and fabrication of large areas of MoS_2 monolayers [59] and the generation of nanopores with homogeneous size distribution. Nevertheless, 2D layer-stacked MoS_2 with few ~7 nm thick layers has been recently tested with promising water permeability of up to 320 $L{\cdot}m^{-2}{\cdot}h^{-1}{\cdot}bar^{-1}$, and stable high ionic sieving capability (> 99%) [12]. This type of membrane allows for water flow in between their grain boundaries and gaps. As for MoS_2, a 300 nm layer-stacked membrane of WS_2 exhibited an even higher water permeance of 730 $L{\cdot}m^{-2}{\cdot}h^{-1}{\cdot}bar^{-1}$ [15] at the cost of rejecting 90% 3 nm Evans blue molecules—a typical solution procedure to evaluate rejection rates. In terms of RO saltwater purposes, it is very important for a membrane to be able to reject 99.5% of salt at standard test conditions [62, 63].

To emulate the natural hydrophilicity present in both MoS_2 and WS_2 nanopores, it is possible to add chemical functional groups in nanoporous graphene to then adjust the pore's chemistry. Using classical MD simulations Cohen-Tanugi and Grossman [64] investigated water flux through hydrogenated (bonded with H) and hydroxylated (bonded with H and OH) graphene nanopores. They reported permeabilities two to three orders of magnitude higher than commercial RO membranes at the same salt rejection rate for some nanopore sizes. By maintaining pore size and using hydroxyl groups, they also discovered water flux enhancement when compared to the hydrogenated case. In addition, Risplendi et al. [65] used quantum (DFT) and classical (MD) simulations to show that functionalized graphene nanopores can reject even neutral solutes such as boric acid (H_3BO_3). In order to guide future membrane designs, it is very important to understand the challenges involved in the commercialization of RO membranes such as chemical and thermal sensitivity, rapid fouling, and cleaning.

An experimental work of Surwade et al. [72] demonstrated the possibility of using this kind of nanoporous graphene monolayer as a desalination membrane. The nanopores, produced through exposure to oxygen plasma, confirmed the indication of previous theoretical studies and presented a permeability of ~252 $L{\cdot}m^{-2}{\cdot}h^{-1}{\cdot}bar^{-1}$—assuming a nanopore density of 10^{12} cm^{-2} and sizes of 1 nm in diameter. Test conditions have further confirmed that graphene oxide membranes remain with excellent water permeance, separation efficiency, chemical and mechanical stability in water, acid, and basic solutions even after months [73]. However, despite the attractive potential improvements in the manufacturing process, a cost-effective graphene-based desalination device and other 2D membranes are still uncertain [66].

The option of layer-stacked graphene membranes is attractive comparing the industrial scale challenges related to the fabrication of large-area monolayer graphene with controlled pore density and size, a process which is intrinsically stochastic [67]. Although the water transport mechanism is a bit different in stacked GO nanosheets, experimental works have demonstrated the possibility of controlling the interlayer spacing in GO membranes and use it as a water transport channel with salt exclusion up to 97% [74], a performance that is comparable to a typical forward osmosis membrane. On the other hand, GO needs some stabilization strategy (embedded in epoxy, for example) once it can disintegrate in aqueous solutions [68].

Long before 2D membranes came to the spotlight of research on desalination, membranes composed of CNTs were extensively studied as attractive materials [71]. The investigation of water transport through CNTs led to the discovery of new and exciting properties, such as flow rate enhancement [1], which was seen as a very useful feature for desalination purposes. In 2D materials, the high transport rate is associated with their extremely low thickness of just a single atom.

Notwithstanding, the high water transport rate observed in CNTs is a consequence of the smooth hydrophobic inner core, which allows for uninterrupted water molecules passage with negligible adsorption and almost no friction [75]. Indeed, the enhancement flow factor—defined as the ratio between measured flow and the ideal no-slip Poiseuille flow—puts classical hydrodynamic theory in check, once the condition of zero interfacial fluid velocity does not necessarily hold at nanoscopic length scales [58]. This means that while water permeability goes down as we increase the thickness of 2D membranes, it is almost independent of the nanotube length, making CNT membranes still commercially attractive. Secchi et al. [17] have confirmed experimentally fast water transport behavior with nearly frictionless interfaces through carbon nanotubes. Besides these elevated speeds, high aspect ratio, and easy functionalization—a prerequisite for desalination purposes, avoiding aggregation that harms ion selectivity and water flux—renders CNTs as a widely explored nanomaterial in water purification research.

Among other categories, we could divide CNT-based membranes into two types based on their configurations: freestanding and mixed with polymeric materials [69]. The former can be produced either with vertically aligned nanotubes—where water is forced through inside them—or as buckypaper membranes—in short, a random network of CNTs with the large specific surface area. This network is structurally similar to the current commercial thin-film RO membrane composite in which CNTs are mixed with the top layer polymer. Buckypaper CNT membrane poses as an excellent alternative for desalination in distillation technology [76]. In contrast, Baek et al. [77] successfully synthesized a vertically aligned CNT membrane with pore diameters of ~ 4.8 nm and a pore density of 6.8×10^{10} cm^{-2}. They presented improved performance (three times higher flux) compared to typical ultrafiltration (UF) membranes. Additionally, vertically aligned CNT membranes can be obtained from thermal and oxygen-plasma treatments of densified outer-wall CNTs [78]. The advantage of the latter is that pore diameter can be readily varied (e.g., from 7 to 40 nm) through simple mechanical compression. CNT wall membranes obtained in this fashion can deliver water permeability that approaches 30,000 $L \cdot m^{-2} \cdot h^{-1} \cdot bar^{-1}$ and still avoid bacterial adhesion and biofilm formation. As in every other nanoscaled membrane, the experimental challenges here lie in the production of well-defined specific nanotube diameters needed for selectivity purposes, pore size homogeneity, alignment, and agglomeration control [62]. MD simulations have suggested that in order to achieve desalination capacity comparable to that of RO membranes the inner diameter of nanotubes should be around 0.6 nm [79], and current state-of-the-art CNT membrane synthesis is not able to meet these requirements. Also, there are concerns about potential nanotoxicity in the aquatic environment [70, 80].

CNTs mixed with polymeric materials are well guided to be applied in RO systems. High-performance RO CNT/Polyamide membranes can be achieved with CNTs dispersion in typical polymeric matrices [81]. The technique of CNT incorporation into polymers is well known and represents an excellent strategy to conceive membranes that are reasonable in terms of water flux (compared to vertically aligned CNTs) but extraordinary in terms of salt permeation, which makes them ideal to improve current RO technology.

Though promising, all of these membrane technologies are still far from reaching the desired production and commercialization stage. The next generation of desalination membranes is about to face the challenge of keeping the same (high) levels of water permeability, salt rejection, and stability while also becoming industrially scalable. In the meantime, there are several pitfalls to overcome such as chlorine tolerance, fouling/scaling, acid/base, and cleaning. The following section is focused on the gaps and possibilities of MoS_2 and WS_2 as viable alternatives for 2D desalination membranes.

5 Research Gaps

(a) ***Nanopore Opening***—The available synthesis of large-scale MoS_2 and WS_2 single layers by CVD makes them suitable to perform as water desalination 2D membranes. However, nanoporous membranes (Fig. 2a) require well-defined and distributed openings with pores of $\leq$ 1 nm radius. This is also true for layered membranes (Fig. 2b), where superficial nanopores lead to improved filtration and selective ion transport channels. From methods relying on plasma treatment or chemical etching to irradiation with energetic particles, i.e., electrons or ions, there are different ways to open nanopores in 2D membranes. Intrinsic defects in CVD-grown MoS_2 and WS_2 membranes stand as a challenge regarding electron irradiation techniques. While high-energy electrons are able to perforate freestanding MoS_2 and WS_2, single or double vacancies are created during the process [82], limiting their application. Exposing the membrane to a high flux of electrons can also result in uncontrollable size distribution and pore density. Nevertheless, a combination of methods can be used to achieve the desired scalability. A recent experiment combining focused electron beams with an in situ heating holder was able to drill nanopores in WS_2 bilayers (See Fig. 4) with precise control over spatial distributions with 5 nm accuracy of patterning and the width of nanowells adjustable by dose-dependent parameters [83]. In contrast to electron beams, we can use ions (or heavy ions) to bombard MoS_2 and WS_2 nanosheets, which involves a wider range of experimental parameters to be explored during the defect creation mechanism. For example, swift heavy ions [84] and highly charged ions [85] have been used to manufacture well-defined openings in freestanding MoS_2. In this way, pores with radii ranging from 0.3 to 3 nm have been created. Some groups have also exploited the electrochemical activity of MoS_2 and WS_2. Debatably, electrochemical reaction (ECR) techniques were used to open well-controlled nanopores in a scalable way [86]. But again, the high density of intrinsic

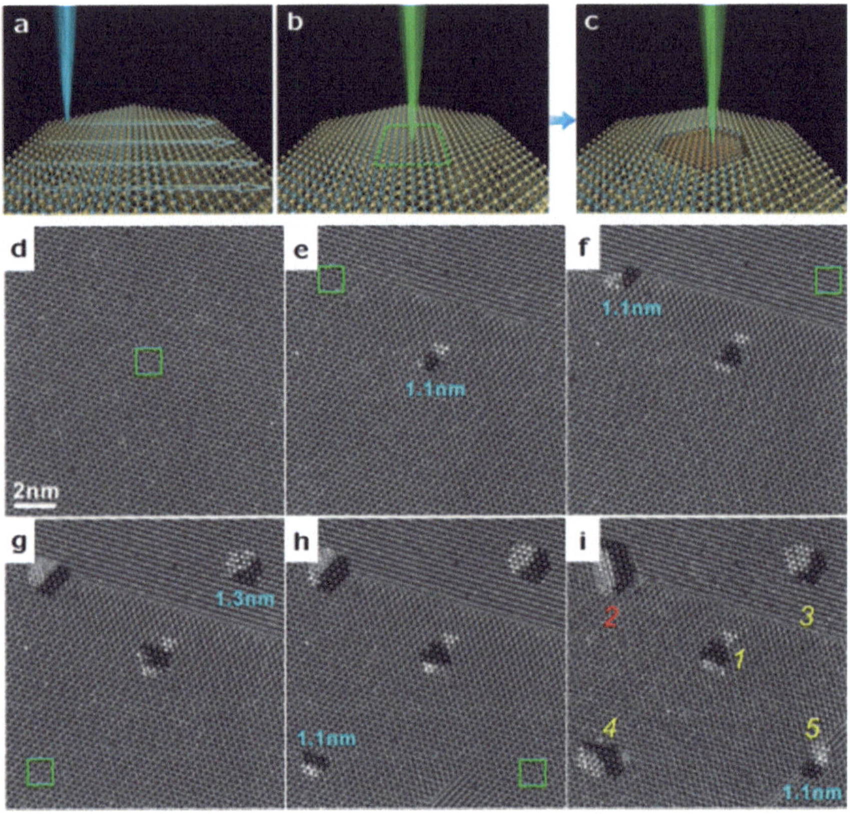

Fig. 4 Schematic illustration of electron beam used to drill nanopores in bilayer WS_2 Adapted with permission from Chen et al. [83]

defects naturally occurring in TMDs can represent a drawback for upscaling the method. The large-scale production of 2D MoS_2 and WS_2 to operate as nanoporous membranes depends on the advancement toward improved synthesis and control of nanopore fabrication.

(b) ***Vacancies***—Due to the imperfect nature of CVD process optimization, intrinsic structural variations such as atomic vacancies and grain boundaries are inherent to CVD-grown 2D MoS_2 and WS_2. In this context, water permeability can be understood in terms of the membrane's intrinsic porosity. Sulfur vacancies of various sizes can yield a large areal density of up to ~10^{13} cm^{-2} [22], rendering CVD-grown 2D MoS_2 layers sulfur-deficient in general [87]. Conversely, the areal density of large-sized (~1 nm) intrinsic sulfur vacancies such as V_{MoS6} is much smaller than that of small-sized vacancies, for example, V_S of <0.3 nm. This competing situation makes it difficult to experimentally quantify and identify the kind of specific vacancies governing water permeability. Additionally, CVD-grown 2D MoS_2 layers contain a large density of intrinsic "nanopores" present along the grain boundaries formed by individually stitching 2D grains of distinct crystallographic orientation. The size of these nanopores is larger than that of the atomic vacancies within basal planes as they are generally composed of multiple uncoordinated atoms. Grain-boundary nanopores play a very important role in the membrane's permeability, governing both water permeation and ion rejection. The uniformly spaced vdW gaps in the CVD-grown 2D MoS_2 also help the membrane to achieve high ionic sieving capability. In an experimental realization of few-layer 2D MoS_2 membranes, Li et al. [12] found the physical sizes of the hydrated ions to be larger than the interlayer vdW MoS_2 gap, which suggests that ion transportation is efficiently impeded. In addition to this geometrical effect, the electrostatic interaction of atomic vacancies is considered another major factor governing both water permeation and ionic sieving. Theoretical works have suggested that salt ions encounter significantly high energetic and steric barriers when approaching sulfur vacancies, while water molecules are relatively unaffected [60]. In this case, cations such as Na^+ will experience a high Coulombic barrier due to the positively charged sulfur vacancies, which in turn expose the hydrophilic Mo-rich sites to ignite a process of water attraction and ion repulsion. Indeed, we have identified in our simulations that nanopores with up to ~1 nm present strong ion rejection rates even at high pressures of ~100 MPa [13]. Although experimentalists have shown good control of the naturally occurring defects in CVD growth of TMDs, we still miss a systematic study on the vacancies distribution where their effect, grain boundary, and vdW gaps would be properly assessed.

(c) ***Imaging***—In order to take advantage of all the prominent desalination and adsorption features of MoS_2 and WS_2 membranes, it is very important to establish a precise, reproducible, rapid, and nondestructive method to effectively image nanopores and grain boundaries. This method should be also independent of the membrane composition, doping, and defect reconstruction. Although 2D nanoporous membranes stand as a new class of materials, the characterization of solid-state 2D nanostructures is an old, always evolving area of research. Therefore, the long-term knowledge we have built around this technology can be used to access atomic-scale information about nanopore geometry and distribution along the membrane. For instance, we have seen that transmission electron microscopy (TEM) directly resolves atomistic details of defects but requires intensive and disruptive sample preparation [19, 88, 89]. Alternatively, we can use photoluminescence or Raman spectroscopy that provides rapid and nondestructive probes of the electronic and vibrational properties of defective regions. Both techniques often manifest as red- or blue-shifted emission with enhanced or suppressed intensities when compared to the response from a pristine crystal, but it depends on multiple factors that affect local electronic properties such as material composition, doping level, defect passivation, grain boundary geometry, or edge terminations [90, 91]. Nonlinear optical spectroscopy can be highly sensitive to imaging imperfections, but their visualization can exhibit weak background contrast [92] or require the use of chemical solvents [93]. We can alternatively boost image contrast using dark-field (DF) microscopy. By comparing a bright-field linear optical image of a TMD monolayer on quartz and a dark-field linear optical image one can clearly see the edges scattered in dark-field light. However, atomically thin grain boundaries cannot be resolved by either imaging technique, because the in-plane linear dielectric response of TMD monolayers is often isotropic, making the scattered light intensity near grain boundaries indistinguishable. To circumvent this limitation, Carvalho et al. [94] have proposed a dark-field nonlinear characterization technique combining a second harmonic generation microscopy and a spatial filter to further enhance the second harmonic contrast of 1D imperfections in 2D systems (MoS_2 or WS_2), thus allowing detailed large-area spatial mapping of grain boundaries and edges regardless of their local atomic and electronic structures. There is still room for technological innovations in 2D materials characterization. New and improved nanometer-scale imaging techniques are going to rise in decades to come, enabling advancements toward both applications and fundamental 2D materials discoveries.

(d) ***High Intensive Energy Use***—Nanostructured membranes represent a breakthrough in membrane technology because they allow to shift the trade-off between permeability and salt rejection [62]. It also makes working with less membrane surface than conventional technology possible. Still, energy consumption at equal operating conditions is very similar to that of traditional saltwater RO (limited to a reduction of almost 15% by theoretical calculations based on phenomenological models [95]). In fact, the RO itself is just one stage among others. Because of that, a 15% reduction does not represent much in the overall cost saving. Improvements in pressure bomb technology and pressure recovery systems allowed to reduce energy demand associated with desalination plants over the last decades, but in the end, the costs are ultimately limited by the osmotic pressure in the system. The benefits of using an ultra-permeable membrane are more significant for brackish water RO: a reduction of 46% in energy consumption could in principle be achieved [95]. Besides that, an ultra-permeable membrane opens the window of possibilities to shift the operating conditions in order to optimize the desalination system. Another breakthrough aimed at minimizing energy demand is challenging and necessary to face both the world's water scarcity and climate emergency [96].

(e) ***Simulations***—Despite a large number of studies, there is still a long way to go when it comes to fully understand the mechanisms behind 2D membrane desalination. Most of the difficulties are associated with the fact that in the vicinity of an environment as complex as a nanoporous membrane, both water and especially ions can assume completely unpredictable behaviors. For instance, Cohen-Tanugi and Grossman [64] investigated the role of chemical functional groups bonded to the edges of graphene pores to suggest that commonly occurring hydroxyl groups can roughly double the water flux thanks to their hydrophilic character. It has been shown that a nozzle-like structure of protein channels and other nanoscale membranes also influences water transport [97]. TMDs such as MoS_2 and WS_2 offer the possibility to craft the pore's edge with Mo(W), S, or both, which allows for the design of functional nanopores. Additionally, the fish-bone structure of these TMDs [14] makes it promising as a nozzle-like sub-nanometer pore to be efficiently used in desalination membranes. In their computational study, Heiranian et al. [14] revealed that pore chemistry can significantly impact fluid transport, hence the ion rejection rate. It is important to note, though, that most of the computational results on desalination membranes come from pure classical dynamics. Despite the widespread agreement of these results with experiments, there is still much to advance in the chemical–physical molecular understanding of the processes that lead to higher or lower water fluxes, as well as the mechanisms and interactions between solution and membrane that provide greater efficiency in solute rejection. This can be achieved by using hybrid simulations that consider explicit charge transfer, for example.

(f) ***Toxicity***—Along with the scalable use of MoS_2 and WS_2 nanomembranes, one of the ultimate challenges is to understand more deeply how these structures are going to affect the environment after release. For instance, nanomaterials may pose a long-term threat if they are persistent and nondegradable [98]. In this regard, both MoS_2 and WS_2 are considered chemically stable materials against environmental stressors because of the absence of dangling bonds in the terminating S atoms [35]. Particularly the solubility of MoS_2 is low under ambient conditions, which leads to long-term persistence in the environment. On the other hand, this scenario can differ for extreme conditions such as high temperature or strong oxidation, where TMD nanomaterials can be oxidized to different oxides [99]. For instance, the oxidation of MoS_2 nanosheets has been shown to occur in aqueous solutions, leading to soluble, low-toxic oxidation products [100]. The oxidation kinetics depend on factors such as the pH and the crystallographic phase of the TMD.

Toxicity varies depending on preparation methods, but both MoS_2 and WS_2 nanosheets generally show high biocompatibility at concentrations up to ~100 ppm and certain cytotoxicity at high concentrations (a few hundred ppm). For instance, low toxicity of exfoliated, well-dispersed MoS_2 nanosheets was observed, but aggregated samples were found to induce acute lung inflammation in mice [101], raising concerns about their size's effects on the toxicity of these nanosheets. When it comes to toxicity we have good reasons to believe that both MoS_2 and WS_2 nanosheets can perform better than graphene-based nanomaterials [102], but there can be significant differences between those TMDs. Polyethylene glycol (PEG)-coated 2H-MoS_2 nanosheets have shown fast degradation and complete excretion within a month, in marked contrast to WS_2, which presented high levels of concentration in the organs for months [103]. Further degradation experiments indicated that the distinctive in vivo excretion behaviors of TMDs can be attributed to their different chemical properties. Still, the peculiarities involved in the degradation and accumulation of each material are yet to be clarified.

6 Conclusions and Perspectives

In this chapter, we analyzed the unusual 2D film properties of the TMDs MoS_2 and WS_2 with a specific focus on their application as adsorbents and membranes for water purification. The study of 2D materials' transport properties, even though inspired by the enhancement flow observed in carbon nanotubes, has a very different origin. While fast flow in nanotubes is due to friction regardless of tube length, in the case of 2D systems fast liquids transport is associated with their extremely low thickness, and this phenomenon disappears as the layer grows beyond a certain limit. Within single-layer surfaces, graphene is the material which not only has more studies but is also more advanced in production. Nevertheless, TMDs exhibit water permeability two to five orders of magnitude greater than the current technology and 70% better than similar-sized graphene nanopores. These results demonstrate how material chemistry (especially in nanopores) leads to exotic relationships with water, which is attracted to the inner pore enhancing both water permeation and the rejection of unwanted substances.

The actual nature of the TMD membranes' higher permeability and ion rejection is not clearly understood. In principle, the mix of hydrophobic and hydrophilic edges are important ingredients, yet they do not explain why WS_2 shows water flux two times greater than MoS_2 laminar membranes when both have a very similar charge distribution.

TMD membranes can be constructed either by making pores in the perfect material or by building up lamellar structures to employ filtration and adsorption qualities. In both cases impact of defect, pore distribution, and thickness of the stacking layers have to be explored together with the stability and toxicity of the material.

So far, what we have are prominent but diffuse experimental results. With the help of theoretical and computational simulations—in addition to some creativity—it is possible that in the near future both MoS_2 and WS_2 membranes can be used in desalination plants, significantly improving their performance.

References and Future Reading

1. Majumder M, Chopra N, Andrews R, Hinds BJ (2005) Enhanced flow in carbon nanotubes. Nature 438:44
2. Novoselov KS, Geim AK, Morozov S, Jiang D, Zhang Y, Dubonos SA, Grigorieva I, Firsov A (2004) Electric field effect in atomically thin carbon films. Science 306:666–669
3. Novoselov KS, Geim AK, Morozov S, Jiang D, Katsnelson M, Grigorieva I, Dubonos S, Firsov A (2005) Two-dimensional gas of massless dirac fermions in graphene. Nature 438:197–200
4. Ortiz-Medina J, Inukai S, Araki T, Morelos-Gomez A, Cruz-Silva R, Takeuchi K, Noguchi T, Kawaguchi T, Terrones M, Endo M (2018) Robust water desalination membranes against degradation using high loads of carbon nanotubes. Sci. Rep 8:2748
5. Jamali SH, Vlugt TJH, Lin L-C (2017) Atomistic understanding of zeolite nanosheets for water desalination. J Phys Chem C 121:11273–11280
6. Dulebohn J, Ahmadiannamini P, Wang T, Kim S-S, Pinnavaia TJ, Tarabara VV (2014) Polymer mesocomposites: ultrafiltration membrane materials with enhanced permeability, selectivity and fouling resistance. J Membr Sci 453:478–488
7. Dong H, Zhao L, Zhang L, Chen H, Gao C, Ho WSW (2015) High-flux reverse osmosis membranes incorporated with NaY zeolite nanoparticles for brackish water desalination. J Membr Sci 476:373–383
8. Mouterde T, Keerthi A, Poggioli AR, Dar SA, Siria A, Geim AK, Bocquet L, Radha B (2019) Molecular streaming and its voltage control in angstrom-scale channels. Nature 567:87–90
9. Bertolazzi S, Brivio J, Kis A (2011) Stretching and breaking of ultrathin MoS_2. ACS Nano 5:9703–9709

10. Toh RJ, Sofer Z, Luxa J, Sedmidubský D, Pumera M (2017) 3R phase of MoS_2 and WS_2 outperforms the corresponding 2H phase for hydrogen evolution. Chem Commun 53:3054–3057
11. Sun L, Huang H, Peng X (2013) Laminar MoS_2 membranes for molecule separation. Chem Commun 49:10718–10720
12. Li H, Ko T-J, Lee M, Chung HS, Han SS, Oh KH, Sadmani A, Kang H, Jung Y (2019) Experimental realization of few layer two-dimensional MoS_2 membranes of near atomic thickness for high efficiency water desalination. Nano Lett 19:5194–5204
13. Köhler MH, Bordin JR, Barbosa MC (2018) 2D nanoporous membrane for cation removal from water: effects of ionic valence, membrane hydrophobicity, and pore size. J Chem Phys 148:
14. Heiranian M, Farimani AB, Aluru NR (2015) Water desalination with a single-layer MoS_2 nanopore. Nat Commun 6:8616
15. Sun L, Ying Y, Huang H, Song Z, Mao Y, Xu Z, Peng X (2014) Ultrafast molecule separation through layered WS_2 nanosheet membranes. ACS Nano 8:6304–6311
16. Liu K, Yan Q, Chen M, Fan W, Sun Y, Suh J, Fu D, Lee S, Zhou J, Tongay S, Ji J, Neaton JB, Wu J (2014) Elastic properties of chemical-vapor-deposited monolayer MoS_2, WS_2, and their Bilayer heterostructures. Nano Lett 14:5097–5103
17. Secchi E, Marbach S, Nigues A, Stein D, Siria A, Bocquet L (2016) Massive radius-dependent flow slippage in carbon nanotubes. Nature 537:210–213
18. Hernández ER (2008) Molecular dynamics: from basic techniques to applications (A molecular dynamics primer). AIP Conf Proc 1077:95–123
19. Zhou W, Zou X, Najmaei S, Liu Z, Shi Y, Kong J, Lou J, Ajayan PM, Yakobson BI, Idrobo J-C (2013) Intrinsic structural defects in monolayer molybdenum disulfide. Nano Lett 13:2615–2622
20. Najmaei S, Liu Z, Zhou W, Zou X, Shi G, Lei S, Yakobson BI, Idrobo J-C, Ajayan PM, Lou J (2013) Vapour phase growth and grain boundary structure of molybdenum disulphide atomic layers. Nat Mater 12:754–759
21. Jeong HY, Jin Y, Yun SJ, Zhao J, Baik J, Keum DH, Lee HS, Lee YH (2017) Heterogeneous defect domains in single-crystalline hexagonal WS_2. Adv Mater 29:1605043
22. Kou J, Yao J, Wu L, Zhou X, Lu H, Wu F, Fan J (2016) Nanoporous two-dimensional MoS_2 membranes for fast saline solution purification. Phys Chem Chem Phys 18:22210–22216
23. Cohen-Tanugi D, Grossman JC (2015) Nanoporous graphene as a reverse osmosis membrane: recent insights from theory and simulation. Desalination 366:59–70
24. Zhu C, Li H, Meng S (2014) Transport behavior of water molecules through two-dimensional nanopores. J Chem Phys 141:18C528
25. Köhler MH, Bordin JR, Barbosa MC (2019) Ion flocculation in water: from bulk to nanoporous membrane desalination. J Mol Liq 277:516–521
26. Pérez MDB, Nicolai A, Delarue P, Meunier V, Drndic M, Senet P (2019) Improved model of ionic transport in 2-D MoS_2 membranes with sub-5 nm pores. Appl Phys Lett 114:
27. Wang Z, Tu Q, Zheng S, Urban JJ, Li S, Mi B (2017) Understanding the aqueous stability and filtration capability of MoS_2 membranes. Nano Lett 17:7289–7298
28. Novoselov KS, Mishchenko A, Carvalho A, Neto AHC (2016) 2D Materials and van der walls heterostructures. Science 353, aac9439
29. Schumann T, Lopes JMJ, Wofford JM Jr, Dubslaff MHO, Hanke MM, Jahn U, Geelhaar L, Riechert H (2015) The impact of substrate selection for the controlled growth of graphene by molecular beam epitaxy. J Cryst Growth 425:274–278
30. Reina A, Jia X, Ho J, Nezich D, Son H, Bulovic V, Dresselhaus MS, Kong J (2009) Large area, few-layer graphene films on arbitrary substrates by chemical vapor deposition. Nano Lett 9:30–35
31. Emtsev KV, Bostwick A, Horn K, Jobst J, Kellogg GL, Ley L, McChesney JL, Ohta T, Reshanov SA, Röhrl J, Rotenberg E, Schmid AK, Waldmann D, Weber HB, Seyller T (2009) Towards wafer-size graphene layers by atmospheric pressure graphitization of silicon carbide. Nat Mater 8:203–207

32. Raman RKS, Banerjee PC, Lobo DE, Gullapalli H, Sumandasa M, Kumar A, Choudhary L, Tkacz R, Ajayan PM, Majumder M (2012) Protecting copper from electrochemical degradation by graphene coating. Carbon 50:4040–4045
33. Xu G-R, Xu J-M, Su H-C, Liu X-Y, Lu-Li, Zhao H-L, Feng H-J, Das R (2019) Two-dimensional (2D) nanoporous membranes with sub-nanopores in reverse osmosis desalination: latest developments and future directions. Desalination 451:18–34
34. Li X, Zheng Z, Liu X, Zhao S, Liu S (2015) Nanostructured photoelectrochemical biosensor for highly sensitive detection of organophosphorous pesticides. Biosens Bioelectron 64:1–5
35. Chhowalla M, Shin HS, Eda G, Li L-J, Loh KP, Zhang H (2014) The chemistry of two dimensional layered transition metal dichalcogenide nanosheets. Nat Chem 5:263–275
36. Kam KK, Parkinson BA (1982) Detailed photocurrent spectroscopy of the semiconducting group VIB transition metal dichalcogenides. J Phys Chem 86:463–467
37. Lopez-Sanchez O, Lembke D, Kayci M, Radenovic A, Kis A (2013) Ultrasensitive photodetectors based on monolayer MoS_2. Nat Nanotech 8:497–501
38. Tongay S, Zhou J, Ataca C, Lo K, Matthews TS, Li J, Grossman JC, Wu J (2012) Thermally driven crossover from indirect toward direct bandgap in 2D semiconductors: MoSe 2 versus MoS_2. Nano Lett 12:5576–5580
39. Ohta T, Bostwick A, Seyller T, Horn K, Rotenberg E (2006) Controlling the electronic structure of bilayer graphene. Science 313:951–954
40. Cao Y, Fatemi V, Fang S, Watanabe K, Taniguchi T, Kaxiras E, Jarillo-Herrero P (2018) Unconventional superconductivity in magic-angle graphene superlattices. Nature 556:43–50
41. Buscema M, Groenendijk DJ, Blanter SI, Steele GA, van der Zant HSJ, Castellanos-Gomez A (2014) Fast and broadband photoresponse of few-layer black phosphorus field-effect transistors. Nano Lett 14:3347–3352
42. Li H, Wu J, Huang X, Lu G, Yang J, Lu X, Xiong Q, Zhang H (2013) Rapid and reliable thickness identification of two-dimensional nanosheets using optical microscopy. ACS Nano 7:10344–10353
43. Lide DR (ed) (2000) CRC handbook of chemistry and physics, 81st edn. CRC Press, Boca Raton, FL
44. Liu K-K, Zhang W, Lee Y-H, Lin Y-C, Chang M-T, Su C-Y, Chang C-S, Li H, Shi Y, Zhang H, Lai C-S, Li L-J (2012) Growth of large-area and highly crystalline MoS_2 thin layers on insulating substrates. Nano Lett 12:1538–1544
45. Yoon J, Park W, Bae G-Y, Kim Y, Jang HS, Hyun Y, Lim SK, Kahng YH, Hong W-K, Lee BH, Ko HC (2013) Highly flexible and transparent multilayer MoS_2 transistors with graphene electrodes. Small 9:3295–3300
46. Salvatore GA, Münzenrieder N, Barraud C, Petti L, Zysset C, Büthe L, Ensslin K, Tröster G (2013) Fabrication and transfer of flexible few-layers MoS_2 thin film transistors to any arbitrary substrate. ACS Nano 7:8809–8815
47. Kalanyan B, Kimes WA, Beams R, Stranick SJ, Garratt E, Kalish I, Davydov AV, Kanjolia RK, Maslar JE (2017) Rapid wafer-scale growth of polycrystalline 2H-MoS2 by pulsed metal-organic chemical vapor deposition. Chem Mater 29:6279–6288
48. Zeng H, Liu G-B, Dai J, Yan Y, Zhu B, He R, Xie L, Xu S, Chen X, Yao W, Cui X (2013) Optical signature of symmetry variations and spin-valley coupling in atomically thin tungsten dichalcogenides. Sci Rep 3:1608
49. Voiry D, Yamaguchi H, Li J, Silva R, Alves DCB, Fujita T, Chen M, Asefa T, Shenoy VB, Eda G, Chhowalla M (2013) Enhanced catalytic activity in strained chemically exfoliated WS_2 nanosheets for hydrogen evolution. Nat Mater 12:850–855
50. Okada M, Sawazaki T, Watanabe K, Taniguch T, Hibino H, Shinohara H, Kitaura R (2014) Direct chemical vapor deposition growth of WS2 atomic layers on hexagonal boron nitride. ACS Nano 8:8273–8277
51. Zhu H, Wang Y, Xiao J, Liu M, Xiong S, Wong ZJ, Ye Z, Ye Y, Yin X, Zhang X (2015) Observation of piezoelectricity in free-standing monolayer MoS_2. Nat Nanotech 10:151–155
52. Liu K, Feng J, Kis A, Radenovic A (2014) Atomically thin molybdenum disulfide nanopores with high sensitivity for DNA translocation. ACS Nano 8:2504–2511

53. Das R, Vecitis CD, Schulze A, Cao B, Ismail AF, Lu X, Chen J, Ramakrishna S (2017) Recent advances in nanomaterials for water protection and monitoring. Chem Soc Rev 46:6946–7020
54. Peimyoo N, Shang J, Yang W, Wang Y, Cong C, Yu T (2015) Thermal conductivity determination of suspended mono- and bilayer WS_2 by Raman spectroscopy. Nano Res 8:1210–1221
55. Kim T, Ding D, Yim J-H, Jho Y-D, Minnich AJ (2017) Elastic and thermal properties of free-standing molybdenum disulfide membranes measured using ultrafast transient grating spectroscopy. APL Mater 5:
56. Waduge P, Bilgin I, Larkin J, Henley RY, Goodfellow K, Graham AC, Bell DC, Vamivakas N, Kar S, Wanunu M (2015) Direct and scalable deposition of atomically thin low-noise MoS_2 membranes on apertures. ACS Nano 9:7352–7359
57. Eda G, Yamaguchi H, Voiry D, Fujita T, Chen M, Chhowalla M (2011) Photoluminescence from chemically exfoliated MoS_2. Nano Lett 11:5111–5116
58. McGaughey AJ, Mattia D (2017) Materials enabling nanofluidic flow enhancement. MRS Bull 42:273–277
59. Wang Z, Mi B (2017) Environmental applications of 2D molybdenum disulfide (MoS_2) nanosheets. Environ Sci Technol 51:8229–8244
60. Li W, Yang Y, Weber JK, Zhang G, Zhou R (2016) Tunable, strain-controlled nanoporous MoS_2 filter for water desalination. ACS Nano 10:1829–1835
61. Zhang H, Taymazov D, Li M-P, Huang Z-H, Liu W-L, Zhang X, Ma X-H, Xu Z-L (2019) Construction of MoS_2 composite membranes on ceramic hollow fibers for efficient water desalination. J Membr Sci 59:
62. Werber JR, Osuji CO, Elimelech M (2016) Materials for next-generation desalination and water purification membranes. Nat Rev Mater 1:16018
63. Voutchkov N (2013) Desalination engineering: planning and design. McGraw-Hill, New York
64. Cohen-Tanugi D, Grossman JC (2012) Water desalination across nanoporous graphene. Nano Lett 12:3602–3608
65. Risplendi F, Raffone F, Lin L-C, Grossman JC, Cicero G (2020) Fundamental insights on hydration environment of boric acid and its role in separation from saline water. J Phys Chem C 124:1438–1445
66. Boretti A, Al-Zubaidy S, Vaclavikova M, Al-Abri M, Castelletto S, Mikhalovsky S (2018) Outlook for graphene-based desalination membranes. npj Clean Water 1, 5
67. Homaeigohar S, Elbahri M (2017) Graphene membranes for water desalination. NPG Asia Mater 9:e427–e427
68. Yeh C-N, Raidongia K, Shao J, Yang Q-H, Huang J (2015) On the origin of the stability of graphene oxide membranes in water. Nat Chem 7:166–170
69. Ihsanullah (2019) Carbon nanotube membranes for water purification: developments, challenges, and prospects for the future. Sep Purif Technol 209:307–337
70. Das R, Leo BF, Murphy F (2018) The toxic truth about carbon nanotubes in water purification: a perspective view. Nanoscale Res Lett 13:183
71. Das R, Ali ME, Hamid SBA, Ramakrishna S, Chowdhury ZZ (2014) Carbon nanotube membranes for water purification: a bright future in water desalination. Desalination 336:97–109
72. Surwade SP, Smirnov SN, Vlassiouk IV, Unocic RR, Veith GM, Dai S, Mahurin SM (2015) Water desalination using nanoporous single-layer graphene. Nat Nanotech 10:459–464
73. Thebo KH, Qian X, Zhang Q, Chen L, Cheng H-M, Ren W (2018) Highly stable graphene-oxide-based membranes with superior permeability. Nat Commun 9:1486
74. Abraham J, Vasu KS, Williams CD, Gopinadhan K, Su Y, Cherian CT, Dix J, Prestat E, Haigh SJ, Grigorieva IV, Carbone P, Geim AK, Nair RR (2017) Tunable sieving of ions using graphene oxide membranes. Nat Nanotech 12:546–550
75. Köhler MH, Bordin JR, de Matos CF, Barbosa MC (2019) Water in nanotubes: the surface effect. Chem Eng Sci 203:54–67
76. Dumée L, Sears K, Schutz J, Finn N, Duke M, Gray S (2010) Carbon nanotube based composite membranes for water desalination by membrane distillation. Desalin Water Treat 17:72–79

77. Baek Y, Kim C, Seo DK, Kim T, Lee JS, Kim YH, Ahn KH, Bae SS, Lee SC, Lim J, Lee K, Yoon J (2014) High performance and antifouling vertically aligned carbon nanotube membrane for water purification. J Membr Sci 460:171–177
78. Lee B, Baek Y, Lee M, Jeong DH, Lee HH, Yoon J, Kim YH (2015) A carbon nanotube wall membrane for water treatment. Nat Commun 6:7109
79. Ahn CH, Baek Y, Lee C, Kim SO, Kim S, Lee S, Kim S-H, Bae SS, Park J, Yoon J (2012) Carbon nanotube-based membranes: fabrication and application to desalination. J Ind Eng Chem 18:1551–1559
80. Francis AP, Devasena T (2018) Toxicity of carbon nanotubes: a review. Toxicol Ind Health 34:200–210
81. Kim HJ, Choi K, Baek Y, Kim D-G, Shim J, Yoon J, Lee J-C (2014) High-performance reverse osmosis CNT/polyamide nanocomposite membrane by controlled interfacial interactions. ACS Appl Mater Interfaces 6:2819–2829
82. Komsa H-P, Kotakoski J, Kurasch S, Lehtinen O, Kaiser U, Krasheninnikov AV (2012) Two-dimensional transition metal dichalcogenides under electron irradiation: defect production and doping. Phys Rev Lett 109:
83. Chen J, Ryu GH, Zhang Q, Wen Y, Tai K-L, Lu Y, Warner JH (2019) Spatially controlled fabrication and mechanisms of atomically thin nanowell patterns in bilayer WS_2 using in situ high temperature electron microscopy. ACS Nano 13:14486–14499
84. Madauß L, Ochedowski O, Lebius H, Ban-d'Etat B, Naylor CH, Johnson ATC, Kotakoski J, Schleberger M (2016) Defect engineering of single- and few-layer MoS_2 by swift heavy ion irradiation. 2D Mater 4:015034
85. Kozubek R, Tripathi M, Ghorbani-Asl M, Kretschmer S, Madauß L, Pollmann E, O'Brien M, McEvoy N, Ludacka U, Susi T, Duesberg GS, Wilhelm RA, Krasheninnikov AV, Kotakoski J, Schleberger M (2019) Perforating freestanding molybdenum disulfide monolayers with highly charged ions. J Phys Chem Lett 10:904–910
86. Feng J, Liu K, Graf M, Lihter M, Bulushev RD, Dumcenco D, Alexander DTL, Krasnozhon D, Vuletic T, Kis A, Radenovic A (2015) Electrochemical reaction in single layer MoS_2: nanopores opened atom by atom. Nano Lett 15:3431–3438
87. Hong J, Hu Z, Probert M, Li K, Lv D, Yang X, Gu L, Mao N, Feng Q, Xie L, Zhang J, Wu D, Zhang Z, Jin C, Ji W, Zhang X, Yuan J, Zhang Z (2015) Exploring atomic defects in molybdenum disulphide monolayers. Nat Commun 6:6293
88. Lehtinen O, Komsa H-P, Pulkin A, Whitwick MB, Chen M-W, Lehnert T, Mohn MJ, Yazyev OV, Kis A, Kaiser U, Krasheninnikov AV (2015) Atomic scale microstructure and properties of Se-deficient two-dimensional MoSe 2. ACS Nano 9:3274–3283
89. Azizi A, Zou X, Ercius P, Zhang Z, Elías AL, Perea-López N, Stone G, Terrones M, Yakobson BI, Alem N (2014) Dislocation motion and grain boundary migration in two-dimensional tungsten disulphide. Nat. Commun. 5:4867
90. Nan H, Wang Z, Wang W, Liang Z, Lu Y, Chen Q, He D, Tan P, Miao F, Wang X, Wang J, Ni Z (2014) Strong photoluminescence enhancement of MoS_2 through defect engineering and oxygen bonding. ACS Nano 8:5738–5745
91. Lin K-I, Ho Y-H, Liu S-B, Ciou J-J, Huang B-T, Chen C, Chang H-C, Tu C-L, Chen C-H (2018) Atom-dependent edge-enhanced second-harmonic generation on MoS_2 monolayers. Nano Lett 18:793–797
92. Yin X, Ye Z, Chenet DA, Ye Y, O'Brien K, Hone JC, Zhang X (2014) Edge nonlinear optics on a MoS_2 atomic monolayer. Science 344:488–490
93. Karvonen L, Säynätjoki A, Huttunen MJ, Autere A, Amirsolaimani B, Li S, Norwood RA, Peyghambarian N, Lipsanen H, Eda G, Kieu K, Sun Z (2017) Rapid visualization of grain boundaries in monolayer MoS_2 by multiphoton microscopy. Nat Commun 8:15714
94. Carvalho BR, Wang Y, Fujisawa K, Zhang T, Kahn E, Bilgin I, Ajayan PM, de Paula AM, Pimenta MA, Kar S, Crespi VH, Terrones M, Malard LM (2020) Nonlinear dark-field imaging of one-dimensional defects in monolayer dichalcogenides. Nano Lett 20:284–291
95. Cohen-Tanugi D, McGovern RK, Dave SH, Lienhard JH, Grossman JC (2014) Quantifying the potential of ultra-permeable membranes for water desalination. Energy Environ Sci 7:1134–1141

96. WWAP and UNESCO (2019) The United Nations World Water Development Report 2019: leaving no one behind
97. Gravelle S, Joly L, Detcheverry F, Ybert C, Cottin-Bizonne C, Bocquet L (2013) Optimizing water permeability through the Hourglass Shape of aquaporins. Proc Natl Acad Sci USA 110:16367–16372
98. Wang Z, Zhu W, Qiu Y, Yi X, von dem Bussche A, Kane A, Gao H, Koski K, Hurt R (2016) Biological and environmental interactions of emerging two-dimensional nanomaterials. Chem Soc Rev 45:1750–1780
99. Gao J, Li B, Tan J, Chow P, Lu T-M, Koratkar N (2016) Aging of transition metal dichalcogenide monolayers. ACS Nano 10:2628–2635
100. Wang Z, von dem Bussche A, Qiu Y, Valentin TM, Gion K, Kane AB, Hurt RH (2016) Chemical dissolution pathways of MoS_2 nanosheets in biological and environmental media. Environ Sci Technol 50:7208–7217
101. Wang X, Mansukhani ND, Guiney LM, Ji Z, Chang CH, Wang M, Liao Y-P, Song T-B, Sun B, Li R, Xia T, Hersam MC, Nel AE (2015) Differences in the toxicological potential of 2D versus aggregated molybdenum disulfide in the lung. Small 11:5079–5087
102. Teo WZ, Chng ELK, Sofer Z, Pumera M (2014) Cytotoxicity of exfoliated transition-metal dichalcogenides (MoS_2, WS_2, and WSe_2) is lower than that of graphene and its analogues. Chem Eur J 20:9627–9632
103. Hao J, Song G, Liu T, Yi X, Yang K, Cheng L, Liu Z (2017) In vivo long-term biodistribution, excretion, and toxicology of PEGylated transition-metal dichalcogenides MS_2 (M = Mo, W, Ti) nanosheets. Adv Sci 4:1600160

Newly Emerging Metal–Organic Frameworks (MOF), MXenes, and Zeolite Nanosheets in Solutes Removal from Water

Guo-Rong Xu

Abstract Freshwater crisis is an ever-increasing severe twenty-first-century problem. Scientists, experts, and engineers are exploring various strategies to meet the freshwater desire. Membranes with high ion rejections and adsorptive capability are playing important roles in desalination and/or wastewater treatment. Many 2D membranes have been demonstrated great potential in water purification, such as metal–organic frameworks (MOF), MXenes, and zeolite nanosheets. In this chapter, applications of these 2D nanosheets in desalting and heavy metal ions removal are discussed from the point of computer simulations, synthesis procedures, and experimental works.

Keywords Desalination · Pollutants · Two-dimensional nanosheets · Metal–organic frameworks · MXene · Zeolites

1 Introduction

Ever-increasing water shortage forces people all over the world to find effective ways for water purification including disinfection, decontamination, reclamation, and desalination. Desalination has developed into the most efficacious strategy and has been widely used globally. Meanwhile, wastewater treatments, such as municipal and industrial wastewater reclamation draw more and more attention in recent years [1]. The specific character of desalination and wastewater treatment is that they do not consume hydrological water resource. Accordingly, there is an important need to develop efficient and sustainable technologies to realize freshwater augmentation without consuming hydrological water cycle. Materials play vital roles in this aspect of fabricating new technology for efficient water purification [2].

Given the fact that reverse osmosis (RO) desalination has become the most important method in seawater desalination, RO membrane (mainly polyamide thin-film

G.-R. Xu (✉)
The Institute of Seawater Desalination and Multipurpose Utilization (ISDMU),
Ministry of Natural Resources (MNR), Tianjin 300192, China
e-mail: labxgr@aliyun.com

R. Das (ed.), *Two-Dimensional (2D) Nanomaterials in Separation Science*,
Springer Series on Polymer and Composite Materials,
https://doi.org/10.1007/978-3-030-72457-3_9

composite membranes, PA-TFC membrane) has gained extensive research in recent years [3]. With the ongoing improvements and adjustments via various strategies, such as surface modifications [4], nanoparticles incorporations [5], and sublayers tuning [6], and interfacial synthesis controlling [7]; PA-TFC-based RO membranes have obtained great advancements in both fundamental research and engineering applications. Nevertheless, despite the comparatively excellent performance, RO membranes are still limited by the low water permeability and membrane fouling [8]. Currently, one of the great challenges for RO membranes is to further increase their permeations in order to decrease the energy consumption, which is the main cost component of RO engineering [9]. For example, Grossman et al. reported that RO membranes with three times higher permeations could reduce the energy cost by 15–46% [8].

Not as that of salty water, wastewater is characterized by more complicated components, sizes, charges, physical, and chemical properties [10]. Traditional methods for the purification of wastewater are flocculation, coagulation, membrane, advanced oxidation processes, filtration, biological processes, adsorption, chemical precipitation, and so forth. Among these methods, because of complex instruments and space-consuming facilities with high maintenance costs of other ones [11], adsorption has always been the most versatile and facile way. Adsorbents are the core of adsorption. Traditional adsorbents, such as ion-exchange resins [12], activated carbon [13], and functional clays [14], are always challenged by their limited functional groups and not satisfied with adsorption ability. It is of great necessity to explore more advanced adsorbents to meet the wastewater treatment desire.

In recent years, two-dimensional (2D) nanosheets and the corresponding assembled layered nanomembranes (2D nanosheet materials) have shown great potential in both desalination and wastewater treatment, especially after the discovery of graphene. Various 2D nanosheet materials such as graphene [15], graphene oxide [16], molybdenum disulfide (MoS_2) [17], boron nitride (BN) [18], silicon carbide [19], MXene, and zeolite [20] have been widely used in desalination mainly theoretically, accompanied by some experimental works. Meanwhile, unique physical and chemical properties (e.g., unsaturated surface atoms, greater surface energy fractions, etc.) enable 2D nanosheet materials to effectively adsorb heavy metal ions [21].

Together with zeolite nanosheets, MXene and MOF nanosheets have attracted great attention in recent years in liquid separation area. Compared with other 2D nanomaterials, these types demonstrate some unique characters. For example, although possessing similar features like GO, MXene shows superiority in its hydrophilic nature [22]. In this chapter, we focus on the applications of MOFs, zeolites, and MXene nanosheets in wastewater treatment by adsorption and filtration (mainly RO). Although some of these 2D nanomaterials also show great potential in removing pollutants via other routes (e.g., organic pollutants removal by photocatalysis with MOFs [23]), herein we do not include these issues because they are totally different disciplines.

2 Metal–Organic Framework, Zeolite and MXene Nanosheets

2.1 MOF Nanosheets

High surface areas and well-ordered porous structures make crystalline porous materials (CPM) that show great potential in separation and adsorption [24]. Among them, MOFs, which are typically constructed from metal ions/clusters and organic ligands via coordination bonds, have attracted much attention [25, 26]. Both the metal ions chemistry and ligand preparation versatility endow MOFs with great tunability and distinguished interesting properties, such as chemical functionality, porosity, and stability, compared with other porous materials. Thus, MOFs have been of particular interest and have been used in multidisciplinary areas since its first discovery in 1995 [27]. Thanks to their facile synthesis, large surface area, and chemical stability; MOFs show great potential applications in purification and separation [28].

Tailorable pore structure and sizes, and porous structures with numerous active sites make MOFs very promising in treating contaminants through adsorption [10]. Compared with other adsorbents, such as zeolites whose pores are limited by small size formed through inorganic crosslinking anions, MOFs display advances owing to their variable choices of metal ions and building blocks. The pore sizes of MOFs could be tuned in the range of few angstroms to several nanometers.

Layered MOFs constitute 2D layers assembled in a vertical direction by stacking via weak interactions, such as Van der Waals forces, hydrogen bonding, and π-π stacking. These weak interactions can be easily disrupted. As a result, the stacked MOFs would be exfoliated to 2D MOF nanosheets. For 2D MOF nanosheets, more accessible active sites are exposed on surface, facilitating the interactions between active sites and substrate molecules and improving the separation performance.

2.2 Zeolite Nanosheets

Zeolite, typically in forms of NaA, MFI, FAU, etc., is a kind of aluminosilicate minerals with a microstructures composed of three-dimensionally (3D) connected nanopores of 0.3–0.8 nm [29]. Considerable attention has been paid to layered zeolites with a thickness varying from one to several unit cells [30]. Moreover, 232 distinct zeolite topologies have been identified and included in the International Zeolite Associations (IZA) database to date, while 48 millions or more have been theoretically predicted [31].

2.3 *MXene Nanosheets*

MXene is a new family of 2D layered membranes based on transition metal carbides and/or nitrides. The general chemical formula of MXene is $M_{n+1}X_nT_z$ (n is in the range of 1–3), M is transition metals, such as Ti, Zr, Hf, V, Nb, Ta, Cr, and Sc, X is C and/or N, T_z is surface groups, such as O^{2-}, OH^-, F^-, NH_3, NH_4^+. Presently, fabrication of MXenes is mainly realized via exfoliation of A element such as Al of its precursor of $M_{n+1}AX_n$ with weak combination by hydrofluoric acid solution or mixed solution of hydrochloric acid and fluoride [32, 33]. MXene has emerged and exhibited promising potential in electrochemical energy storage area via controlling ions insertion between the atomically thin layers [34].

Besides, it should be noted that excellent characters, such as high flexibility, mechanical strength, and hydrophilicity, enable MXenes with controllable thickness from nanometers to micrometers, which are very promising in separation science. The most studied MXene is $Ti_3C_2T_x$. $Ti_3C_2T_x$ displays comparable mechanical strength with GO. For example, $Ti_3C_2T_x$ film with thickness of 3 μm exhibits a tensile strength of 22 MPa [35], while GO membrane with thickness of 2.5 μm shows 55 MPa [36]. Moreover, hydroxyl groups on alkalization interacted MXenes make them suitable for removing heavy metal ions from wastewater [37]. For instance, high surface area and hydrophilic surface endow $Ti_3C_2T_x$ great potential in removing various heavy metal ions and organic pollutants [38–40].

3 Computer Simulation

3.1 *MOF Nanosheets*

Hg_O and $HgCl_2$ adsorption on UiO-66 was investigated using computer simulation [41]. A reliable force fields (FFs) stemmed from plane-wave-density functional theory (DFT) calculations was introduced to describe interactions between Hg_O, $HgCl_2$, and UiO-66. Grand Canonical Monte Carlo (GCMC) calculations were used to indicate the single and/or multi adsorption of Hg_O, $HgCl_2$, and other components into UiO-66. The interaction energies between mercury species and UiO-66 were researched. $HgCl_2$ has a stronger affinity than Hg0 for UiO-66 with a heat of adsorption of ~ 60 kJ/mol. Considering that MOFs materials are always limited by the low chemical stability compared with that of zeolites [42]. Tb with higher chemical stability and facile functionality with abundant functional groups was used to construct MOFs. Adsorption of Pb^{2+} onto Tb-MOFs was theoretically studied recently (Fig. 1) [43]. Carboxylate-based ligands of H3TATAB (4,4′,4″-(1,3,5-triazine-2,4,6-triyltriimino) tris-benzoic acid, containing amino, imine, and carboxylic acid groups and Tb^{3+} were assembled into Tb-MOFs with nanotubular shapes via solvothermal method. N groups seem to be the most prospective ligands to remove heavy metal ions, while they did not react with constitutive metal ions.

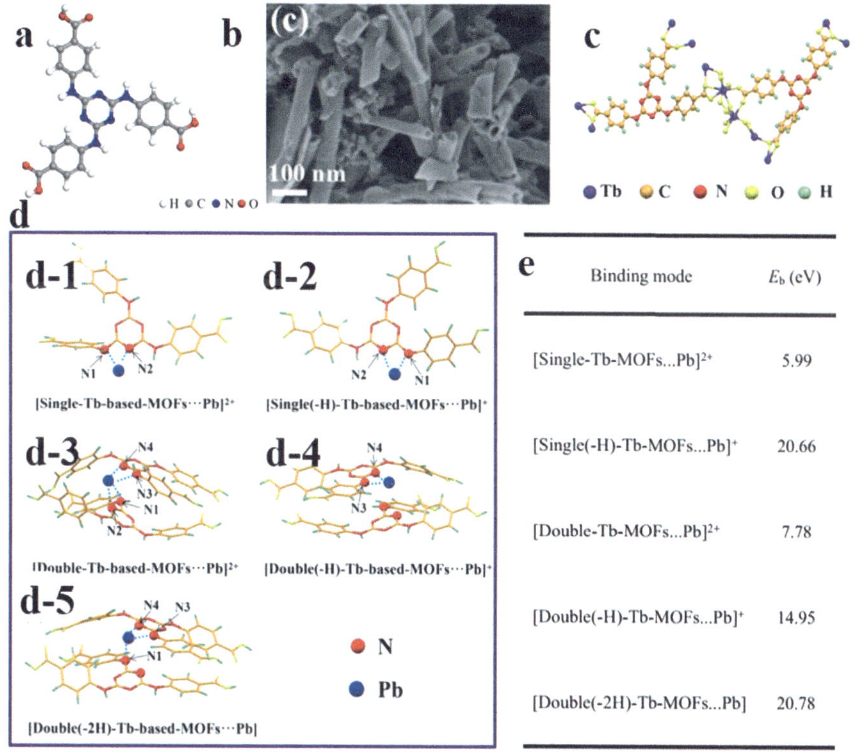

Binding mode	E_b (eV)
$[Single-Tb-MOFs...Pb]^{2+}$	5.99
$[Single(-H)-Tb-MOFs...Pb]^{+}$	20.66
$[Double-Tb-MOFs...Pb]^{2+}$	7.78
$[Double(-H)-Tb-MOFs...Pb]^{+}$	14.95
[Double(-2H)-Tb-MOFs...Pb]	20.78

Fig. 1 **a** The structure of H3TATAB, **b** SEM images for Tb-MOFs nanotubular structures, **c** the coordinated environments of Tb^{3+}, **d** The optimized geometries for Pb^{2+} ion adsorbed on Tb-MOFs: (d-1) the single ligand binding with Pb^{2+}, (d-2) the single ligand binding (N1 removed a proton) with Pb^{2+}, (d-3) double ligands binding with Pb^{2+}, (d-4) double ligands binding (N3 removed a proton) with Pb^{2+}, **e** double ligands binding (N1 and N3 removed two protons) with Pb^{2+}, and **e** The calculated binding energies (Eb) between Pb^{2+} and Tb-MOFs by DFT calculations. Figures are reproduced with permissions from (Ref. 43)

H3TATAB has good affinity for Pb^{2+} through complexation and electrostatic interactions [44]. Tb-MOFs displayed 547 mg/g of adsorption capacity toward Pb. Besides the evidence of Pb–N– and Pb–N = bonding was confirmed by X-ray photoelectron spectroscopy (XPS). DFT calculations revealed that the formation of the inner-sphere complex between the carbon/nitrogen and Pb^{2+} followed the primary adsorption mechanism. Pb^{2+} can bind with N groups of Tb-MOFs through the formation of Pb-N- and Pb-N = to form inner-sphere complexes. The binding energy E_b which indicates the interaction between the heavy metal ions and the adsorbents was calculated. All the E_b values were higher than 1.0 eV, indicating the chemical adsorption of Pb^{2+} on Tb-MOFs. The higher E_b indicates the more favorable state. Deprotonation increased the E_b energy for both single- and double-Tb-MOFs…Pb complexes, suggesting the improved adsorption.

2D MOF nanosheets see little applications in solute separations. Zr-MOF nanosheets were used as photocatalysts because of their ultrathin thickness of ~1.5 nm and high exposed active sites [45].

3.2 *Zeolite Nanosheets*

The potential of zeolitic imidazolate framework (ZIF) in desalination was firstly reported by Jiang et al. [46]. They further extended the research to five ZIFs including ZIF-25 (dimethyl imidazolate), ZIF-71 (dichloro imidazolate), ZIF-93 (aldehydemethyl imidazolate), ZIF-96 (cyanide imidazolate), and ZIF-97 (hydroxymethyl imidazolate) with varied functionalization and diameters of 0.51, 0.54, 0.37, 0.55, and 0.35 nm [47]. Compared to ZIF-93 and 97, ZIF-25, 71, and 96 exhibited higher water permeation because of larger aperture size (da). Meanwhile, water flux difference for ZIF-25, 71, and 96 themselves was ascribed to functional groups polarity rather than da. In spite of larger da, hydrophilic CH_3 group of ZIF-25 enabled them with higher water flux compared to ZIF-71 and 96. As for the salt rejections, ZIF-25 exhibited 97% and the others showed 100%. ZIF-25 was considered comprehensively the most potential alternative.

In spite of great potential, desalination performance of zeolite nanosheets is still currently limited in computer simulations with little experimental study. For example, the structures and desalination performance of zeolite nanosheets were researched by MD simulation [20]. It was believed that ultrathin-film nature and the versatile pore structures enabled zeolite nanosheets to have a great opportunity in desalination. Influence of pore density, free energy barrier, and cages inclusion was studied. It is indicated that desalination performance of zeolite nanosheets was derived from their one-atom-thin thickness and unique water channels including one-dimensional (1D), multi-dimensional, and cage-containing channels. Moreover, appropriate pore diameter of 0.4–0.75 nm was also important. MD simulation results indicated that zeolite nanosheets could exhibit salt rejection of 100% and water permeability one time higher than that of currently used PA-TFC membranes. Water permeability of zeolite nanosheet with one layer was as high as 40 L/(cm^2·day·MPa), similar as that of other novel ultrathin membranes such as graphene. Even with thickness of 100 nm, zeolite nanosheets still demonstrated water permeability of 1.3 L/(cm^2·day·MPa), much higher compared to conventional membranes in the range of 0.03–0.2 L/(cm^2·day·MPa).

Generally, heavy metal ions are always removed by adsorption. But membrane filtration using nanosheets could also be used to realize heavy metal ions removal. Potential of nanomembranes in removing heavy metal ions was Initially studied by DFT [48]. The graphene structure was optimized to obtain the highest Cu^{2+} and Hg^{2+} removal. Moreover, they also investigate the ability of BN and graphene sheets in removing Zn^{2+} and indicated that pore structure played key roles [49]. Furthermore, they found that only ions with energy barrier difference could be effectively separated. Besides the nanosheets above, MoS_2 nanosheets were also applied to separate

Hg^{2+} [50] and As^{2+} [51] from wastewater. In spite of these achievements, great challenges still exist in large-scale purification and desalination for all the nanosheets above because of deficiency in long-term stable, multifunctional, and cost-effective fabrication procedures [15].

The perpendicular pores in MFI zeolite enable water transport in all directions and thus endow them with great permeation in desalination. Besides, the appropriate pore size provides highs salt rejection [52]. More importantly, MFI zeolites could be derived from mineral resources, which is abundant in natural and the cost is very low [53]. Actually, zeolite is considered to be more advantageous in liquid separation [54]. Recently, molecular dynamics simulation was used to study the desalination performance of MFI nanosheets [55]. Non-equilibrium MD in a constructed RO system was applied to investigate the $HgCl_2$ and $CuCl_2$ removal by MFI nanosheets under 10–200 MPa (Fig. 2a and b).

Figure 2c shows Cu^{2+}, Hg^{2+}, and Cl^- rejection by MFI nanosheets under various pressures. Time-dependent simulated separation process under pressure of 50 and 100 MPa is displayed in Fig. 2d. $Cu^{2+}$and Hg^{2+} could be totally rejected under all pressures, while the pressure was increased to 100 MPa some Cl^- could permeate because of the overcoming of electrostatic interactions and Van der Waals. Figure 2d indicated that all the salt ions were totally rejected under 50 MPa even in 10 ns. But when the pressure was increased to 100 MPa, permeation of some Cl^- occurred. In spite of the comparatively low salt rejections under high pressures, extremely high water flux could make some compensation. For instance, water flux dramatically increased from 25.6 ns^{-1} to 173.6 ns^{-1} from 10 to 200 MPa.

3.3 MXene Nanosheet

It was reported that the surrounding groups of MXene could significantly affect the heavy metal ions adsorption. For example, F and H sites weakened the adsorption, while the intercalation of X^+ ($X = Li, Na, K$) could strengthen the adsorption behavior of $Ti_3C_2(OH)_2$. And alkalization intercalation could act as an effective method to accelerate the heavy metal ions adsorption [39]. Different transition metal ions of M and X endow MXenes with different properties because of various electronic structures [56]. Taking this under consideration, based on their research on the lead adsorption on alkalization-intercalated MXene ($Ti_3C_2(OH)_2$) [40], Guo et al.systematically studied the adsorption of Pb on various MXenes ($M_2X(OH)_2$, M = Sc, Ti, Cr, Zr, Nb, Mo, Hf, Ta, and X = C or N) using DFT by first principle calculations [57].

The adopted M_2X type MXene is the thinnest in MXene family. The alkalization treatment after the exfoliating by HF always leads to the formation of terminated OH groups at their outer layers [58]. It is composed of tri-layer sheets characterized by a hexagonal-like unit in which the M is nearly closely packed, and X fills in octahedral sites [59]. The OH groups are always added to enhance their stability. Three kinds of $M_2X(OH)_2$ were adopted in the research (Fig. 3a–c). For Model A, two OH groups sited on the top of M hollow site. For model B, one OH group was

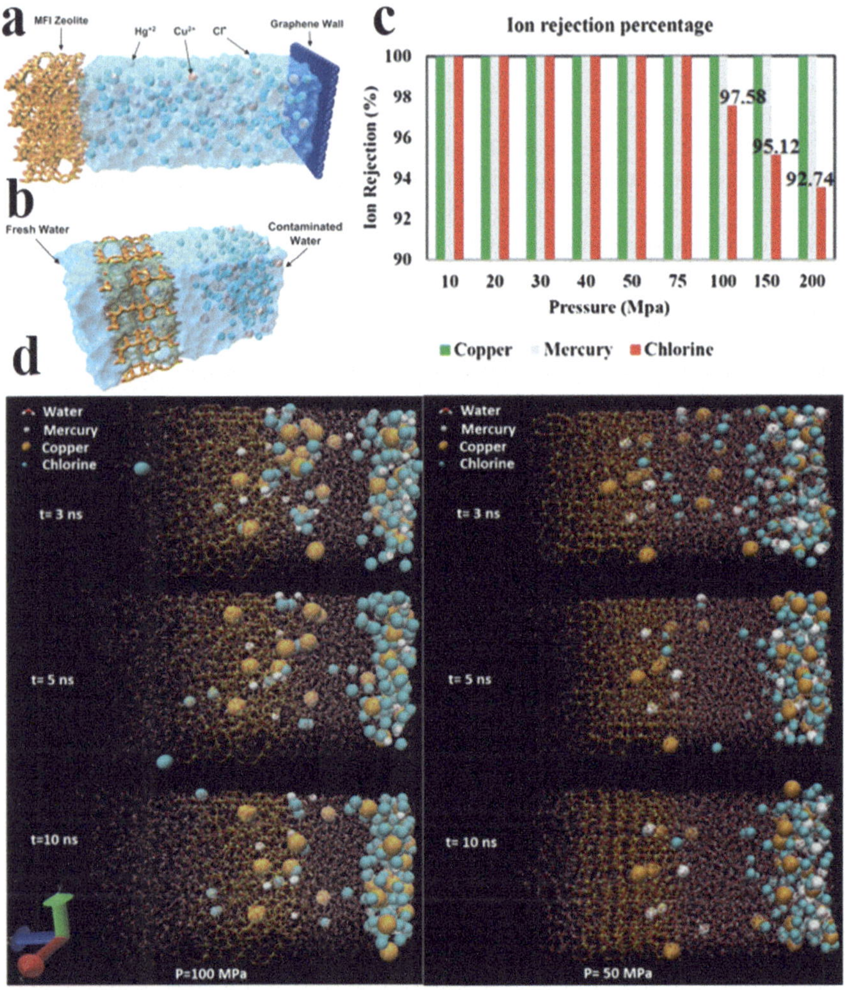

Fig. 2 **a** Schematic of the problem before ion separation at 50 MPa, **b** Ion separation at 50 MPa after 10 ns MD simulation, **c** Ion rejection versus pressure using Zeolite MFI nanosheet, and **d** Separation process at different times and pressures using MFI zeolite. Figures are reproduced with permissions from (Ref. 55)

located on top of X atom and the other was located on top of M. For model C, two OH groups were on top of X atom. Three modes exhibited different stabilities with varying transition metal ions and C/N compositions. Except Zr2C, Nb2N, Hf2C, TaC, and Ta2N; model A was energetically most stable. For Zr2C and Hf2C, model B is more suitable. For Nb2N, Ta2C, and Ta2N, model C was most stable. Pb could be adsorbed onto MXenes by replacing H atoms [60]. In this research, Pb could also bond with O atoms but was restrained by the adjacent OH groups (Fig. 3d and e). Changing M sites induced small difference for the three modes. Types of M played

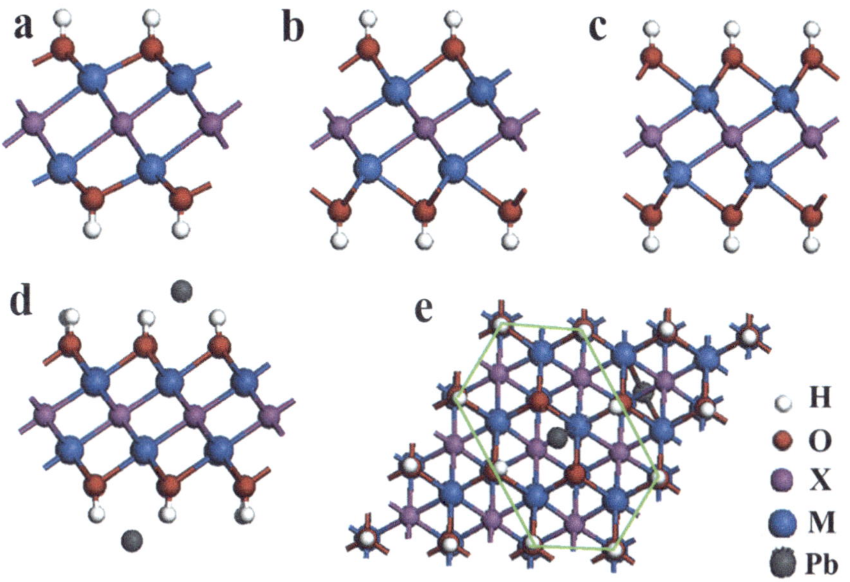

Fig. 3 The sketch maps of three different modes of $M_2X(OH)_2$ MXene **a** Mode A-$M_2X(OH)_2$, **b** Mode B-$M_2X(OH)_2$, **c** Mode C-$M_2X(OH)_2$; the side **d** and **e** top view of $M_2X(O_2H_{2-2x}Pb_x)$. Figures are reproduced with permissions from (Ref. 57)

important roles in Pb adsorption. For all the $M_2C(OH)_2$ MXenes, only $Sc_2C(OH)_2$ and $Zr_2C(OH)_2$ could not be able to adsorb Pb due to their positive formation energy values. $M_2N(OH)_2$ had stronger affinity to Pb than $M_2C(OH)_2$, which was evidenced by the lower formation energy of $Ti_2N(O_2H_{2-2X}Pb_X)$ than that of $Ti_2C(O_2H_{2-2X}Pb_X)$. Overall, $Ti_2C(OH)_2$ is most ideal MXene nanosheets for heavy metal ions removal.

Another research has also confirmed the roles of surface groups in affecting the heavy metal ions adsorption on MXenen nanosheets using MD simulations [38]. Pb and Cu ions adsorption on Ti_2C_2, V_2C_1, and T_2C_1 was investigated. Various groups including H, OH, and F were decorated on the surface of MXene. It was found that both adsorption capacities and binding energies were sensitive to the surface functional groups. All the MXenes exhibited excellent adsorption ability to Pb and Cu. Especially, the Pb adsorption capability could be as high as 2560 mg/g.

4 Syntheses Methods

4.1 MOF Nanosheets

Top-down and bottom-up methods are currently mainly used methods to fabricate MOF nanosheets. In top-down process, bulk MOFs are exfoliated into nanosheets via the weak interaction such as Van der Waals forces and hydrogen bonding [61, 62]. Sonication [62], mechanical [63], intercalation [64], and chemical exfoliations [65] are all belong to top-down methods. Zamora et al. firstly reported the synthesis of MOF nanosheets ($[Cu_2Br(IN)_2]_n$) by exfoliation [66]. This kind of nanosheets is composed of one Cu dimer with ligands of bromine and isonicotinato. The layers isonicotinato aromatic rings were stacked via π-π interactions, which could be deteriorated by mechanical force, and nanosheets with thickness of ~0.5 nm were fabricated. MOF-2 (ZnBDC, BDC = 1, 4-benzenedicarboxylate) nanosheet was synthesized by the similar strategy. The thickness and lateral dimensions were 1.5–6.0 nm and 100 μm, respectively. In this work, the effect of various solvents was paid a special attention [67].

$Zn_2(bim)_4$ (bim = benzimidazole) was exfoliated by combination technique of sonication and well ball-milling [62]. 2D layers with interlayer thickness of 0.988 nm were stacked with weak Van der Waals interactions. Exfoliation of stacked crystals was realized with treatment by wet ball-milled and the consequent sonication, and thus fabricated nanosheets exhibited thickness of 1.12 nm. Recently, $Ni_8(5\text{-BBDC})_6(m\text{-OH})_4$ (denoted as MAMS-1, BBDC = 5-tert-butyl- 1,3-benzenedicarboxylic acid) nanosheets were fabricated by freeze–thaw process using MAMS-1 as the raw materials [68]. Dispersion of MAMS-1 powder in hexane was firstly carried out and then they were frozen in liquid nitrogen of 196 °C. Finally, thawing in hot water of 80 °Cwas executed. At the solid/liquid interface, nanosheets were formed by the shear force induced by volume change of hexane. Thus fabricated nanosheets are with thickness and lateral size of ~4 nm and 10.7 mm, respectively.

A combined intercalation/chemical strategy to fabricate MOF nanosheets via exfoliation using organic ligands as intercalants was reported by Zhou et al. [65]. Chemically labile dipyridyl ligand (i.e., 4,40- dipyridyl disulfide (DPDS)) was inserted into layered MOF (Zn_2(PdTCPP) (TCPP = tetrakis(4-carboxyphenyl)-porphyrin)). The interactions among the MOFs interlayer could be weakened by the coordination between DPDS and metal nodes. Single-layer MOF nanosheet with yield of 57% was fabricated by exfoliation through the reduction of disulfide bonds with trimethylphosphine.

Different from top-down method, in bottom-up process, MOFs nanosheets are directly synthesized from organic precursors and metals. During bottom-up process, lateral directional growth is not affected but the growth in vertical direction is restricted. Interfacial synthesis [69], three-layer synthesis [70], surfactant-assisted synthesis [71], modulated synthesis [72], and sonication synthesis [73] are all belong to bottom-up fabrication methods.

Interfacial synthesis is the most widely used method among all the bottom-up methods. The main interfacial synthesis methods include liquid/liquid, liquid/air, and liquid/solid processes. Generally, two immiscible liquids such as water/dichloromethane and water/ethyl acetate were used to dissolve organic ligands and metal ions in liquid/liquid process [74, 75]. Synthesis of some MOF nanosheets (e.g., $Fe(py)_2[Pt(CN)_4]$ (py = pyridine)) is an example of interface synthesis [76]. Particularly, liquid/air synthesis is the most widely used interfacial synthesis. In liquid/air process, to control MOF nucleation and growth was controlled by reaction occurs at solvent interface. The thickness could be controlled by organic ligands dispersion. For example, Nano-a (nickel bis(dithiolene) nanosheets) nanosheet was fabricated by coordination reaction that occurred at the interface of benzenehexathiol (BHT) and nickel acetate [69]. Ethyl acetate thin layer containing BHT spread on the surface of aqueous solution of $Ni(OAc)_2/NaBr$. By this method, nano-1 nanosheet with thickness of 0.6 nm was formed.

Although with the wide application, interfacial synthesis is faced with a challenge in the low yield. In order to deal with this, a three-layer synthesis strategy was explored [70]. In this process, DMF and acetonitrile were used as miscible solvents. Because of the different densities, acetonitrile was on the top and DMF was at the bottom, and middle layer composed of equal DMF and acetonitrile was formed. $Cu(NO_3)_2$ was dissolved in the top layer and BDC ligand was in the bottom layer. With the continuous diffusion of Cu^{2+} and BDC ligands into the middle layer, MOF nanosheets are formed. MOF nanosheets synthesis by this strategy had a very high yield. CuBDS nanosheets fabricated in this process are characterized by the thickness of 4–25 nm and lateral side of 0.5–4.0 μm, respectively. This method also displayed great practicability in synthesis of many other MOF nanosheets such as ZnBDC, CoBDC, Cu(1,4-NDC), and Cu(2,6-NDC) (NDC = naphthalenedicarboxylate).

Surfactant-assisted method has been used to fabricate MOF nanosheets with a thickness below 10 nm [77]. The MOF products formation difference in the process with and without surfactant was compared. Without surfactant, only the bulk crystals were formed. When the surfactant of (Polyvinylpyrrolidone, PVP) was introduced into the synthesis system, nanosheets with a thickness of ~8 nm were obtained because of the restricted growth in the vertical direction. In addition to the surfactants, many other small molecules (e.g., acetic acid and pyridine) could also play the same role [78]. They can coordinate with specific planes of MOF crystals and induce the anisotropic MOFs growth [79]. In addition to the methods mentioned above, many other ones (e.g., chemical vapor deposition (CVD)) could also be used to synthesize MOF nanosheets [80, 81].

Besides the intrinsic synthesis, functionalization of MOF nanosheets is also an important aspect [82]. Recently, various functional materials such as noble metals [83], metal sulfide [84], and carbon nanomaterials [85] were incorporated into MOF nanosheets to obtain the maximized synergetic effect.

4.2 Zeolite Nanosheet

Zeolites nanosheets with various interlayer linkers have been synthesized [86–89] with the purpose to tune various catalytic systems, for example, isomerization and hydrocracking [90], selective formation of molecules [91], and epoxidation reaction [92]. Moreover, high-aspect-ratio zeolite nanosheets could further assemble into zeolite films, which shows great potential in molecular sieving. The currently used method to synthesize zeolite nanosheets was exfoliation. Nevertheless, because of morphological damage (aggregation, curling, and fragmentation) and structural deterioration, it is still faced a great challenge to synthesize well-dispersed zeolite nanosheets with high-aspect ratio. Recently, Varoon et al. reported the synthesis of highly crystallized MWW and MFI nanosheets derived from ITQ-1 and multilamellar silicalite-1/polystyrene (PS) blending [93]. Thus fabricated nanosheets, which exhibited a highly crystallized flake-like structure with a thickness of 2.6 ± 0.3–3.4 ± 0.3 nm, could be further deposited on the surface of the porous substrate to obtain sieving membranes.

Zeolite nanosheets could be significantly affected by the thickness and wrinkling. Taking this under consideration, a 3D mapping method was exploited to precisely control the wrinkles and thickness of zeolite nanosheets [94]. Thus synthesized MFI nanosheets exhibited a thickness of 3.0 nm and a surface roughness of 0.8 nm.

In order to decrease the production cost, many other strategies in addition to the exfoliation with expensive cost and low yielding were explored to synthesize zeolite nanosheets. For example, MFI, sodalite (SOD), and Linde Type A (LTA) zeolite membranes have been fabricated by in-situ microwave synthesis and/or microwave-assisted secondary growth methods (SGM) considering its rapid yield [95]. Nonetheless, these methods always require multiple steps including lengthy hydrothermal synthesis, drying, and calcination for template removal procedure such as drying and calcination. Thus they are actually labor and energy intensive. Consequently, optimization is needed to make these methods more acceptable. Just recently, tetrapropyl ammonium cation (TPA) was used as a structure-directing agent (SDA) to fabricate MFI nanosheets with high-aspect ratio by a bottom-up seeded growth method [96]. Thus fabricated nanosheets displayed thickness of 5 nm and the pores are highly oriented. This method displayed obvious advantages compared to exfoliation. In general, the fabrication methods of zeolite membranes include.

(a) lodging of zeolite onto matrices (e.g., silica and polymers).
(b) layering zeolite onto a support (e.g., ceramic, clay, and carbon).
(c) self-supporting film.

Mobil five (MFI) zeolite membrane has been widely prepared by direct in-situ crystallization.

4.3 MXene Nanosheet

The most usual method to synthesis MXene is etching of A element layers (Al or Si) from MAX phase by HF (Fig. 4) [32]. In this process, MAX powders are dissolved in HF aqueous solution, followed by the centrifugation and filtration to separate solid from suspensions. The solids are then washed till the pH value of suspension reaches 4–6. Various types of MXenes, such as M2X, M3X2, M4X3 MXenes, have been synthesized via this method [34]. Especially, Ti_3AlC_2 is the most commonly used MAX phase for the preparation of Ti_3C_2Tx MXene.

MXenes obtained using etching method are always terminated with -F, -OH, and -O- (Tx) with multilayer structures [97]. Multilayered MXenes can be treated by

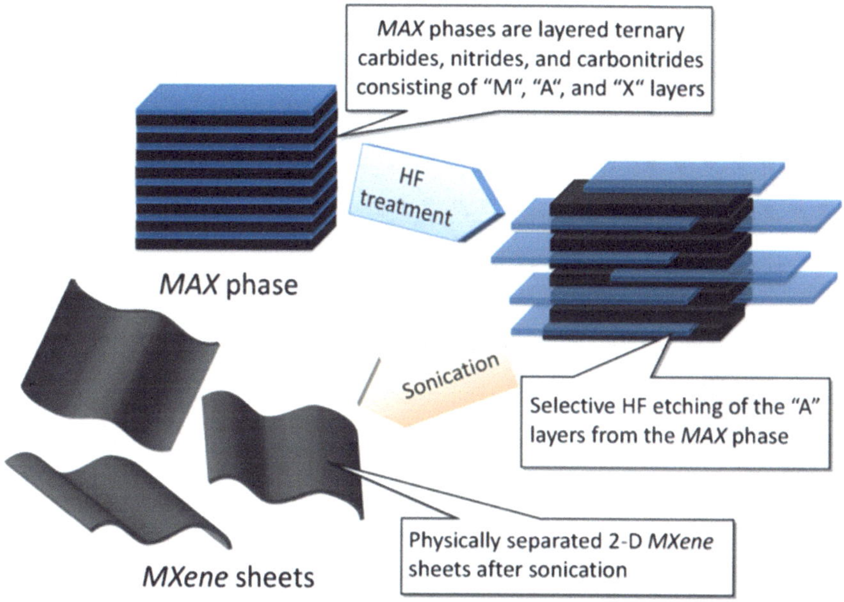

Fig. 4 MXene synthesis by etching A layers from MAX phase using HF (a) processes. Figures are reproduced with permissions from (Ref. 32)

intercalation of small molecules (DMSO) or ions (Li^+) [98] to increase the interlayer distance by weak Van der Waals forces. The intercalated MXenes could be further delaminated by sonication (Fig. 5a) [99]. Direct use of HF always involves multiple hazardous materials [100]. Thus, besides HF, other etchants such as in-situ HF formed by the reaction of HCl and fluoride salt (e.g., LiF) are also used (Fig. 5b and c).

5 Experimental Work

5.1 *MOF Nanosheets*

Various heavy metal ions including arsenic (As), cadmium (Cd), mercury (Hg), and Uranium (U) adsorption on diverse MOFs were removed by adsorption using various MOFs. As exists in forms of As^{5+} or As^{3+}. As adsorption on MOFs was firstly studied by Wang et al. using UiO-66. High surface area and numerous adsorption sites endowed UiO-66 with high adsorption capacity of 303 mg/g at pH of 2 [101]. High adsorption ability was ascribed to the hydroxyl group and benzene dicarboxylate ligand. Strong Zr–O bonding and hydrolysis stability render arsenic removal ability to MOF-808. Li et al. fabricated monodispersed MOF-808 octahedral nanoparticles using microwave-assisted method and used them in removing

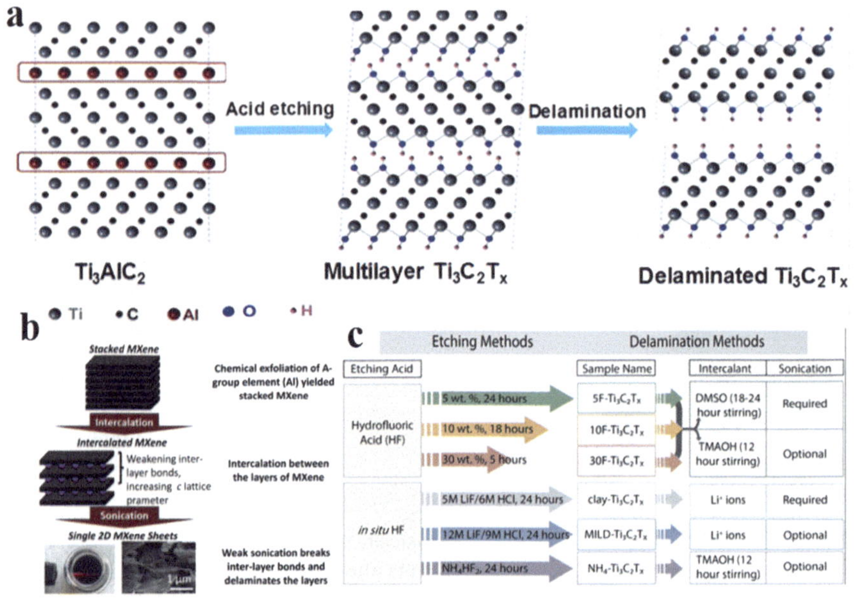

Fig. 5 **a** and **b** Multilayered and delaminated MXenes obtained during the process of acid etching; **c** various etching and delamination methods. Figures are reproduced with permissions from (Ref. 99)

arsenic from wastewater with an initial concentration of 5 mg/L [102]. The adsorption capacity was 24.8 mg/g, which was higher than other porous materials such as silica and aluminum oxide. Audu et al. reported both As^{5+} and As^{3+} on UiO–MOFs involving Zr^{6+} node with missing-linker sites and a thiolate organic ligand [103]. The binding ability of missing-linker Zr^{6+} node and capture capability of thiolate linkers, UiO-66 could adsorb both As^{5+} and As^{3+}. Ma et al. anchored $-SO_3H$ group on $Cu_3(BTC)_2$ by post-modification and oxidation. The resulted $Cu_3(BTC)_2$–SO_3H displayed un adsorption capacity of 88.7 mg/g for Cd, which was higher than that of benchmark adsorbents [104]. Hong et al. investigated the role of MOF cavity for Cd adsorption [105]. They used FJI-H9 to adsorb Cd and found that the high adsorption was attributed to the synergetic effect of suitable cavity and strong hydrogen bonding interaction. Chromium exists in wastewater in the form of either CrO_4 or Cr_2O_7. Due to severe detrimental effect of Cr on both the environment and the human beings, extensive research has focused on the Cr removal by various adsorbents, among which MOFs were a great alternative [106–108]. Zhang et al. used two cationic MOFs (FIR-53 and FIR-54) fabricated through ion-exchange technique to adsorb Cr, and both MOFs showed high capacity of over 100 mg/g [109]. Chen et al. modified MOF-867 using methyl groups to obtain a cationic Zr-MOF (ZJU-101) and used them to adsorb Cr. The uptake capacity was as high as 245 mg/g, higher than any known porous solids [110]. Xu et al. reported Hg removal by various MOFs containing sulfur-containing ligands [111–113]. They fabricated a kind of Zr-DMBN by introducing thiol groups into MOFs. Zr–DMBN could lower Hg concentration to below 0.01 mg/L after 12 h adsorption. Chem et al. reported a kind of sulfur functionalized MOF (FJI-H12) starting from SCN, Co^{2+}, and 2,4,6-tri(1-imidazolyl)-1,3,5-trizaine (Timt) [114]. Adsorption capability of FJI-H12 could be as high as 439.8 mg/g. Li and co-workers synthesized isoreticular luminescent MOFs (LMOF-261, 262, and 263) by introducing fluorescent molecules and sulfone functionalized co-linkers into the framework. The obtained LMOFs were highly water stable and sensitive to Hg, and the detection limit was as low as 3.3 ppb. Wu et al. reported a kind of rod-like MOF-5 nanomaterials for U removal. The maximum sorption capability could be 237 mg/g at pH = 5 and 298 K [115]. The dominant sorption mechanism between U and rod-like MOF-5 was the synergy of surface complexation and electrostatic interaction.

Besides heavy metal ions, organic pollutants remove by MOFs have also been extensively investigated. In 2010, Jhung et al. executed the pioneering work on the applications of MOFs in dyes removal. They used MIL-101, MIL-53, and some modified MOFs for adsorbing MO [116]. High porosity and large pore sizes contribute to the high adsorption capacity. Specific affinities, such as electrostatic interactions, also play important roles. Afterward, adsorption of dyes on other MOFs such as MIL-100(Fe, Cr) was also reported [117]. Apart from the central metals influence on the adsorption ability of MOFs, Bibi et al. investigated the role of surface conditions [118]. MOFs have also shown adsorption ability to agrochemicals, such as 2,4-D [119], MCPP [120], and herbicides [121].

MOF nanosheets have also shown applications in heavy metal ions removal. A facile operating method was adopted to exfoliate few-layered CoCNSP nanosheets from the bulk MOFs (Fig. 6a) [122]. Periodically aligned sulfur species were distributed on the nanosheets. The nanosheets exhibited lateral area of 0.5 μm (Fig. 6b) and thickness of sub-4.0 nm (Fig. 6c). S···S distance of 0.9476 nm verified that the obtained CoCNSP nanosheets constituted few layers. CoCNSP nanosheets displayed excellent heavy metal ions removal in the sequence of Co(II), Zn(II), Cd(II), Ni(II) $\ll$ Cu(II) < Pb(II) < U(VI) < Hg(II) regardless of single-component and/or multi-component adsorption backgrounds. The uptake capacities of the nanosheets toward Hg(II), U(VI), Pb(II), and Cu(II) were 716, 661, 534, and 325 mg/g, respectively. Even with the ionic strength up to 100 mM, CoCNSP nanosheets could also maintain high adsorption efficiency in a wide pH range with good recycling performance. Xu et al. sreported a kind of 2D zinc-based MOF nanosheets of Zn(Bim)(OAc) with high heavy metals capture (Fig. 6d) [123]. Zn(Bim)(OAc) nanosheets showed an ultrathin 2D structure with lateral length of more than 2 μm thickness of 7.05 nm (Fig. 6e and f) with highly exposed active sites. The adsorption capacity for Pb(II) and Cu(II) was 253.8 and 335.57 mg/g, respectively.

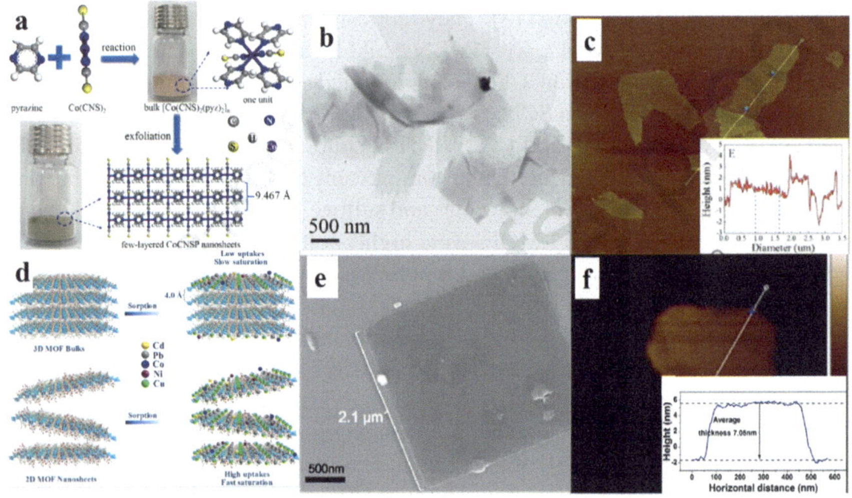

Fig. 6 CoCNSP nanosheets: **a** synthesis routes, **b** TEM and **c** AFM images; and Zn(Bim)(OAc) nanosheets: **d** complex with Pb and Cu, **e** SEM and **f** AFM images. (the inset in c and f shows the corresponding height profiles). Figures are reproduced with permissions from (Refs. [122, 123])

5.2 Zeolite Nanosheets

Thanks to their sub-nanopores and high porosity (30–40%), zeolites have seen lots of applications in adsorption and catalysis [123]. In RO desalination they have also displayed great potential [124]. While bulk zeolite membranes could be directly used as a membrane in desalination [125, 126], zeolite nanoparticles can be used as functional materials in tuning the structures and performance of RO membranes such as introduction into surface polyamide layer [127–129] and/or sublayers [130] to fabricate nanocomposite membranes. In spite of the great potential, direct application of zeolite nanosheets in desalination have not seen in experimental or computational investigations [86] although with reports in liquid adsorption [131] and membranes [132]. Nevertheless, the ultrathin-film nature and versatile pore structures of zeolite nanosheet make them very promising and in future it is speculated to be seen more advance.

5.3 MXene Nanosheets

Although their desalination applications have not been seen, MXenes displayed some unique potential, such as ultrafast water flux and size-selectivity, which has been studied by $Ti_3C_2T_x$ MXene membranes [133]. $Ti_3C_2T_x$ MXene membranes were fabricated by combined exfoliation, delamination, and vacuum filtration. They displayed similar structures as that of GO. $Ti_3C_2T_x$ MXene membranes with interlayer spacing of ~0.6 nm demonstrated ultrafast water flux of 37.4 L/(m^2·h·bar)), which was attributed to the hydrophilic nature and the presence of H2O between the layers. $Ti_3C_2T_x$ showed high selectivity toward single-, double-, and triple-charged metal cations and dye cations of different sizes. For cationic dyes (MB) with large size, the rejection was 100%. For double-charged Mg^{2+}, interlayer space shrank and the permeation was lowered. Inversely, single charged Na^+ expanded the interlayer spacing and enhance the permeation. Such character enables $Ti_3C_2T_x$ to be effectively used in separating higher charged cations.

Ag nanoparticles were used to modify MXene (Ti_3C_2Tx) to enhance the permeations and improve the antifouling capability (Fig. 7a–c) [134]. Ag@MXene with variable Ag nanoparticles loadings were fabricated by in-situ production of Ag via the self-reduction of $Ag(NO_3)_2$ on MXene surface. The intriguing thing was that MXene acted as both the membrane and reducing agent. The optimized Ag loading of 21% enhanced the water flux of MXene significantly from 118 to 420 LMH under the same experiment conditions while the organic molecules rejections were well maintained (Fig. 7d and e). In addition, Ag@MXene displayed better water flux recovery than pristine MXene (Fig. 7f). Moreover, Ag@MXenen exhibited prominently increased antifouling ability toward *Escherichia coli* with more than 99% of inhibition, while the MXene manifested only 60% of bacterial growth inhibition and hydrophilic PVDF as a reference sample exhibited no bacterial inhibition. This kind of membrane can be a good candidate for NF applications (Fig. 7g).

Recently, MXene was intercalated into GO to synthesize MXene/GO lamellar membranes to be used in molecular separations (Fig. 8) [135]. Because GO lamellar membranes are not permeable for organic solvents, MXene nanosheets were intercalated into the interlayers of GO membranes to reduce the spacing distance via vacuum filtration method. The composite membranes showed excellent water and organic solvent permeations. The flux of pure solvents of water, acetone, methanol, ethanol, and IPA was 21.02, 48.32, 25.03, 10.76, and 6.18 $L{\cdot}m^{-2}{\cdot}h^{-1}$ (LMH), respectively, and the rejections for all the dyes were over 990% in aqueous and organic solvents.

MXene-based 2D nanosheets have also shown usage in heavy metal ions removal [136]. Considering that Ti_3C_2Tx MXene might be oxidized in aqueous media, magnetic Fe_2O_3 nanoparticles were deposited onto the surface of MXene (MGMX) by solvothermal method using $Fe(CH_3COO)_2$ as a precursor to enhance their stability (Fig. 9) [137]. MGMX could adsorb 99.9% OF Hg^{2+} with capacity of 1128.41 mg/g even in the existence of background metal ions. The high adsorption was attributed to the presence of large amount of O- and OH- groups together with negatively charged surfaces. Magnetic properties of MGMX made them easily separated from aqueous solutions after the adsorption.

And recently, a kind of $Ti_3C_2T_x$ MXene-based core–shell spheres (MX-SA) containing MXene core and alginate (SA) shell were fabricated and used as adsorbents for Hg^{2+} (Fig. 10) [138]. The spheres were fabricated using varied concentrations of $Ti_3C_2T_x$ MXene and SA via $CaCl_2$ as the crosslinking agent. The optimized MX/SA ratio was 4:20. Under this condition, the spheres exhibited the inner core

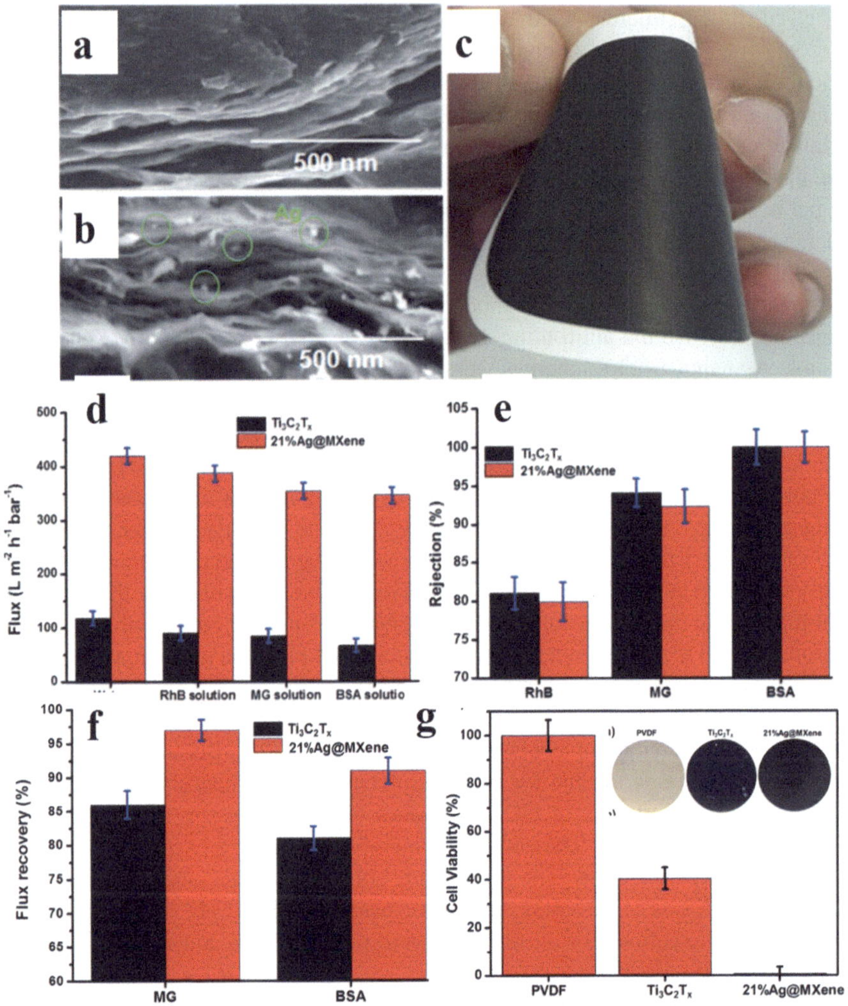

Fig. 7 Cross-sectional SEM for **a** MXene, and **b** 21% Ag@MXene, **c** photograph of 21% Ag@MXene membrane, comparison of the filtration performance of MXene and 21 Ag@MXene for RhB, MG, and BSA molecules **d** flux and **e** rejection. Figures are reproduced with permissions from (Ref. 134)

with thickness of ~1.5 mm and outer shell wall of ~20 μm. Unique inside structures including high porosities, large specific surface areas, and oxygenated functional groups of MXenes enables MX-SA to effectively adsorb Hg^{2+}. Especially, binding group of [Ti–O]–H^+ had strong metal–ligand interaction with Hg^{2+}. Additionally, MX-SA spheres could further provide cage-like spherical structures that can entrap Hg^{2+}. Adsorption capacity of Hg^{2+} was 932 mg/g with removal efficiency of 100% and other five heavy metal ions removal were all >90%. The binding of Hg^{2+} with

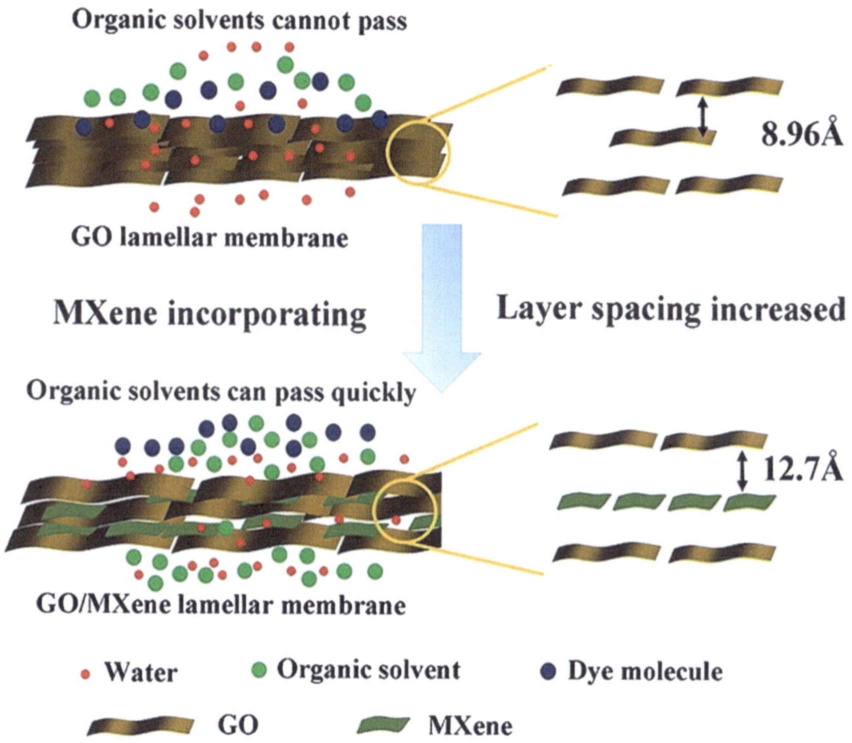

Fig. 8 Intercalation of MXene into GO to fabricated lamellar GO/MXene membrane for solvent permeation and molecular separations. Figures are reproduced with permissions from (Ref. 135)

MX-SA spheres was so strong that only >8 M HCl could desorb the adsorbed Hg^{2+}, which was similar as that of MOFs [139]. It should be noted that with MX-SA treatment, Hg^{2+} concentration could be reduced to be as low as 1.8 ppb.

6 Research Gaps

Bringing 2D nanosheets-based water purification technology from the lab-scale experiments and simulations to the industrial and/or field-level applications have remained as one of the major challenges. For MOFs, MXene and zeolite nanosheets, although their separation performance has been successfully evidenced in lab scale, even now it is still difficult to realize their practical usage. The first challenge is the sophisticated synthesis procedures which always involve the usage of abundant organic solvents. Besides, the large-scale synthesis of nanosheets is also challenging, which hinders their widespread applications in adsorption. In addition, when being used in desalination, the interlayer spacing distance is hard to maintain under the

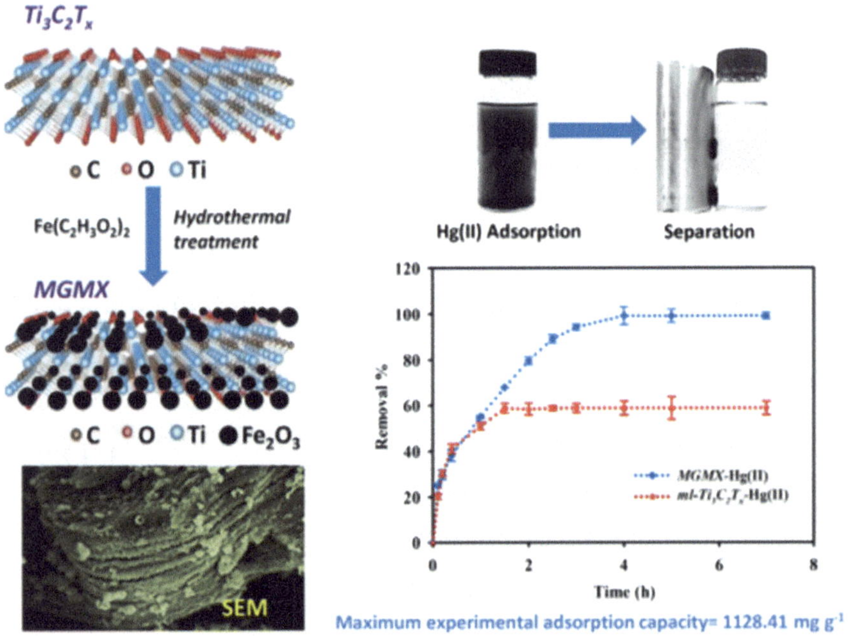

Fig. 9 Schematic illustration for the synthesis of MGMX, SEM images and the adsorption ability for Hg. Figures are reproduced with permissions from (Ref. 137)

harsh ionic environment. The removal efficiency toward single valent ions, such as Na^+ is usually very low, seriously limiting their desalination performance. The harsh surroundings also reduce the stability of the nanosheets.

7 Conclusions and Perspectives

(a) **More advanced fabrication strategies**
Despite the promising research advances, challenges still exist in the synthesis of 2D nanosheets. For example, the fabrication of MOF nanosheet using top-down method is limited by the low yielding. MOF nanosheets obtained by exfoliation method face the problem of stabilization and that obtained via interfacial synthesis are encountered by the low crystallinity. It is urgent to find more effective and facile method for the synthesis of 2D MOF nanosheets. In addition, how to improve the yield should be further investigated. Furthermore, the green synthesis procedures using environmentally friendly solvents should also be taken into considerations. For example, in the fabrication of MXene via exfoliation method, abundant fluorine containing solvents are used, which might bring severe environmental problems.

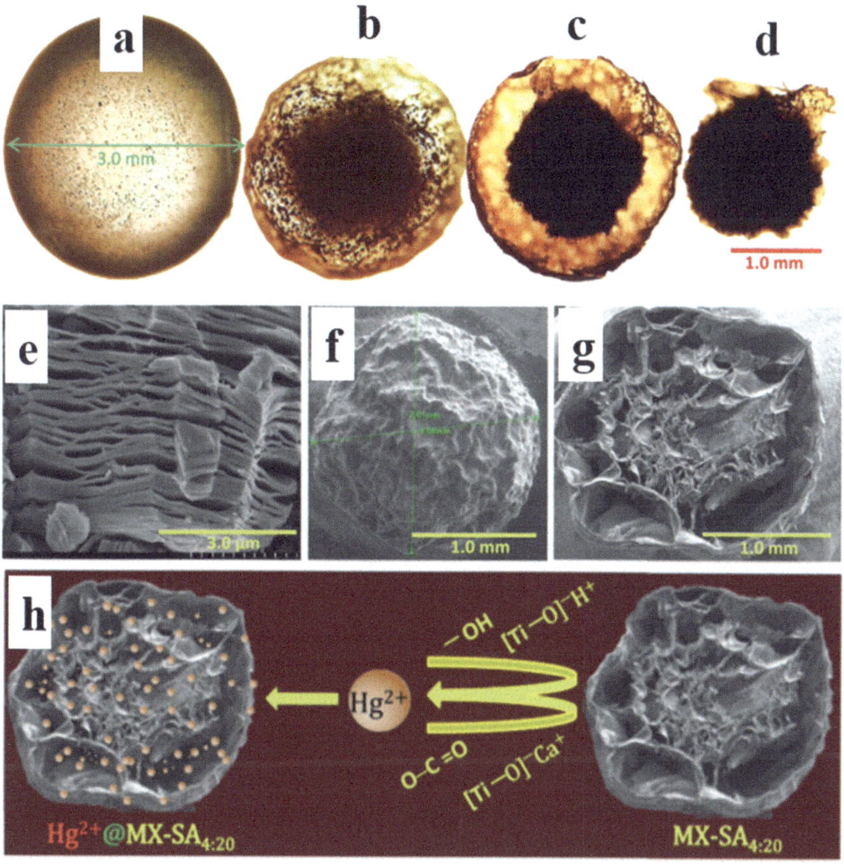

Fig. 10 Optical microscopic images for MX–$SA_{4:20}$: **a** hydrogel spheres (3 mm), **b** following vacuum freeze-drying (~2 mm), **c** cross-sectional view, and **d** core of MX-$SA_{4:20}$ spheres. SEM images for **e** Ti_3C_2Tx MXene, **f** dried MX–$SA_{4:20}$ sphere, and **g** internal structure of sphere cross section. **h** Schematic illustration of Hg^{2+} ions uptake onto MX–$SA_{4:20}$ sphere. Figures are reproduced with permissions from (Ref. 138)

(b) **Precise control and tuning of the interlayer spacing distances**
When 2D nanosheets are used in desalination, the critical point is to find effective method to control the interlayer spacing more precisely. Meanwhile, their stability should also be strengthened. Only by doing this it can be possible to design the membranes constructed by 2D nanosheets with specific purpose toward the unique ions removal.

(c) **Available adsorption sites strengthen**
For 2D nanosheets used as adsorbents, two aspects can be considered in the future research. Firstly, explore an efficacious method to distribute the functional groups on the surface as much as possible to enhance the interactions between the adsorbents and the pollutants. Furthermore, from the point of

the industrial application feasibility and the recycle utilization, perhaps it is desirable to load the nanosheets onto the traditional supports, such as ceramic membranes, polymeric hollow fiber membranes, and electrospun nanofibrous membranes that has been successfully used in water separation,

(d) **Propel the feasibility in industrial scale**
The majority of 2D nanosheets in desalination are based on the theoretical studies. There is still a long way to realize the realistic large-scale desalination and solutes separation using 2D nanosheets. Combined with the research gap analysis, further works should be devoted to the large-scale synthesis of stable membranes. Spacing distance between the 2D nanosheets in membranes should be precisely controlled via greener and more environmentally friendly synthesis routes.

Overall, MOFs, MXene and zeolite nanosheets exhibit great potential in water remediation and desalination owing to their unique structures. With the ongoing efforts, their practical applications and roles in water resource remediation could be promisingly anticipated.

References and Future Readings

1. Van Loosdrecht MC, Brdjanovic D (2014) Anticipating the next century of wastewater treatment. Science 344:1452–1453
2. Werber JR, Osuji CO, Elimelech M (2016) Materials for next-generation desalination and water purification membranes. Nat. Rev. Mater. 1:16018
3. Ismail A, Padaki M, Hilal N, Matsuura T, Lau W (2015) Thin film composite membrane-recent development and future potential. Desalination 356:140–148
4. Yang Z, Saeki D, Matsuyama H (2018) Zwitterionic polymer modification of polyamide reverse-osmosis membranes via surface amination and atom transfer radical polymerization for anti-biofouling. J. Membr. Sci. 550:332–339
5. Goh P, Ismail A, Hilal N (2016) Nano-enabled membranes technology: sustainable and revolutionary solutions for membrane desalination? Desalination 380:100–104
6. Xu GR, Xu JM, Feng HJ, Zhao HL, Wu SB (2017) Tailoring structures and performance of polyamide thin film composite (PA-TFC) desalination membranes via sublayers adjustment-a review. Desalination 417:19–35
7. Thür R, Corvilain M, Klaysom C, Hartanto Y, Vankelecom IF (2019) Tuning the selectivity of thin film composite forward osmosis membranes: effect of co-solvent and different interfacial polymerization synthesis routes. Sep. Purif. Technol. 227:115671
8. Cohen-Tanugi D, McGovern RK, Dave SH, Lienhard JH, Grossman JC (2014) Quantifying the potential of ultra-permeable membranes for water desalination. Energy Environ. Sci. 7:1134–1141
9. Karabelas AJ, Koutsou CP, Kostoglou M, Sioutopoulos DC (2018) Analysis of specific energy consumption in reverse osmosis desalination processes. Desalination 431:15–21
10. Gao Q, Xu J, Bu XH (2019) Recent advances about metal–organic frameworks in the removal of pollutants from wastewater. Coordin. Chem. Rev. 378:17–31
11. Montes-Hernandez G, Concha-Lozano N, Renard F, Quirico E (2009) Removal of oxyanions from synthetic wastewater via carbonation process of calcium hydroxide: Applied and fundamental aspects. J. Hazard. Mater. 166:788–795

12. Dujardin MC, Cazé C, Vroman I (2000) Ion-exchange resins bearing thiol groups to remove mercury: Part 1: synthesis and use of polymers prepared from thioester supported resin. React. Funct. Polym. 43:123–132
13. Hadi P, To MH, Hui CW, Lin CSK, McKay G (2015) Aqueous mercury adsorption by activated carbons. Water Res. 73:37–55
14. Tchinda AJ, Ngameni E, Kenfack IT, Walcarius A (2009) One-step preparation of thiol-functionalized porous clay heterostructures: application to hg(ii) binding and characterization of mass transport issues. Chem. Mater. 21:4111–4121
15. Homaeigohar S, Elbahri M (2017) Graphene membranes for water desalination. NPG Asia Mater. 9:e427
16. Abraham J, Vasu KS, Williams CD, Gopinadhan K, Su Y, Cherian CT, Dix J, Prestat E, Haigh SJ, Grigorieva IV (2017) Tunable sieving of ions using graphene oxide membranes. Nat. Nanotechnol. 12:546–550
17. Heiranian M, Farimani AB, Aluru NR (2015) Water desalination with a single-layer MoS2 nanopore. Nat. Commun. 6:8616
18. Gao H, Shi Q, Rao D, Zhang Y, Su J, Liu Y, Wang Y, Deng K, Lu R (2017) Rational design and strain engineering of nanoporous boron nitride nanosheet membranes for water desalination. J. Phys. Chem. C. 121:22105–22113
19. Khataee A, Bayat G, Azamat J (2017) Molecular dynamics simulation of salt rejection through silicon carbide nanotubes as a nanostructure membrane. J. Molecul. Graph. Modell. 71:176–183
20. Jamali SH, Vlugt TJ, Lin LC (2017) Atomistic understanding of zeolite nanosheets for water desalination. J. Phys. Chem. C. 121:11273–11280
21. Khajeh M, Laurent S, Dastafkan K (2013) Nanoadsorbents: classification, preparation, and applications (with emphasis on aqueous media). Chem. Rev. 113:7728–7768
22. Manos MJ, Petkov VG, Kanatzidis MG (2009) $H_{2x}Mn_xSn_{3-x}S_6$ (x= 0.11–0.25): a novel reusable sorbent for highly specific mercury capture under extreme ph conditions. Adv. Funct. Mater. 19:1087–1092
23. Pi Y, Li X, Xia Q, Wu J, Li Y, Xiao J, Li Z (2018) Adsorptive and photocatalytic removal of persistent organic pollutants (pops) in water by metal-organic frameworks (MOFs). Chem. Eng. J. 337:351–371
24. Zhou J, Wang B (2017) Emerging crystalline porous materials as a multifunctional platform for electrochemical energy storage. Chem. Soc. Rev. 46:6927–6945
25. Eddaoudi M, Kim J, Rosi N, Vodak D, Wachter J, O'Keeffe M, Yaghi OM (2002) Systematic design of pore size and functionality in isoreticular mofs and their application in methane storage. Science 295:469–472
26. Li H, Eddaoudi M, O'Keeffe M, Yaghi OM (1999) Design and synthesis of an exceptionally stable and highly porous metal-organic framework. Nature 402:276–279
27. Silva P, Vilela SMF, Tomé JPC, Almeida Paz FA (2015) Multifunctional metal–organic frameworks: from academia to industrial applications. Chem. Soc. Rev. 44:6774–6803
28. Li J, Wang X, Zhao G, Chen C, Chai Z, Alsaedi A, Hayat T, Wang X (2018) Metal-organic framework-based materials: superior adsorbents for the capture of toxic and radioactive metal ions. Chem. Soc. Rev. 47:2322–2356
29. Humplik T, Lee J, O'Hern SC, Fellman BA, Baig MA, Hassan SF, Atieh MA, Rahman F, Laoui T, Karnik R (2011) Nanostructured materials for water desalination. Nanotechnology 22:292001
30. Xu L, Sun J (2006) Recent advances in the synthesis and application of two-dimensional zeolites. Adv. Energy Mater. 6:1600441
31. Pophale R, Cheeseman PA, Deem MW (2011) A database of new zeolite-like materials. Phys. Chem. Chem. Phys. 13:12407–12412
32. Naguib M, Mashtalir O, Carle J, Presser V, Lu J, Hultman L, Gogotsi Y, Barsoum MW (2012) Two-dimensional transition metal carbides. ACS Nano 6:1322–1331
33. Anasori B, Xie Y, Beidaghi M, Lu J, Hosler BC, Hultman L, Kent PRC, Gogotsi Y, Barsoum MW (2015) Two-dimensional, ordered, double transition metals carbides (mxenes). ACS Nano 9:9507–9516

34. Naguib M, Mochalin VN, Barsoum MW, Gogotsi Y (2014) MXenes: a new family of two-dimensional materials. Adv. Mater. 26:992–1005
35. Ling Z, Ren CE, Zhao MQ, Yang J, Giammarco JM, Qiu J, Barsoum MW, Gogotsi Y (2014) Flexible and conductive MXene films and nanocomposites with high capacitance. P. Natl. Acad. Sci. 111:16676–16681
36. Dikin DA, Stankovich S, Zimney EJ, Piner RD, Dommett GHB, Evmenenko G, Nguyen ST, Ruoff RS (2007) Preparation and characterization of graphene oxide paper. Nature 448:457–460
37. Zhang Q, Du Q, Hua M, Jiao T, Gao F, Pan B (2013) Sorption enhancement of lead ions from water by surface charged polystyrene-supported nano-zirconium oxide composites. Environ. Sci. Technol. 47:6536–6544
38. Guo X, Zhang X, Zhao S, Huang Q, Xue J (2016) High adsorption capacity of heavy metals on two-dimensional MXenes: an ab initio study with molecular dynamics simulation. Phys. Chem. Chem. Phys. 18:228–233
39. Guo J, Peng Q, Fu H, Zou G, Zhang Q (2015) Heavy-metal adsorption behavior of two-dimensional alkalization-intercalated mxene by first-principles calculations. J. Phys. Chem. C. 119:20923–20930
40. Peng Q, Guo J, Zhang Q, Xiang J, Liu B, Zhou A, Liu R, Tian Y (2014) Unique lead adsorption behavior of activated hydroxyl group in two-dimensional titanium carbide. J. Am. Chem. Soc. 136:4113–4116
41. Tang H, Fang H, Duan Y, Sholl DS (2019) Predictions of hg0 and hgcl2 adsorption properties in uio-66 from flue gas using molecular simulations. J. Phys. Chem. C. 123:5972–5979
42. Xiao C, Silver MA, Wang S (2017) Metal-organic frameworks for radionuclide sequestration from aqueous solution: a brief overview and outlook. Dal Trans. 46:16381–16386
43. Zhu H, Yuan J, Tan X, Zhang W, Fang M, Wang X (2019) Efficient removal of Pb^{2+} by Tb-MOFs: identifying the adsorption mechanism through experimental and theoretical investigations. Environ Sci Nano 6:261–272
44. Zhu H, Wu J, Fang M, Tan L, Chen C, Alharbi NS, Hayat T, Tan X (2017) Synthesis of a core-shell magnetic Fe_3O_4–NH_2@PmPD nanocomposite for efficient removal of Cr(vi) from aqueous media. RSC Adv. 7:36231–36241
45. He T, Ni B, Zhang S, Gong Y, Wang H, Gu L, Zhuang J, Hu W, Wang X (2018) Ultra-thin 2D zirconium metal-organic framework nanosheets: preparation and application in photocatalysis. Small 14:1703929
46. Hu Z, Chen Y, Jiang J (2011) Zeolitic imidazolate framework-8 as a reverse osmosis membrane for water desalination: insight from molecular simulation. J. Chem. Phys. 134:134705
47. Gupta KM, Zhang K, Jiang J (2015) Water desalination through zeolitic imidazolate framework membranes: significant role of functional groups. Langmuir 31:13230–13237
48. Azamat J, Khataee A, Joo SW (2014) Functionalized graphene as a nanostructured membrane for removal of copper and mercury from aqueous solution: a molecular dynamics simulation study. J. Mol. Graph. Modell. 53:112–117
49. Azamat J, Sattary BS, Khataee A, Joo SW (2015) Removal of a hazardous heavy metal from aqueous solution using functionalized graphene and boron nitride nanosheets: insights from simulations. J. Mol. Graph. Modell. 61:13–20
50. Azamat J, Khataee A (2017) Improving the performance of heavy metal separation from water using MoS_2 membrane: molecular dynamics simulation. Computation. Mater. Sci. 137:201–207
51. Azamat J, Khataee A, Sadikoglu F (2018) Computational study on the efficiency of MoS_2 membrane for removing arsenic from contaminated water. J. Mol. Liq. 249:110–116
52. Roth WJ, Čejka J (2011) Two-dimensional zeolites: dream or reality? Catal. Sci. Technol. 1:43–53
53. Wei R, Yang H, Scott JA, Aguey-Zinsou KF, Zhang D (2018) Synthesis of 2D MFI zeolites in the form of self-interlocked nanosheet stacks with tuneable structural and chemical properties for catalysis. Appl. Mater. Today 11:22–33

54. Foo KY, Hameed BH (2011) The environmental applications of activated carbon/zeolite composite materials. Adv. Colloidal Interface Sci. 162:22–28
55. Rassoulinejad-Mousavi SM, Azamat J, Khataee A, Zhang Y (2020) Molecular dynamics simulation of water purification using zeolite MFI nanosheets. Sep. Purif. Technol. 234:116080
56. Kurtoglu M, Naguib M, Gogotsi Y, Barsoum MW (2012) First principles study of two-dimensional early transition metal carbides. MRS Commun. 2:133–137
57. Guo J, Fu H, Zou G, Zhang Q, Zhang Z, Peng Q (2016) Theoretical interpretation on lead adsorption behavior of new two-dimensional transition metal carbides and nitrides. J. Alloys Compd. 684:504–509
58. Mashtalir O, Naguib M, Mochalin VN, Dall'Agnese Y, Heon M, Barsoum MW, Gogotsi Y (2013) Intercalation and delamination of layered carbides and carbonitrides. Nat. Commun. 4:1716
59. Khazaei M, Arai M, Sasaki T, Chung CY, Venkataramanan NS, Estili M, Sakka Y, Kawazoe Y (2013) Novel electronic and magnetic properties of two-dimensional transition metal carbides and nitrides. Adv. Funct. Mater. 23:2185–2192
60. Xie Y, Naguib M, Mochalin VN, Barsoum MW, Gogotsi Y, Yu X, Nam KW, Yang XQ, Kent KAI, PR, (2014) Role of surface structure on Li-ion energy storage capacity of two-dimensional transition-metal carbides. J. Am. Chem. Soc. 136:6385–6394
61. Nicolosi V, Chhowalla M, Kanatzidis MG, Strano MS, Coleman JN (2013) Liquid exfoliation of layered materials. Science 340:1226419
62. Peng Y, Li Y, Ban Y, Jin H, Jiao W, Liu X, Yang W (2014) Metal-organic framework nanosheets as building blocks for molecular sieving membranes. Science 346:1356–1359
63. Abhervé A, Manas-Valero S, Clemente-León M, Coronado E (2015) Graphene related magnetic materials: micromechanical exfoliation of 2D layered magnets based on bimetallic anilate complexes with inserted [Fe III (acac 2-trien)]+ and [Fe III (sal 2-trien)]$^+$ molecules. Chem. Sci. 6:4665–4673
64. Wang HS, Li J, Li JY, Wang K, Ding Y, Xia XH (2017) Lanthanide-based metal-organic framework nanosheets with unique fluorescence quenching properties for two-color intracellular adenosine imaging in living cells. NPG Asia Mater. 9:e354
65. Ding Y, Chen YP, Zhang X, Chen L, Dong Z, Jiang HL, Xu H, Zhou HC (2017) Controlled intercalation and chemical exfoliation of layered metal–organic frameworks using a chemically labile intercalating agent. J. Am. Chem. Soc. 139:9136–9139
66. Amo-Ochoa P, Welte L, González-Prieto R, Miguel PJS, Gómez-García CJ, Mateo-Martí E, Delgado S, Gómez-Herrero J, Zamora F (2010) Single layers of a multifunctional laminar Cu (I, II) coordination polymer. Chem. Commun. 46:3262–3264
67. Li PZ, Maeda Y, Xu Q (2011) Top-down fabrication of crystalline metal–organic framework nanosheets. Chem. Commun. 47:8436–8438
68. Wang X, Chi C, Zhang K, Qian Y, Gupta KM, Kang Z, Jiang J, Zhao D (2017) Reversed thermo-switchable molecular sieving membranes composed of two-dimensional metal-organic nanosheets for gas separation. Nat. Commun. 8:14460
69. Kambe T, Sakamoto R, Hoshiko K, Takada K, Miyachi M, Ryu JH, Sasaki S, Kim J, Nakazato K, Takata M (2013) π-Conjugated nickel bis (dithiolene) complex nanosheet. J. Am. Chem. Soc. 135:2462–2465
70. Rodenas T, Luz I, Prieto G, Seoane B, Miro H, Corma A, Kapteijn F, Xamena FXL, Gascon J (2015) Metal-organic framework nanosheets in polymer composite materials for gas separation. Nat. Mater. 14:48–55
71. Junggeburth SC, Diehl L, Werner S, Duppel V, Sigle W, Lotsch BV (2013) Ultrathin 2D coordination polymer nanosheets by surfactant-mediated synthesis. J. Am. Chem. Soc. 135:6157–6164
72. Sakata Y, Furukawa S, Kondo M, Hirai K, Horike N, Takashima Y, Uehara H, Louvain N, Meilikhov M, Tsuruoka T (2013) Shape-memory nanopores induced in coordination frameworks by crystal downsizing. Science 339:193–196
73. Zhao S, Wang Y, Dong J, He CT, Yin H, An P, Zhao K, Zhang X, Gao C, Zhang L (2016) Ultrathin metal-organic framework nanosheets for electrocatalytic oxygen evolution. Nat. Energy 1:16184

74. Clough AJ, Skelton JM, Downes CA, Rosa AA, Yoo JW, Walsh A, Melot BC, Marinescu SC (2017) Metallic conductivity in a two-dimensional cobalt dithiolene metal-organic framework. J. Am. Chem. Soc. 139:10863–10867
75. Clough AJ, Yoo JW, Mecklenburg MH, Marinescu SC (2015) Two-dimensional metal-organic surfaces for efficient hydrogen evolution from water. J. Am. Chem. Soc. 137:118–121
76. Zhao M, Huang Y, Peng Y, Huang Z, Ma Q, Zhang H (2018) Two-dimensional metal-organic framework nanosheets: synthesis and applications. Chem. Soc. Rev. 47:6267–6295
77. Zhao M, Wang Y, Ma Q, Huang Y, Zhang X, Ping J, Zhang Z, Lu Q, Yu Y, Xu H (2015) Ultrathin 2D metal-organic framework nanosheets. Adv. Mater. 27:7372–7378
78. Hermes S, Witte T, Hikov T, Zacher D, Bahnmüller S, Langstein G, Huber K, Fischer RA (2007) Trapping metal-organic framework nanocrystals: an in-situ time-resolved light scattering study on the crystal growth of mof-5 in solution. J. Am. Chem. Soc. 129:5324–5325
79. Tsuruoka T, Furukawa S, Takashima Y, Yoshida K, Isoda S, Kitagawa S (2009) Nanoporous nanorods fabricated by coordination modulation and oriented attachment growth. Angew. Chem. Int. Ed. 48:4739–4743
80. Dmitriev A, Spillmann H, Lin N, Barth JV, Kern K (2003) Modular assembly of two-dimensional metal-organic coordination networks at a metal surface. Angew. Chem. Int. Ed. 115:2774–2777
81. Wurster B, Grumelli D, Hötger D, Gutzler R, Kern K (2016) Driving the oxygen evolution reaction by nonlinear cooperativity in bimetallic coordination catalysts. J. Am. Chem. Soc. 138:3623–3626
82. Zhu QL, Xu Q (2014) Metal-organic framework composites. Chem. Soc. Rev. 3:5468–5512
83. Zhan G, Zeng HC (2016) Synthesis and functionalization of oriented metal–organic-framework nanosheets: toward a series of 2D catalysts. Adv. Funct. Mater. 26:3268–3281
84. Lu Q, Zhao M, Chen J, Chen B, Tan C, Zhang X, Huang Y, Yang J, Cao F, Yu Y (2006) In situ synthesis of metal sulfide nanoparticles based on 2D metal-organic framework nanosheets. Small 12:4669–4674
85. Kumar R, Jayaramulu K, Maji TK, Rao CNR (2014) Growth of 2D sheets of a MOF on graphene surfaces to yield composites with novel gas adsorption characteristics. Dalton T 43:7383–7386
86. Roth WJ, Nachtigall P, Morris RE, Cejka J (2014) Two-dimensional zeolites: current status and perspectives. Chem. Rev. 114:4807–4837
87. Luo Y, Wang Z, Jin S, Zhang B, Sun H, Yuan X, Yang W (2016) Synthesis and crystal growth mechanism of ZSM-22 zeolite nanosheets. Cryst. Eng. Comm. 18:5611–5615
88. Chen HL, Li SW, Wang YM (2015) Synthesis and catalytic properties of multilayered MEL-type titanosilicate nanosheets. J. Mater. Chem. A 3:5889–5990
89. Choi M, Na K, Kim J, Sakamoto Y, Terasaki O, Ryoo R (2009) Stable single-unit-cell nanosheets of zeolite MFI as active and long-lived catalysts. Nature 461:246–249
90. Ren L, Guo Q, Kumar P, Orazov M, Xu D, Alhassan SM, Mkhoyan KA, Davis ME, Tsapatsis M (2015) Self-pillared, single-unit-cell sn-mfi zeolite nanosheets and their use for glucose and lactose isomerization. Angew Chem Int Ed 54:10848–10851
91. Hu S, Shan J, Zhang Q, Wang Y, Liu Y, Gong Y, Wu Z, Dou T (2012) Selective formation of propylene from methanol over high-silica nanosheets of MFI zeolite. Appl. Catal. A. 445:215–220
92. Na K, Jo C, Kim J, Ahn WS, Ryoo R (2011) MFI titanosilicate nanosheets with single-unit-cell thickness as an oxidation catalyst using peroxides. ACS Catal. 1:901–907
93. Varoon K, Zhang X, Elyassi B, Brewer DD, Gettel M, Kumar S, Lee JA, Maheshwari S, Mittal A, Sung CY (2011) Dispersible exfoliated zeolite nanosheets and their application as a selective membrane. Science 334:72–75
94. Kumar P, Agrawal KV, Tsapatsis M, Mkhoyan KA (2015) Quantification of thickness and wrinkling of exfoliated two-dimensional zeolite nanosheets. Nat. Commun. 6:7128
95. Zhu B, Li B, Zou L, Hill AJ, Zhao D, Lin JY, Duke M (2013) Wiley Online Library
96. Jeon MY, Kim D, Kumar P, Lee PS, Rangnekar N, Bai P, Shete M, Elyassi B, Lee HS, Narasimharao K (2017) Ultra-selective high-flux membranes from directly synthesized zeolite nanosheets. Nature 543:690–694

97. Anasori B, Lukatskaya MR, Gogotsi Y (2017) 2D metal carbides and nitrides (MXenes) for energy storage. Nat. Rev. Mater. 2:16098
98. Ma R, Sasaki T (2015) Two-dimensional oxide and hydroxide nanosheets: controllable high-quality exfoliation, molecular assembly, and exploration of functionality. Acc. Chem. Res. 48:136–143
99. Rasool K, Pandey RP, Rasheed PA, Buczek S, Gogotsi Y, Mahmoud KA (2019) Water treatment and environmental remediation applications of two-dimensional metal carbides (MXenes). Mater. Today 30:80–102
100. Ghidiu M, Lukatskaya MR, Zhao MQ, Gogotsi Y, Barsoum MW (2014) Conductive two-dimensional titanium carbide 'clay' with high volumetric capacitance. Nature 516:78–81
101. Wang C, Liu X, Chen JP, Li K (2015) Superior removal of arsenic from water with zirconium metal-organic framework UiO-66. Sci. Rep. 5:16613
102. Li ZQ, Yang JC, Sui KW, Yin N (2015) Facile synthesis of metal-organic framework MOF-808 for arsenic removal. Mater. Lett. 160:412–414
103. Audu CO, Nguyen HG, CY, Katz MJ, Mao L, Farha OK, Hupp JT, Nguyen ST (2016) The dual capture of As V and As III by UiO-66 and analogues. Chem. Sci. 7:6492–6498
104. Wang Y, Ye G, Chen H, Hu X, Niu Z, Ma S (2015) Functionalized metal-organic framework as a new platform for efficient and selective removal of cadmium (II) from aqueous solution. J. Mater. Chem. A 3:15292–15298
105. Xue H, Chen Q, Jiang F, Yuan D, Lv G, Liang L, Liu L, Hong MA (2016) A regenerative metal-organic framework for reversible uptake of Cd (II): from effective adsorption to in situ detection. Chem. Sci. 7:5983–5988
106. Rapti S, Pournara A, Sarma D, Papadas IT, Armatas GS, Hassan YS, Alkordi MH, Kanatzidis MG, Manos MJ (2016) Rapid, green and inexpensive synthesis of high quality UiO-66 amino-functionalized materials with exceptional capability for removal of hexavalent chromium from industrial waste. Inorg. Chem. Front. 3:635–644
107. Desai AV, Manna B, Karmakar A, Sahu A, Ghosh SK (2016) A water-stable cationic metal-organic framework as a dual adsorbent of oxoanion pollutants. Angew. Chem. Int. Ed. 55:7811–7815
108. Lv XX, Shi LL, Li K, Li BL, Li HY (2017) An unusual porous cationic metal–organic framework based on a tetranuclear hydroxyl-copper (II) cluster for fast and highly efficient dichromate trapping through a single-crystal to single-crystal process. Chem. Commun 53:1860–1863
109. Fu HR, Xu ZX, Zhang J (2014) Water-stable metal-organic frameworks for fast and high dichromate trapping via single-crystal-to-single-crystal ion exchange. Chem. Mater. 27:205–210
110. Zhang Q, Yu J, Cai J, Zhang L, Cui Y, Yang Y, Chen B, Qian G (2015) A porous Zr-cluster-based cationic metal-organic framework for highly efficient Cr 2 O 7 2—removal from water. Chem. Commun. 51:14732–14734
111. Zhou XP, Xu Z, Zeller M, Hunter AD, Chui SSY, Che CM (2008) Coordination networks from a bifunctional molecule containing carboxyl and thioether groups. Inorg Chem 47:7459–7461
112. Zhou XP, Xu Z, Zeller M, Hunter AD (2009) Reversible uptake of $HgCl_2$ in a porous coordination polymer based on the dual functions of carboxylate and thioether. Chem Commun 5439–5441
113. He J, Yee KK, Xu Z, Zeller M, Hunter AD, Chui SSY, Che CM (2011) Thioether side chains improve the stability, fluorescence, and metal uptake of a metal-organic framework. Chem. Mater. 23:2940–2947
114. Liang L, Chen Q, Jiang F, Yuan D, Qian J, Lv G, Xue H, Liu L, Jiang HL, Hong M (2016) In situ large-scale construction of sulfur-functionalized metal-organic framework and its efficient removal of Hg (II) from water. J. Mater. Chem. A 4:15370–15374
115. Wu Y, Pang H, Yao W, Wang X, Yu S, Yu Z, Wang X (2018) Synthesis of rod-like metal-organic framework (MOF-5) nanomaterial for efficient removal of U (VI): batch experiments and spectroscopy study. Sci. Bull. 63:831–839

116. Haque E, Lee JE, Jang IT, Hwang YK, Chang JS, Jegal J, Jhung SH (2010) Adsorptive removal of methyl orange from aqueous solution with metal-organic frameworks, porous chromium-benzenedicarboxylates. J. Hazard. Mater. 181:535–542
117. Tong M, Liu D, Yang Q, Devautour-Vinot S, Maurin G, Zhong C (2013) Influence of framework metal ions on the dye capture behavior of MIL-100 (Fe, Cr) MOF type solids. J Mater Chem A 1:8534–8537
118. Bibi R, Wei L, Shen Q, Tian W, Oderinde O, Li N, Zhou J (2017) Effect of amino functionality on the uptake of cationic dye by titanium-based metal organic frameworks. J. Chem. Eng. Data 62:1615–1622
119. Jung BK, Hasan Z, Jhung SH (2013) Adsorptive removal of 2, 4-dichlorophenoxyacetic acid (2, 4-D) from water with a metal-organic framework. Chem. Eng. J. 234:99–105
120. Seo YS, Khan NA, Jhung SH (2015) Adsorptive removal of methylchlorophenoxypropionic acid from water with a metal-organic framework. Chem. Eng. J. 270:22–27
121. Jia YY, Zhang YH, Xu J, Feng R, Zhang MS, Bu XH (2015) A high-performance "sweeper" for toxic cationic herbicides: an anionic metal-organic framework with a tetrapodal cage. Chem. Commun. 51:17439–17442
122. Li J, Duan Q, Wu Z, Li X, Chen K, Song G, Alsaedi A, Hayat T, Chen C (2020) Few-layered metal-organic framework nanosheets as a highly selective and efficient scavenger for heavy metal pollution treatment. Chem. Eng. J. 383:123189
123. Xu R, Jian M, Ji Q, Hu C, Tang C, Liu R, Zhang X, Qu J (2020) 2D water-stable zinc-benzimidazole framework nanosheets for ultrafast and selective removal of heavy metals. Chem. Eng. J. 382:122658
124. Cundy CS, Cox PA (2003) The hydrothermal synthesis of zeolites: history and development from the earliest days to the present time. Chem. Rev. 103:663–702
125. Duke MC, Zhu B, Doherty CM, Hill MR, Hill AJ, Carreon MA (2016) Structural effects on SAPO-34 and ZIF-8 materials exposed to seawater solutions, and their potential as desalination membranes. Desalination 377:128–137
126. Zhu B, Myat DT, Shin JW, Na YH, Moon IS, Connor G, Maeda S, Morris G, Gray S, Duke M (2015) Application of robust MFI-type zeolite membrane for desalination of saline wastewater. J. Membr. Sci. 475:167–174
127. Safarpour M, Vatanpour V, Khataee A, Zarrabi H, Gholami P, Yekavalangi ME (2017) High flux and fouling resistant reverse osmosis membrane modified with plasma treated natural zeolite. Desalination 411:89–100
128. Dong H, Zhao L, Zhang L, Chen H, Gao C, Ho WW (2015) High-flux reverse osmosis membranes incorporated with NaY zeolite nanoparticles for brackish water desalination. J. Membr. Sci. 476:373–383
129. Ma N, Wei J, Liao R, Tang CY (2012) Zeolite-polyamide thin film nanocomposite membranes: towards enhanced performance for forward osmosis. J. Membr. Sci. 405:149–157
130. Pendergast MM, Ghosh AK, Hoek E (2013) Separation performance and interfacial properties of nanocomposite reverse osmosis membranes. Desalination 308:180–185
131. Kabalan I, Rioland G, Nouali H, Lebeau B, Rigolet S, Fadlallah MB, Toufaily J, Hamiyeh T, Daou T (2014) Synthesis of purely silica MFI-type nanosheets for molecular decontamination. RSC Adv. 4:37353–37358
132. Hu Y, Wei J, Liang Y, Zhang H, Zhang X, Shen W, Wang H (2016) Zeolitic imidazolate framework/graphene oxide hybrid nanosheets as seeds for the growth of ultrathin molecular sieving membranes. Angew. Chem. Int. Ed. 55:2048–2052
133. Ren CE, Hatzell KB, Alhabeb M, Ling Z, Mahmoud KA, Gogotsi Y (2015) Charge-and size-selective ion sieving through $Ti_3C_2T_x$ MXene membranes. J. Phys. Chem. Lett. 6:4026–4031
134. Pandey RP, Rasool K, Madhavan VE, Aïssa B, Gogotsi Y, Mahmoud KA (2018) Ultrahigh-flux and fouling-resistant membranes based on layered silver/MXene ($Ti_3C_2T_x$) nanosheets. J. Mater. Chem. A 6:3522–3533
135. Wei S, Xie Y, Xing Y, Wang L, Ye H, Xiong X, Wang S, Han K (2019) Two-dimensional graphene Oxide/MXene composite lamellar membranes for efficient solvent permeation and molecular separation. J. Membr. Sci. 582:414–422

136. Shahzad A, Rasool K, Miran W, Nawaz M, Jang J, Mahmoud KA, Lee DS (2017) Two-dimensional ti3c2t x mxene nanosheets for efficient copper removal from water. ACS Sustain. Chem. Eng. 5:11481–11488
137. Shahzad A, Rasool K, Miran W, Nawaz M, Jang J, Mahmoud KA, Lee DS (2018) Mercuric ion capturing by recoverable titanium carbide magnetic nanocomposite. J. Hazard. Mater. 344:811–818
138. Shahzad A, Nawaz M, Moztahida M, Jang J, Tahir K, Kim J, Lim Y, Vassiliadis VS, Woo SH, Lee DS (2019) $Ti_3C_2T_x$ MXene core-shell spheres for ultrahigh removal of mercuric ions. Chem. Eng. J. 368:400–408
139. Shin Y, Fryxell GE, Um W, Parker K, Mattigod SV, Skaggs R (2007) Sulfur-functionalized mesoporous carbon. Adv. Funct. Mater. 17:2897–2901

The manufacturer's authorised representative in the EU is Springer Nature Customer Service Centre GmbH, Europaplatz 3, 69115 Heidelberg, Germany. If you have any concerns regarding our products, please contact ProductSafety@springernature.com

Printed and bound by CPI Group (UK) Ltd, Croydon, CR0 4YY
30/09/2026
02225692-0005